The Geomagnetic Field and Life

Geomagnetobiology

The Geomagnetic Field and Life

Geomagnetobiology

A.P. Dubrov
Academy of Sciences of the USSR

Translated from Russian by
Frank L. Sinclair

Translation editor
Frank A. Brown, Jr.
Northwestern University
Evanston, Illinois

PLENUM PRESS • NEW YORK AND LONDON

Library of Congress Cataloging in Publication Data

Dubrov, Aleksandr Petrovich.
The geomagnetic field and life.

Translation of Geomagnitnoe pole i zhizn'.
Bibliography: p.
1. Magnetic fields – Physiological effect. 2. Life (Biology) 3. Magnetism, Terrestrial.
I. Title QP82.2.M3D813 574.1'91 78-1705
ISBN 0-306-31072-4

The original Russian text, published by Gidrometeoizdat in Leningrad in 1974, has been substantially expanded and revised by the author for the present edition. This translation is published under an agreement with the Copyright Agency of the USSR (VAAP).

Геомагнитное поле и жизнь
А. П. Дубров

GEOMAGNITNOE POLE I ZHIZN'
A.P. Dubrov

A Division of Plenum Publishing Corporation
227 West 17th Street, New York, N.Y. 10011

Printed in the United States of America

To the memory
of my
father and **mother**

Foreword to the American Edition

Dubrov's treatise is the first general review of the biological effects of the geomagnetic field to be published in English, itself a very valuable contribution. Dubrov appears to have done a fine job of understanding and summarizing the non-Russian literature and an even larger mass of Russian literature, of the interesting nature of much of which I, for one, was unaware. This last is, to my mind, an especially worthwhile aspect of the book for western biologists unfamiliar with the Russian language.

While a number of excellent contributions have appeared in Western journals since Dubrov completed his manuscript, and presumably some others in Russian journals, nothing with which I am familiar gainsays any of the facts, interpretations, or hypotheses presented in the book. It is remarkably complete in its coverage of the literature to the beginning of 1976. Indeed, later reports simply strengthen themes that are presented and defended by Dubrov.

The book carries a nice flavor of the current state of the subject, a subject that undoubtedly will become increasingly appreciated as we move down from the crest of the reductionist binge biology has been on over the past decade or two and commence asking broader questions about regulation and interrelations. And this book, coming along now as it is, should have a tremendous impact on the transition. Biologists are now in the process of learning that something that fairly recently was regarded as absurd and even "theoretically impossible" is now becoming a well-established fact. However, its flip-flop properties render it difficult to evaluate though the overall phenomenon is clearly real. The subject will be urged upon science and scientists for resolution by an enlightened public that is being bombarded with more and more semipopular articles on biological influences of microwaves, radio waves, powerline frequencies, electric and magnetic fields, airborne ions, etc. People will recall the sad episodes of the recent past when ignorance of the latent effects of ionizing radiation permitted deleterious exposure to it.

My only points of ambivalence with respect to Dubrov's book are his speculations concerning possible roles of the geomagnetic field beyond those already demonstrated and concerning its mechanisms of action. And yet, for many persons more theoretically or speculatively inclined than I am, these could be among the most stimulating parts of the book. Anyway, every author has the right to express his own views, and Dubrov does so with an apparent wealth of general background.

In brief, in my opinion this volume will be an excellent, much needed addition to the scientific literature.

Frank A. Brown, Jr.

Morrison Professor of Biological Sciences
Northwestern University
Evanston, Illinois

Preface to the American Edition

I am very pleased that my book *The Geomagnetic Field and Life* is being published in English in the United States. Thanks to the initiative of Plenum Press, a publishing house that is widely known in all countries, I have a great new opportunity to make direct contact with friends throughout the world.

My book on the geomagnetic field can be regarded as an abstraction, whose purpose is to provide a better picture and understanding of the world around us, its main driving forces, and factors, to help us to know ourselves, and to proceed further. The essence of the abstraction is that in treating the problem I have deliberately ignored the diverse effects of various external factors on living organisms and have confined myself to an analysis of the effect of the GMF. This approach allows me to go one step further—to draw various conclusions and propose theories that might bring us closer to a proper understanding of the true nature of the phenomena. Philosophers have long been aware that by such abstract thinking we can determine the nature of phenomena more reliably, completely, and comprehensively, penetrate to the very core of the observed effects, and perceive the depth of their interrelations.

Hence, although I deal only with the GMF and the entire book is devoted to its decisive role in all the processes occurring in the biosphere, it should be borne in mind that the GMF never acts alone anywhere, but is always combined in an intricate complex with the action of other important physical environmental factors that affect the vital activity of living creatures—temperature and light, atmospheric and terrestrial electricity, ionization and pressure, gravitation and cosmic rays, and many other, as yet unknown, fields and radiations.

Many scientists have contributed to the investigation of this problem, but the first names to be noted must surely be those of the Russian scientist A. L. Chizhevskii and the Italian scientist G. Piccardi, who have earned worldwide fame by their titanic research on heliobiology and heliophysicochemistry over many years.

A great contribution toward the clarification of this complex problem was made by the fundamental discoveries in 1965–1967 by Frank Brown and Y. Park (Evanston, Illinois), who showed the important role of the GMF horizontal component in biological rhythmicity and spatial orientation, and also the ability of living organisms to form relations between the GMF and the surrounding spatial distribution of light.

After we discovered in 1969 that the permeability of biological membranes depends on and is regulated by diurnal variations of the GMF, the immense class of rhythmic phenomena studied in chronobiology received an explanation. The theory of functional dissymmetry, which we developed in 1973, deepened our knowledge of this relation between living objects and the GMF.

An important milestone in the forward march of biogeophysics was the establishment of the effect of slow rotation of biological objects and the effect of slowly rotating magnetic fields (1–2 rpm) on biological function (T. Hoshizaki and K. Hamner, 1962–1968; F. Brown and C. Chow, 1973–1976). This opened up new pathways to the discovery of new geophysical factors of great ecological significance and to the study of the interrelations of living organisms.

Our discovery of the role of the GMF in genetic homeostasis from an analysis of available research (1969–1975) provided the basis not only for a new outlook on evolution and understanding of the past history of development of the organic world, but also (and this was most important) showed the role of the GMF in evolutionary processes taking place in the biosphere at present. We shall obtain an even deeper and greater understanding of this close relation between the GMF and biological objects when scientists in the near future discover the intimate mechanism of interaction of biological fields at different levels (molecular and cellular, organismal, and populational) with the GMF, and the essential way by which the GMF acts on organic molecules and water molecules—the basis of our existence on earth.

Some of my future readers, especially physicists, who have a skeptical attitude toward the biological effect of the GMF and who, of course, cannot have a full and detailed knowledge of the geomagnetobiological research cited, will find it difficult to understand the solidity and intensity of my belief in the role of the GMF in the development of the biosphere and all the processes occurring in it. On this point I think it apt to quote the very eminent evolutionist Ernst Mayr: "In each period of time a particular group of facts and some prevailing theories so occupy the minds that to any other viewpoint it is extremely difficult to give any objective attention. Hence, we must show caution in the assessment of our present-day convictions" (982, p.21). This appeal of one of the greatest evolutionists in the world makes us confident that our aspirations, approaches, and outlooks in relation to the role of the GMF in the biosphere will be correctly understood by scientists of the world and the broad mass of readers.

I am deeply grateful for the exceptional assistance given to me by Mr.

Christopher Bird (Washington, D.C.) and by Lidya Zhdanova and Tamilla Nedoshivina (Leningrad), editors in the Soviet publishing house ''Gidrometeoizdat,'' for preparation of the Russian and American editions of my book. It is only by their efforts that this book has been published in the form in which I present it now for the judgment of my numerous friends and readers throughout the world.

In conclusion, I would like to mention that the person who first suggested this book was my great and wonderful friend Dr. Lazar' Vitel's (Leningrad), whose faith in my competence and ability to write such a book has inspired me.

Aleksandr P. Dubrov

Acknowledgments

The author sincerely thanks the following persons, whose help contributed to the successful completion of the work on the book:

Docent Dr. V. V. Abros'kin, Voronezh, USSR
Dr. Matyas Banyai, Estergom, Aranyhegy, Hungary
Prof. Dr. G. Becker, Berlin—Dahlem, FRG
Dr. C. J. Beer, Tucson, Arizona
Prof. F. A. Brown, Jr., Evanston, Illinois
Dr. Mrs. C. Capel-Boute, Brussels, Belgium
Prof. Dr. P. P. Chuvaev, Minsk, Belorussian SSR
Prof. Mrs. Eleonora Francini Corti, Florence, Italy
Prof. T. A. Davis, Calcutta
Mr. Ira Einhorn, Philadelphia, Pennsylvania
Mr. V. V. Fedulov, Moscow
Dr. Mrs. L. E. Fisher, Moscow
Mr. Michel and Mrs. Francoise Gauquelin, Paris
Dr. S. I. Gleizer, Kaliningrad
Dr. Carlo R. Lenzi Grillini, Florence, Italy
Prof. Roberto Gaultierotti, Milan, Italy
Prof. Franz Halberg, Minneapolis, Minnesota
Dr. Ernst Hartmann, Eberbach am Neckar, FRG
Dr. Benson Herbert, Downton, Wiltshire, England
Dr. Eugen Jonas, Mana, Czechoslovakia
Dr. B. N. Kazak, Moscow
Prof. C. Louis Kervran, Paris
Dr. A. Korschunoff, Fürstenfeldbruck, FRG
Dr. A. V. Koval'chuk, Uzhgorod, Ukrainian SSR
Dr. S. M. Krylov, Moscow

Prof. V. Laursen, Charlottenlund, Denmark
Dr. Mrs. L. M. Luk'yanova, Apatity, Karelian SSR
Prof. S. A. Mamaev, Sverdlovsk, USSR
Dr. I. A. Maslov, Moscow
Dr. E. Stanton Maxey, Stuart, Florida
Prof. Dr. H. von Mayersbach, Hanover, FRG
Dr. L. D. Meshalkin, Moscow
Mr. Juri and Mrs. Ludmila Molchanov, Moscow
Prof. Hideo Moriyama, Oiso-machi, Kanagawa-ken, Japan
Dr. E. N. Moskalyanova, Voronezh, USSR
Miss Olga Muzyleva, Moscow
Dr. T. G. Neeme, Tallinn, Estonian SSR
Dr. Yu. I. Novitskii, Moscow
Dr. W. Paulishak, Rockville, Maryland
Prof. Mrs. A. T. Platonova, Irkutsk, USSR
Dr. Mrs. E. D. Rogacheva, Gorki, USSR
Dr. Mrs. G. Shelepina, Moscow
Dr. A. D. Shevnin, Moscow
Dr. O. V. Starovskii, Moscow
Docent Dr. S. Sh. Ter-Kazar'yan, Erevan, Armenian SSR
Prof. Dr. Solco W. Tromp, Oegstgeest (Leiden), Netherlands
Dr. R. Vasilik, Kiev, Ukrainian SSR
Dr. B. M. Vladimirskii, Nauchnyi, Crimea, USSR
Dr. Mrs. Violetta S. Zabelina, Kharkov, Ukrainian SSR

I am grateful to Mrs. Irina Petukhova, Mrs. Marina Lotvina, and Miss Svetlana Dolinskaya for technical assistance.

A.P.D.

Contents

Foreword to the Russian Edition

It is now generally acknowledged that many processes in the biosphere depend on cosmic conditions, primarily the state of the magnetosphere. In this respect the appearance of A. P. Dubrov's *The Geomagnetic Field and Life* is timely and essential.

This book, however, can be regarded from different standpoints and will undoubtedly give rise to a great deal of scientific controversy and discussion. For instance, consideration from the viewpoint of geophysics will certainly give rise to the question: Is it possible that slight variations of the geomagnetic field, measured in gammas, can have such a pronounced effect on living organisms when the biosphere is filled with artificial electromagnetic fields of much greater strength? Biologists, on the other hand, who know the great importance of such climatic factors as light intensity, temperature, and air humidity for the vital activity of living organisms, will quite rightly pose the question: Can the effect of the geomagnetic field and its variations be significant? A similar question can certainly also be posed by medical researchers, geneticists, and ecologists, since in this book the author gives an account of observed correlations between the variation of the geomagnetic field and biological processes.

I would like to point out, however, that, irrespective of the standpoint of the reader, this book, which contains a considerable amount of biological and geophysical data, cannot fail to be of interest.

By correlating numerous and diverse investigations the author shows for the first time the role of the geomagnetic field as a very important *environmental factor* affecting man and living organisms on the earth. This I regard as the important service rendered by the author and the great value of the book as a whole.

To the author and all those who are studying the biological role and importance of natural magnetic fields we wish great success in this complex and difficult work.

V. A. Troitskaya

President of International Association of Geomagnetism and Aeronomy
Chairman of Scientific Council on Geomagnetism
Academy of Sciences of the USSR

Editor's Note

Ten to twenty years ago very few papers devoted to the biological action of magnetic fields could be found in the world scientific literature. Some scientists completely rejected the idea that any magnetic field could affect any biological system. Yet now we have a book on the effect of the geomagnetic field (GMF) on life processes.

While this book was being prepared for publication there occurred two events in the scientific life of our country, which have brought out even more the topicality of the questions raised by the author.

In Belgorod in September, 1973 the Council on the Complex Problem "Cybernetics," Academy of Sciences of the USSR, the Belgorod Pedagogic Institute, the Institute of Problems of the Kursk Magnetic Anomaly, and the Belgorod Regional Health Department held the Second All-Union Symposium on the Effect of Natural and Weak Artificial Magnetic Fields on Biological Objects.

The interest in this effect is dictated by life itself. Alteration of environmental conditions and the modern tempo of life have made man more and more sensitive to the stimuli always present in the earth's biosphere and to those which have appeared only in this country, e.g., electromagnetic fields (emfs), to which all life on earth is exposed.

The problems raised were of a complex nature, involving careful measurements of the physical parameters of emfs, the detection of fine biological reactions, and the assessment of the overall response of the complex human organism. The solution of these problems requires the creative collaboration of different specialists. Hence, biologists, doctors, biophysicists, physicists, chemists, geologists, and representatives of other specialties—a total of 70 delegates from 18 towns in the Soviet Union—took part in the work of the symposium.

In Moscow in March, 1974, there was a conference on the theme "Cosmic

Factors and the Evolution of the Organic World,'' convened on the initiative of the Scientific Council, Academy of Sciences of the USSR, on Pathways and Features of the Historical Development of Animal and Plant Organisms; the Paleontological Institute, Academy of Sciences of the USSR; and the Moscow Society of Naturalists. Several papers presented at this conference showed that reversals of the GMF could affect the evolution of the organic world.

Thus, from complete rejection of any biological effect of magnetic fields to the recognition of them as a factor affecting the evolution of the organic world is the path traversed by magnetobiology in recent years. To provide a picture of this arduous path in a small book is a difficult task. Dubrov, however, who has had great personal experience in dealing with some problems of magnetobiology, has managed to cope with this task.

Why have biologists and medical researchers hitherto shown hardly any interest in the GMF and, with a clear conscience, ignored this environmental factor in their accounts of all life processes?

The fact is that the birth of the space age, which has made an impact on many purely terrestrial matters, has also had a considerable effect on the development of magnetobiology. As the author observes, the development of magnetobiology has been closely correlated with progress in the conquest of space.

The flights of the first space ships heightened interest in the problem under discussion. On being separated from earth, the cosmonaut is deprived not only of the earth's gravitational attraction, its usual atmosphere, and many other conditions essential for life, but also of the GMF. On the other hand, space-ship designers hope to provide protection for the vehicle from harmful ionizing radiation by a powerful artificial magnetic field. Thus the cosmonaut and other biological occupants of a spaceship will, more than anyone else, be subjected for a long period to magnetic fields differing from those on earth. Can such changes affect vital activity?

Affirmative answers to this question are given in this book. The author shows that with each step in the investigation of the biosphere man has become more and more convinced that terrestrial life is extremely closely ''fitted'' to terrestrial conditions of existence. The linkup of different sciences, sometimes quite remote from one another, has now reached the stage where biophysics and geophysics unite. Geophysicists have hitherto characterized magnetic storms and earthquakes only in technical terms, whereas they are now beginning to show interest in, for instance, the number of heart attacks among people during a magnetic storm and the behavior of animals before an earthquake. Biophysicists, in their turn, though not abandoning their favorite laboratory factors—electric current, ionizing radiation, light, etc.—have become interested in the possible biological effect of the GMF.

The author, a biologist by training, has quite rightly devoted the first chapter of the book to a detailed account of the geophysical data relating to the GMF. This is followed by a description and a critical analysis of diverse biological data indicating an effect of natural and weak artificial magnetic fields on various living objects.

Of particular importance is the evidence that artificial weakened magnetic fields can produce responses in man, animals, plants, and microorganisms. The few investigations made in this area so far indicate a possible role of the GMF in life processes.

A comparison of normal and abnormal (with respect to the constant component of the GMF) parts of Belgorod Region shows that they differ as to crop yield, occurrence of relict plants, and incidence of disease among the population. In addition, the motor activity of insects and birds is altered in the area of the Kursk magnetic anomaly. This initial information indicates that a biological analysis of the conditions in the regions of different magnetic anomalies will be worthwhile.

Finally, the vectorial nature (the directivity of the lines of force) of the GMF also affects biological processes. We are not referring here merely to the use of the GMF, among other orienting factors, by migrating birds and fishes.

It should be noted that since artificially produced "magnetic anomalies" also affect very diverse biological objects, ecological magnetobiology has become a recognized branch of science. The orbit of interest of this science includes not only the spatial variations of the GMF, but also the temporal variations of this geophysical factor. To the union of geophysics and biophysics we must add astrophysics, since changes in the GMF are closely correlated with solar activity.

This branch of ecological magnetobiology is interwoven with heliobiology, which is concerned with the effect of solar activity on the biosphere. The founder of heliobiology, A. L. Chizhevskii, did not believe that solar activity affects the biosphere only through alteration of the GMF. The biologically active factor could be radio waves, air ions, ionizing radiation, or solar *Z* emission, which is still not understood by physicists.

Thus, the hypothesis of an orienting (in time and space) effect of the GMF on biological objects has been formulated. It is gratifying to find comparisons of the results of field and laboratory experiments that can resolve the main problem of magnetobiology—the explanation of the intimate mechanism of action of magnetic fields on biological objects.

There are probably several such mechanisms, including those mentioned by Dubrov—the effect of natural magnetic fields on the intrinsic magnetic fields of biological objects, on the permeability of biological membranes, and on the properties of aqueous systems of biological objects. In the last case, problems of magnetobiology are intertwined with problems of physics and engineering, since magnetic treatment of aqueous solutions is finding wide application as a solution to some technological problems.

Yu. A. Kholodov

Introduction

A remarkable feature of recent decades has been the development of space biology. Space research has had a far-reaching and revolutionizing effect on every branch of scientific knowledge. It has confronted scientists with many new problems and has necessitated a critical examination of many of the standpoints of biology and medicine. In particular, the question of the biological significance of natural electromagnetic fields and of the role of the geomagnetic field (GMF) in the functioning of living organisms has again been raised.

It is extremely important for scientists engaged in space research to know how closely living organisms are adapted to terrestrial conditions of existence and whether there is any hazard for them in a prolonged separation from the earth where they have evolved over many centuries. By terrestrial conditions here we mean not only the usual combination of climatic factors (temperature, air humidity, precipitation, etc.), but also the factors whose important role in the vital activity of living organisms has become apparent only relatively recently—these are the gravitational, magnetic, electric, and radiation fields of the earth.

In this book the author gives an account of the present state of research on the biological role of such an important environmental factor as geomagnetism.

It is quite obvious that today, when scientific disciplines are becoming more specialized, the collation of the numerous and diverse data on the biological effect of the GMF is a difficult task. Being well aware of this, the author recognizes that the book is not free of faults, but he has nevertheless decided to submit it for judgment. It is appropriate here to recall the words of Gauss when he developed the general theory of the earth's magnetic field: "The bricks are merely assembled, but there will be no building until the confused phenomena conform to a single universal principle." These words apply equally well to the problem forming the subject of this volume. The author examines the investigations in various areas of medicine and biology, and from the analysis of the various experimental facts he attempts to indicate the state and level of our knowledge of the biological role of the GMF.

Progress in physics has led to a rapid development of the area of magnetobiology, which is concerned, in particular, with the effect of artificial magnetic fields. A great contribution to magnetobiology was made by the physicists and biologists who initiated systematic investigations of the biological effect of the GMF and artificial magnetic fields (241, 676, 678, 751, 752).

Important work has been done by heliobiologists, who have clearly shown the existence of a biological effect of solar activity associated with changes in the earth's magnetic field (92, 178, 516, 721, 766, 767).*

Research in the field of magnetobiology in our country at present is being directed by the Scientific Council on Geomagnetism, Academy of Sciences of the USSR, and the Scientific Council on the Complex Problem "Cybernetics," Academy of Sciences of the USSR. The USSR has a well-developed network of geophysical observatories and scientific-research institutes, and also has one of the international centers for geophysical data, which collects worldwide information on solar activity, the ionosphere, and geomagnetism.

The author expresses his deep gratitude to Academician E. M. Kreps, Academician V. N. Chernigovskii, Doctor of Physicomathematical Sciences M. M. Ivanov, Candidate of Medical Sciences B. A. Ryvkin, Doctor of Biological Sciences Yu. A. Kholodov, Candidate of Physicomathematical Sciences A. I. Ol', and Candidate of Geographical Sciences L. A. Vitel's for useful advice, consultation, and assistance in the compilation of the book. The author sincerely thanks Prof. V. I. Afanas'eva, Candidates of Physicomathematical Sciences K. G. Ivanov, A. D. Shevnin, V. I. Bobrov, and T. P. Puolokainen, and also all the staff of the Terrestrial Electromagnetic Field Sector of the Institute of Earth Physics, Academy of Sciences of the USSR, for friendly assistance.

*In the consideration of the indirect effect of solar activity via the earth's magnetic field the direct effect of wave radiation and corpuscles must not be overlooked. *This and the subsequent notes are due to B. A. Ryvkin.*

CHAPTER 1

General Account of the Geomagnetic Field

The existence of the earth's magnetic field has been known since ancient times. The unusual properties of a magnet gave rise to legends and beliefs in a supernatural force. This applied equally to the earth's magnetic field.

Progressive scientists in their time tried to determine the true physical causes of this exceptional natural phenomenon—magnetism. In 1600 the English scientist W. Gilbert in his book *De Magnete, Magneticisque Corporibus, et de Magno Magnete Tellure* put forward the hypothesis that the earth is a uniformly magnetized sphere with two magnetic poles, and that the cause of the magnetism is hidden inside the sphere.

As long ago as 1769, however, Lomonosov suggested in his investigation *Discourse on Greater Accuracy of Marine Navigation,* that the terrestrial globe consists of differently magnetized bodies and hence is magnetized nonuniformly.

An analytical expression for the earth's magnetic field was first obtained by the scientists I. M. Simonov in 1835 and K. Gauss in 1838.

The physicists' view on the role and significance of the geomagnetic field (GMF) was followed by those of biologists, doctors, and naturalists, who were keenly interested in the discoveries of physics (148, 149). There accumulated a fairly large body of data indicating that the GMF has a very significant effect on processes occurring in the upper and lower atmosphere, on natural regions of the earth, and on climate and atmospheric circulation (151).

The GMF, however, is of special significance for the living organisms inhabiting the earth. Before discussing this very important question we shall give a brief account of the GMF itself, its origin, structure, and some of its manifested features (56, 429, 651, 995). This will lead to a better understanding of cyclic changes in the biosphere, their nature, seasonal and latitudinal variations, and also of the correlation between biological rhythms and variations of the GMF.

Elements of the Geomagnetic Field

It was established experimentally long ago that the terrestrial globe has a magnetic field. The greatest strength of the permanent magnetic field (modulus of the total vector) at the earth's surface is 0.7 Oe (oersteds). The magnetic moment of the earth is an immense quantity—8.1×10^{25} cg s m. If one assumes that this moment is concentrated in a magnet situated at a distance of approximately 400 km from the earth's center, the direction of the axis of magnetization of this magnet makes an angle of 11.5° with the earth's axis of rotation. The field produced by it is called a dipole field.

The earth's magnetic field (like a vector field) is characterized by an intensity vector $\mathbf{H}_T$, which can be resolved into two components in the direction of the magnetic meridian: a horizontal component H acting in a horizontal plane, and a vertical component Z perpendicular to H. The horizontal component can be resolved in turn into a force directed along the geographic meridian—the north component X, and a force perpendicular to the meridian—the east component Y. A compass needle at any point on the earth lies along the magnetic meridian, i.e., along an arbitrary plane of the earth's surface coinciding with the direction of the GMF.

The total vector of the earth's magnetic field undergoes changes in space as well as changes in absolute magnitude. The position of the vector in space can be characterized by two angular quantities—the *declination* and the *dip*. The magnetic declination D is the angle in the horizontal plane, i.e., the angle between the geographic meridian (north–south line) and the magnetic meridian of the

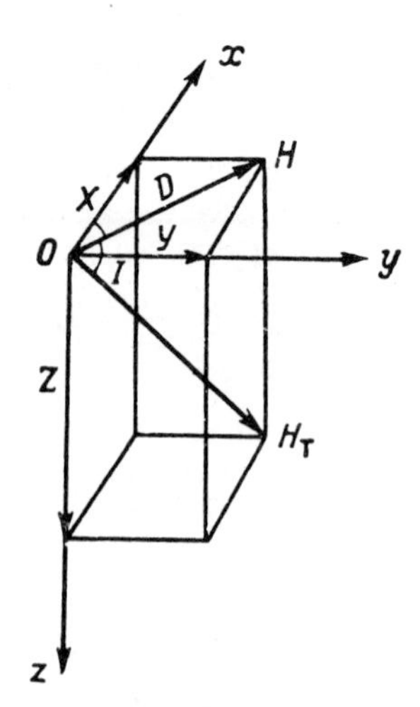

Fig. 1. Elements of terrestrial magnetism (651).

particular point on the earth. The magnetic dip I is the angle between the horizontal plane and the direction of the intensity of the total GMF vector (Fig. 1).

The total GMF vector and its elements are connected with one another by the following relations:

$$T = (H^2 + Z^2)^{1/2}, \qquad \tan I = Z/H, \qquad \tan D = Y/X$$

$$T = H \sec I, \qquad H = (X^2 + Y^2)^{1/2}, \qquad X = H \cos D, \qquad Y = H \sin D$$

The total vector and its vertical and horizontal components are measured in oersteds and gammas (1 Oe $= 10^5\ \gamma$); the declination and dip are measured in angular degrees and minutes. During magnetically quiet and magnetically disturbed days the elements of the magnetic field undergo various changes.

Constant Magnetic Field

The geomagnetic field varies in structure and dynamic properties. On the one hand, the GMF can be subdivided according to its degree of variation and its dynamics into a constant (main) and a variable field. The constant magnetic field is that which is free from all types of variations with a period of up to a year. Even in this case, however, it is not strictly constant, since it is subject to secular variations. In any case, the period of variation of the earth's constant magnetic field is very large and is many hundreds of years. The variable magnetic field of the earth consists of the changes in field with various periods within a year (from fractions of a second to months).

On the other hand, if one considers the structure or causes underlying the origin of the GMF, the GMF can be subdivided into a uniform magnetic field, a continental field, an anomalous field, an external field, and a variation field.

According to Yanovskii's classification (651), the GMF is the sum of several fields: $\mathbf{H}_0$, the field due to uniform magnetization of the globe; $\mathbf{H}_m$, the field due to inhomogeneity of the deep layers of the globe, i.e., the continental field; $\mathbf{H}_a$, the field due to different magnetization of the upper parts of the crust, i.e., the anomalous field: $\mathbf{H}_e$, the field whose source lies outside the earth, i.e., the external field; and $\delta\mathbf{H}$, the variation field due to extraterrestrial causes. Thus, the total field is given by

$$\mathbf{H}_T = \mathbf{H}_0 + \mathbf{H}_m + \mathbf{H}_e + \mathbf{H}_a + \delta\mathbf{H}$$

The anomalous field, i.e., the distorted GMF, is regarded as consisting of fields of three kinds. The first of these is *continental* anomalies, whose area is commensurable with the continents. The cause of continental anomalies has not been conclusively established. It has been suggested that powerful eddy currents play an important role in producing these anomalies. Yanovskii indicates that there are six continental anomalies, the most intense of which is the East Asian.

Another kind of anomaly is represented by *regional* anomalies, which occupy an area of tens or hundreds of square kilometers. These anomalies are due to tectonic and structural causes. The regional anomalies are not necessarily distinguished by high magnetization (like, for instance, the Kursk magnetic anomaly), but the magnetic properties of the rocks of the particular region always differ from those of other regions.

Finally, the third kind of anomaly consists of *local* anomalies. They occur where magnetic rocks lie at the earth's surface. The area of local anomalies is small and usually does not exceed ten or twenty square kilometers.

We must also discuss the earth's magnetic poles. It is generally accepted that the geomagnetic poles are situated where the magnetic axis intersects the earth's surface. These poles do not coincide with the geographic poles and lie at some distance from the corresponding geographic poles—800 km from the geographic north pole, and 1000 km from the geographic south pole. The present coordinates of the geomagnetic north pole are $\varphi = 78°30'$ N and $\lambda = 68°48'$ W; those of the geomagnetic south pole are $\varphi = 78°30'$ S and $\lambda = 111°30'$ E.

It should be noted that the geomagnetic south pole is situated in the northern hemisphere, and the geomagnetic north pole in the southern hemisphere, but in everyday speech they have been named to correspond with the geographic poles—which is incorrect.

The lines of force of the earth's magnetic field issue from approximately the center of the earth (since the magnetic dipole is eccentrically situated) through the southern hemisphere and, enveloping the earth, are again directed toward the center through the northern hemisphere (Fig. 2). At the geomagnetic north pole the field intensity is 0.6 Oe, at the south pole 0.7 Oe, and on the magnetic equator 0.35 Oe. It should be noted that the above-cited coordinates of the geomagnetic poles relate only to a uniformly magnetized sphere of perfect form. In reality, owing to diverse factors, the structure of the geomagnetic field is much more complex. Hence, the actual magnetic poles, where a magnetic needle assumes a vertical position, have the following coordinates (1970 epoch): $\varphi = 75°$ N, $\lambda = 101°$ W for the north pole, and $\varphi = 70°$ S, $\lambda = 140°$ E for the south pole.

As already mentioned, the magnetic field observed at the earth's surface consists of constant and variable fields. The magnitude of the variable field is less than about 2% of the constant field, but its biological significance, as will be shown later, is very great. Researchers have established that the GMF is almost entirely due to internal causes associated with deep layers of the earth (core, mantle, crust) and complex induction currents in them, while a much smaller part of the GMF is due to external causes, the most important of which are currents in the ionosphere and magnetosphere. Contemporary theoretical studies suggest that the GMF is due mainly to eddy electric currents in the earth's liquid core.

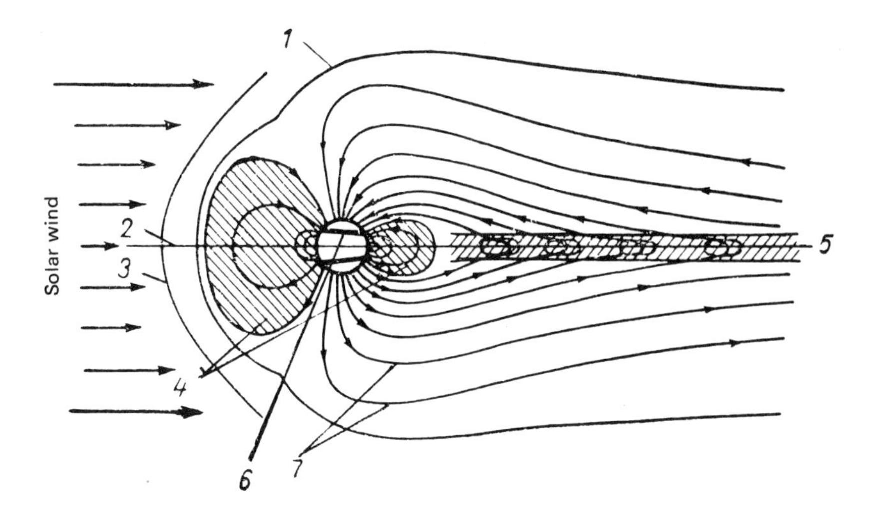

Fig. 2. Schematic meridional section of earth's magnetosphere (402). (1) Magnetopause, (2) geomagnetic equator, (3) shock wave, (4) trapped-radiation regions, (5) neutral layer, (6) zone of auroras, (7) magnetic field lines.

Reversals of Constant Geomagnetic Field

Investigations of the magnetism of rocks of diverse origin have revealed the magnitude and direction of the GMF in various geological epochs (246, 902, 1008). It has been established that the earth's constant magnetic field in remote epochs has periodically altered its direction through 180°, i.e., it has become directly opposite to that in the preceding period. These sharp changes in the direction of the earth's magnetic field, leading to a change in sign of the magnetic poles, are called polar reversals, or simply reversals.

The precise causes of geomagnetic reversals, as of the GMF itself, have not been determined, but the generally accepted theory at present is that of the geomagnetic dynamo, developed by Bullard, Runcorn, Rikitake, and Braginskii and based on eddy motions of self-induction currents originating in the earth's core (651). Other hypotheses regarding the reasons for reversals have been advanced, particularly the effect of cosmic matter on this process (554), changes in the period of rotation of the earth around the center of our galaxy (300, 882), and changes in the rate of the earth's axial rotation (225).

Irrespective of the viewpoints on the reason for polar reversals, their general planetary nature is an established fact. On the basis of paleomagnetic data geochronological scales of GMF reversals have now been devised, and their main features and the cyclicity of their appearance in ancient epochs have been determined (225, 245, 246, 419, 784, 902, 923). As a whole, briefly summing up the work of the cited authors, we can infer that the main changes in GMF direction in the last 600 million years are characterized by cyclicity of different orders: the periods of relatively frequent changes in polarity or relatively stable state of the GMF of the same polarity have a duration of about 150 million years; within this period the frequency of reversals also has a periodicity of approximately 50–60 million years, and the rhythms of the stable state of the GMF, i.e., with practically no reversals, sometimes last 20–50 million years as, for instance, in the late Paleozoic and Mesozoic. During the frequent alternations of reversals the duration of epochs of the same polarity has been, on the average, approximately 0.5 million years. In recent years there have been shorter intervals of reversed polarity—about 10^4–10^5 years (these are called "events"). Events have been most reliably distinguished and investigated for the last 5 million years. There are shorter variations of the GMF, covering a period of 10^2–10^3 years, in which the major part of the GMF does not undergo reversals—these are the secular variations.

The change in polarity of the GMF, i.e., reversal, lasts approximately 10^4 years and, what is exceptionally important, is characterized by a significant reduction of intensity, to a fraction of its former value, for 5000 years.

After an appreciable change in the GMF intensity we would naturally expect associated changes in the organic world, and we shall show this in what follows. Sharp changes in the evolution of fauna and flora should be correlated with GMF reversals, and even greater evolutionary "catastrophes" should be correlated with epochs of frequent reversals. For instance, there have been periods in the history of the earth when eight or more reversals have taken place in 10 million years (783, 784) and in the Neogene and Paleogene reversals occurred every 0.5 million years (245). Investigations connecting evolutionary changes in the organic world and sharp changes in the GMF are still far from adequate but, as will be shown later, such a correlation has been obtained for many periods in the development of the earth.

Variable Magnetic Field and Its Variations

Thus, as the preceding account shows, there are internal and external causes of formation of the GMF. The internal causes, associated directly with the earth (mainly current systems at the core–mantle boundary and other factors), are responsible for the secular variations of the earth's permanent field. The external factors (primarily electric current systems in the ionosphere and the space surrounding the earth) cause rapid changes in the GMF and are largely responsible for its variable part.

Changes in the variable GMF (variations) are of different types. In some cases they have a definite and smooth course (magnetically quiet days), and in other cases they have an irregular course, when the amplitudes, phases, and periods of variations change rapidly and continuously (magnetically disturbed days). We shall briefly discuss the variations of these types.

Quiet and Disturbed Variations of Magnetic Field

The quiet variations include (1) solar diurnal variations (Sq) with a period equal to the solar day, (2) lunar diurnal variations (L) due to the moon, with a period equal to the lunar day, and (3) annual variations. The special feature of solar diurnal variations is that they occur according to local time and depend on the magnetic activity on the particular day. Hence, this activity (u) is determined from them. The amplitudes and phases of these variations change during the day and during the year: The maximum values are observed in the period of the summer solstice, and the minimum ones in the period of the winter solstice (Fig. 3). These variations also depend on solar activity. In a period of maximum solar activity, the amplitude of the variations can be 100% greater than in a minimum period, and the correlation with the sunspot number* is 0.98.

Solar diurnal variations at different points on the earth are of diverse nature. It is of interest to note that the phases of solar diurnal variation undergo inversion at a latitude of approximately 31° (Fig. 4). The source of the magnetic field of solar diurnal variations is the closed eddy-current systems located in the ionospheric E layer (100–120 km from the earth's surface), within which the globe rotates.

The distinctive feature of lunar diurnal variations (L) is their semidiurnal period: During the day there are two minima and two maxima, whose times of onset change daily during the lunar month. We mention these features because they are implicated in the biological rhythmicity of living organisms. The amplitude of lunar diurnal variations is only 10–15% of the solar variations. They depend on the distance between the moon and earth and undergo pronounced annual fluctuations (1143).

As already mentioned, the GMF undergoes nonperiodic changes as well as periodic variations. These changes are due mainly to sharp disturbances of the geomagnetic field, called magnetic storms. Magnetic storms are a very complex phenomenon. They consist of various phases (variations). One of them—the aperiodic disturbed variation (D_{st})—is manifested in a marked change in the horizontal component and a slightly smaller change in the vertical component of the geomagnetic field. There are other kinds of variations—baylike disturbances, irregular disturbances, etc.

*The sunspot number is a solar activity index given by the sum of the spots on the visible solar disk added to ten times the number of groups of spots.

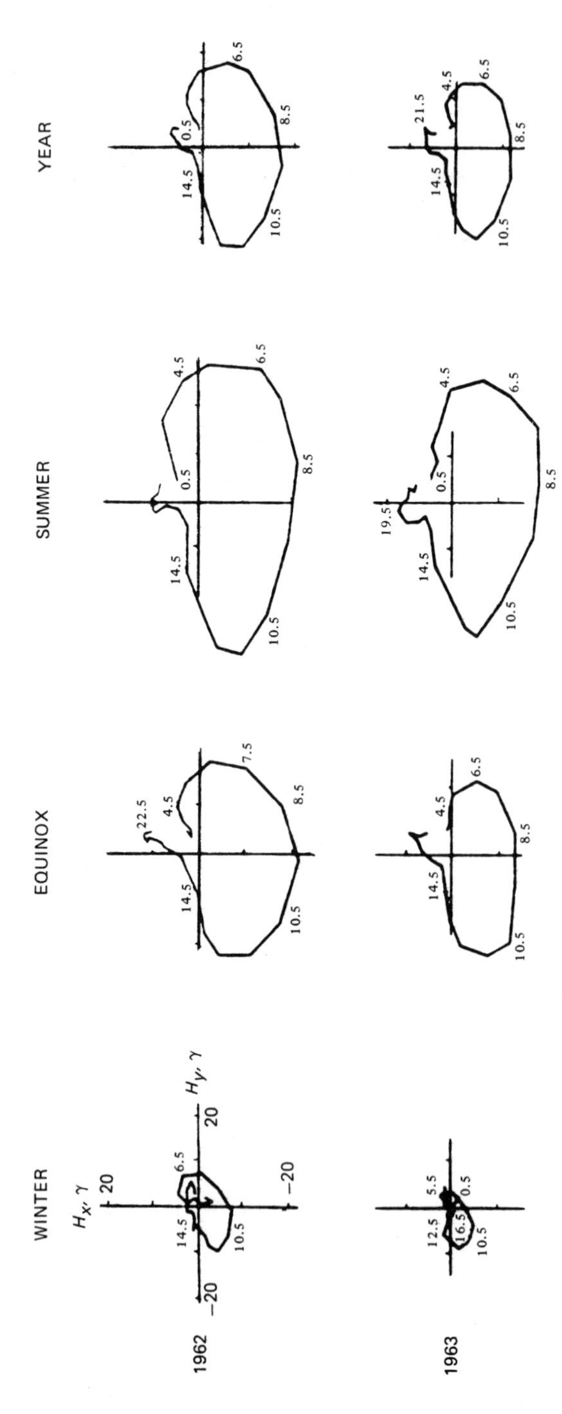
(a) GEOMAGNETIC FIELD
WINTER
EQUINOX
SUMMER
YEAR
1962
1963
H_x, γ
H_y, γ
20
−20
20
−20
14.5
6.5
10.5
22.5
4.5
7.5
8.5
0.5
21.5
19.5
5.5
12.5
16.5

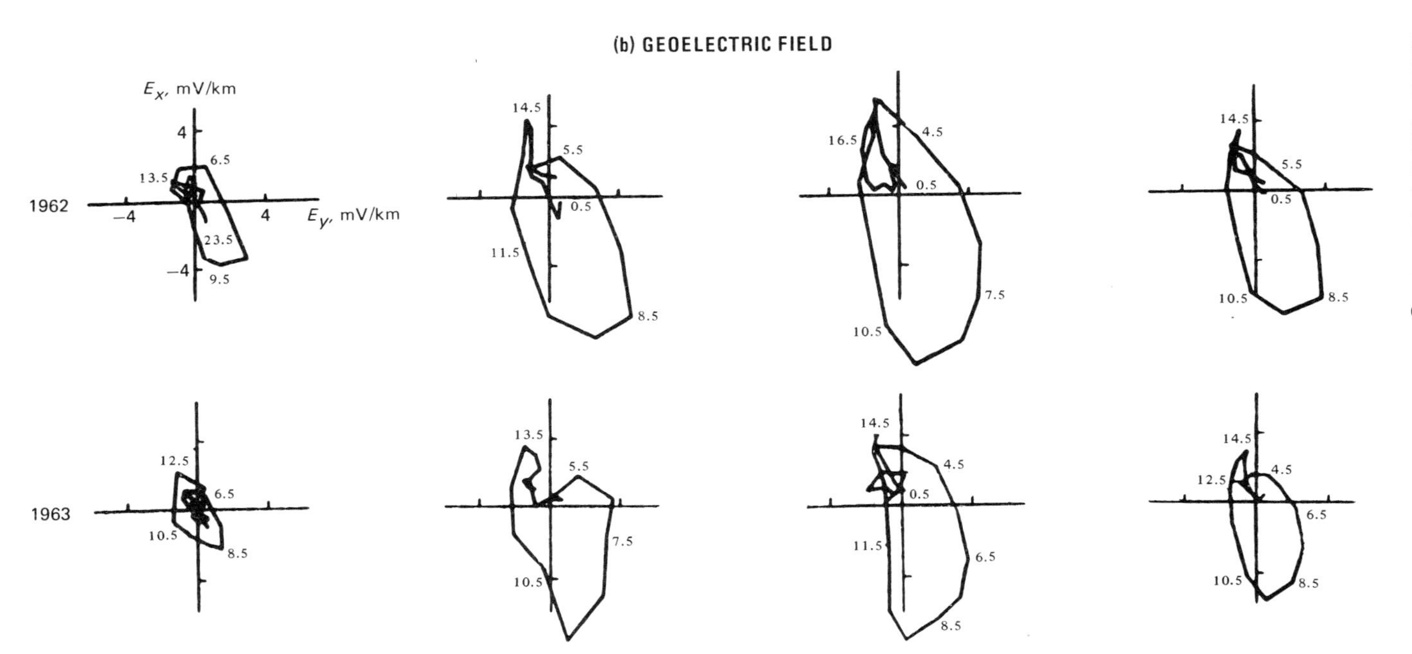

Fig. 3. Hodographs of diurnal variations of (a) GMF (b) geoelectric field for 1962–1963 in horizontal plane at Pleshchenitsy Observatory (BSSR). From (318).

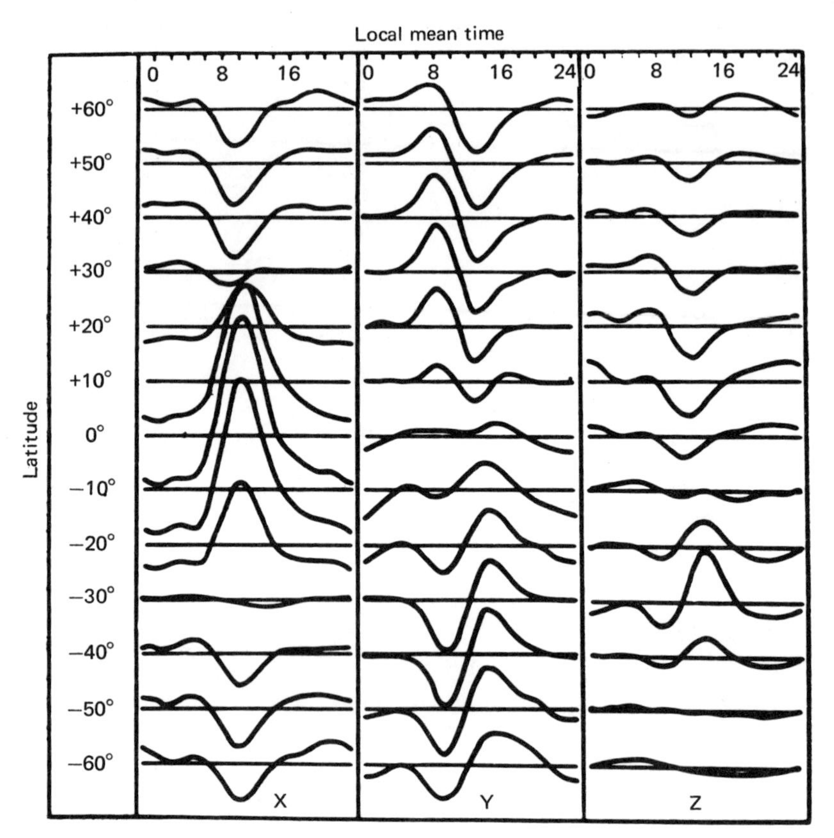

Fig.4. Planetary distribution of diurnal variation of *X, Y,* and *Z* elements of GMF. From (318), given in (651).

Geomagnetic Pulsations

A special kind of variation of the geomagnetic field is represented by the geomagnetic pulsations—electromagnetic waves of very low frequency observed at the earth's surface. Geomagnetic pulsations (we shall denote them by the abbreviation SPF—short-period fluctuations) are due to the interaction of the plasma of hydromagnetic waves, coming from the sun, with the earth's magnetosphere. Geomagnetic pulsations can be divided into two classes—regular (*pc*) and irregular (*pi*). Regular pulsations can be subdivided into five subclasses, which include persistent pulsations with a period of 0.2–600 sec. Depending on their period and particular physical nature, the pulsations can be subdivided into types *pc*1, *pc*2, *pc*3, *pc*4, and *pc*5 (186). Irregular pulsations are divided into two groups with periods of 1–40 and 40–150 sec and are denoted by *pi*1 and *pi*2, respectively.

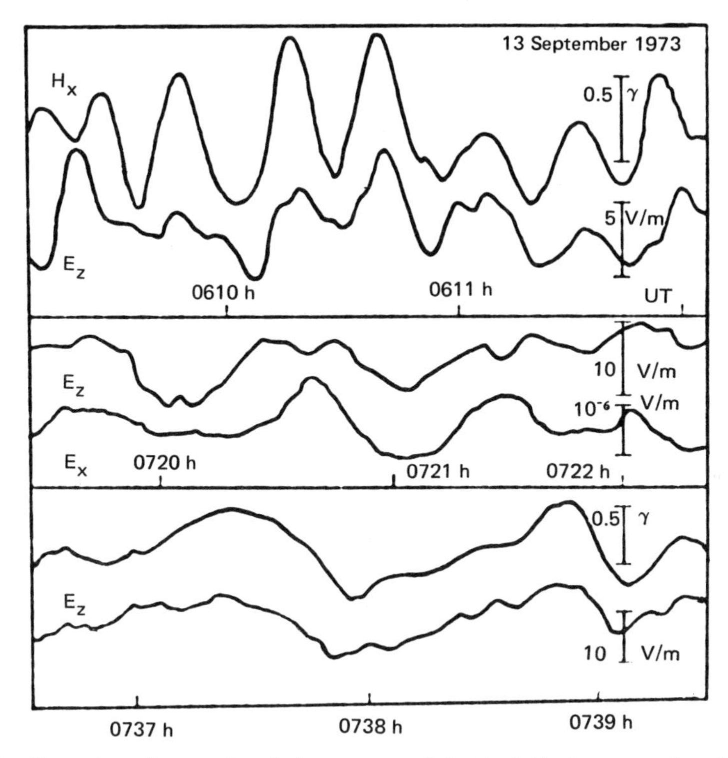

Fig. 5. Comparison of trace of vertical component of electric field of geomagnetic pulsations ($\mathbf{E}_Z{}^{\text{air}}$) in air with geomagnetic component $\mathbf{H}_X$ and geoelectric component $\mathbf{E}_X$ in earth in various intervals of time on September 13, 1973. From (765).

The duration of the SPF during a day varies from a few minutes to several hours, and their amplitude, depending on the class, varies from tenths of a gamma to tens of gammas ($pi2, pc5$). The biological activity of a geomagnetic field of this frequency range will be discussed later, but here we would like to mention that it has recently been discovered that the vertical component of the electric field of the SPF is of the order of fractions of a volt or even tens of volts per meter (86). This fact helps to explain many effects reported by biologists who have investigated the effect of the natural variable electromagnetic field and atmospheric electricity on living organisms (Fig. 5).

Magnetic Activity

Hence, the earth's magnetic field undergoes continual changes, whose complexity is reflected in changes in various parameters of the GMF.

Changes in the GMF are associated mainly with solar activity. This association, however, is not a strict functional relation, since it is the result of the mutual superposition of processes of different scale and different physical nature, i.e., processes occurring on the sun, in interplanetary space, and in the earth's atmosphere (618, 619). The degree of magnetic activity can be characterized by various indices.

Indices of magnetic disturbance are divided into two classes: local (C, K, ω_k, etc.), characterizing the disturbance of the magnetic field mainly close to the particular observatory, and planetary (C_i, K_p, A_p, etc.), characterizing the disturbance of the magnetic field of the whole earth (401, 402).

The C system of geomagnetic characteristics is as follows: Days with quiet magnetic conditions (the changes in the elements H, Z, D, and I are smooth and regular) are denoted by 0, days when these elements are disturbed are denoted by 1, and days when these elements are highly disturbed are denoted by 2. In the international center, which receives information from all observatories, the mean magnetic activity is calculated for each day and is called the international geomagnetic characteristic C_i. We dwell on this index because it is used by the international center for the selection of five quiet and five disturbed days for analysis, standardization of calculation, and comparison of geophysical data (see Appendix). Another index (the measure of the magnetic activity u—the change in mean value of H from the given day to a subsequent day) is based on evaluation of the degree of change of the horizontal component of the geomagnetic field, which is always reduced in absolute value when a magnetic disturbance occurs. This index is closely correlated with the spot-forming activity of the sun and is used to characterize the disturbance for longer periods.

The magnetic indices C and u, however, give only an overall estimate of the magnetic activity for the day. For a more accurate and better assessment of the magnetic environment prevailing in the course of a day, the index K, calculated for three-hour intervals, is used. The index K provides an estimate of the amplitude of the most disturbed element of the GMF for each three-hour interval, beginning from 0 hours. This index reflects the geomagnetic changes due to chromospheric flares.

The index K includes the regular changes in the elements of the magnetic field in the period of disturbance D, the mean value of the disturbed field D_u, and the diurnal disturbed variations S_D. The index K has gradations from 0 to 9. It is selected in accordance with a particular scale corresponding to each observatory, since the degree of disturbance at different latitudes is different for the same solar activity. As an example we give the index K scale used in the Izmiran observatory (Krasnaya Pakhra):

K:	0	1	2	3	4	5	6	7	8	9
R_γ:	5	10	20	40	70	120	200	330	550	>551

The quantity $R\gamma$ is the equivalent amplitude of the scale of index K, expressed in gammas.

In the international center the information obtained from the 11 largest observatories in the world is used to derive a "standardized" index K_s, and data providing a measure of the magnetic disturbance for the planet as a whole are published in the form of an index K_p. It should be noted that for a more accurate and more thorough assessment of the GMF other indices—A, Q, R, etc.—are used.

Magnetic Disturbances

Disturbances of the GMF can be of a periodic or sporadic nature and are also due to solar activity.

The special feature of the cyclic changes in magnetic disturbance is the coincidence of the maximum disturbance with the minimum of solar activity (measured by the sunspot number), or a year later. Near the epoch of the maximum of the 11-year cycle, when the main role is played by corpuscular streams ejected by the sun during powerful chromospheric flares, there usually occur strong magnetic storms with a sudden commencement and a well-expressed main phase.

The minimum of solar activity is closely preceded by an increase in intensity of another kind of solar corpuscular emission, which emanates from the so-called M regions. The M regions are not associated with any visible solar formations, but sometimes last for many revolutions of the sun. Storms caused by the M regions are characterized for the most part by moderate intensity, a gradual commencement, and a weak main phase (401). Periodic disturbances of the GMF (S_D) occur in correspondence with local time, whereas sporadic disturbances (magnetic storms) occur simultaneously over all or part of the planet (D_{st}). Magnetic storms can be classed according to intensity as very strong (more than 200 γ), strong (100–200 γ), and weak (50 γ). According to their distribution over the planet we can distinguish *synphasic* disturbances, which occur simultaneously over the entire surface of the earth and, hence, are in phase, *local* disturbances, occurring within a particular limited region, and *permanent,* i.e., continuous disturbances, occurring mainly in the region of the geomagnetic poles.

According to the nature of their onset magnetic disturbances are divided into storms with a sudden or gradual commencement. Storms with a sudden commencement occur when there is a sharp change in all the elements of the magnetic field over the whole planet.

The appearance of magnetic storms is correlated with solar activity. There is also a 27-day recurrence of magnetic storms due to the persistence of active

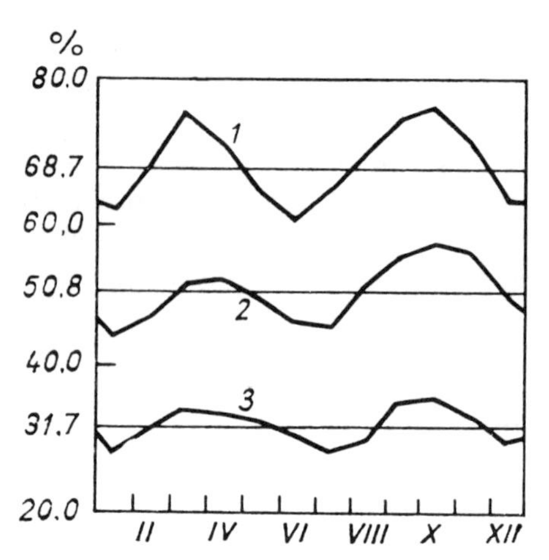

Fig. 6. Mean annual magnetic activity and deviations from it (%) in the course of a year (651). Highly disturbed (1), moderately disturbed (2), and weakly disturbed years (3).

regions on the sun during the 27-day period of its revolution. The number of magnetic storms also varies in a very definite manner over the course of a year: There are considerably more at the equinoxes and fewer during the solstices, which depends on the position of the earth in the plane of the ecliptic relative to the sun (Fig. 6). It is of interest that a 27-day and a characteristic yearly recurrence are also observed in biological processes, which indicates that they may depend on the rhythmicity of natural electromagnetic factors.

Interplanetary Magnetic Field and Geomagnetic Activity

The earth is closely linked with the cosmic space surrounding it. Since the sun is the central star of our planetary system, it is obvious that its influence is decisive for the dynamics of the geophysical and biological processes occurring on the earth (117, 118, 120, 136, 301, 588).

The relative proximity of the earth to the sun means that the interplanetary magnetic field (IMF) in circumterrestrial space can be regarded as the average magnetic field of the solar photosphere (1124). The sun's magnetic field has a complex structure and conforms to specific laws (301). One of the most important of these is Hale's law, according to which the changeover from one 11-year solar cycle to another is accompanied by a reversal of the magnetic polarity of sunspots. The polarity of the IMF varies with the change in the photospheric magnetic field (1124) and has a sectoral structure (1212).

The concept of a sectoral structure of the IMF was introduced on the basis of measurements made by artificial earth satellites (1210–1212). Investigations

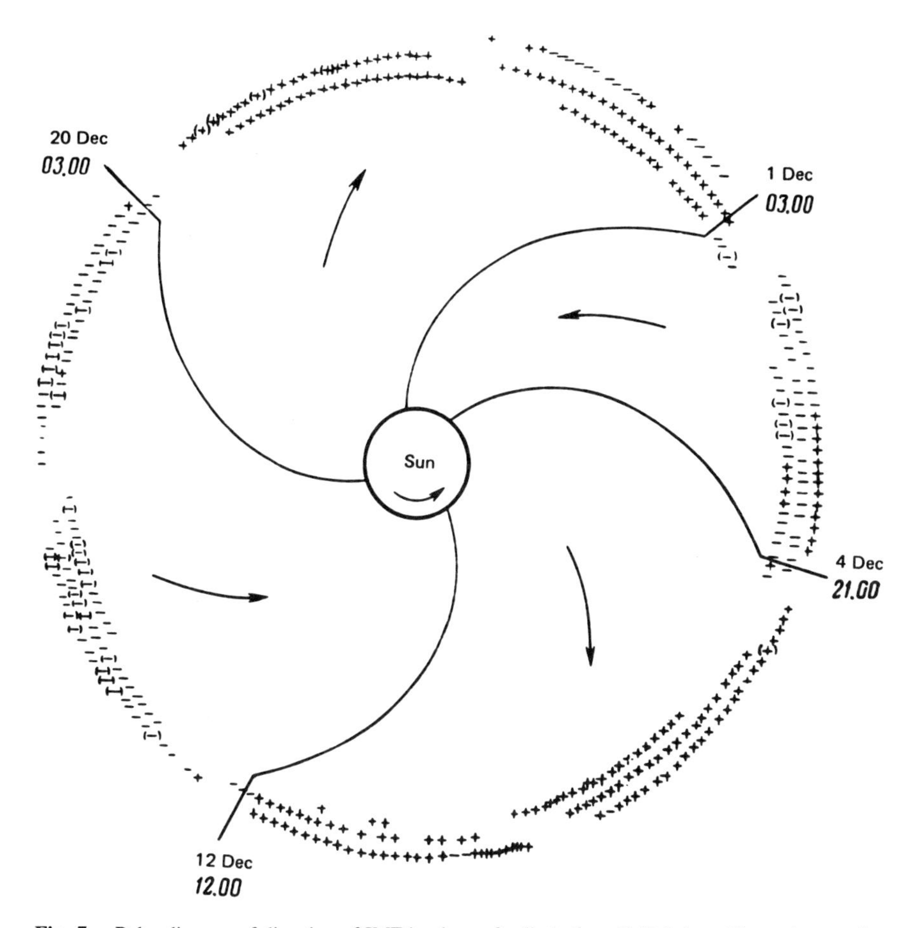

Fig. 7. Polar diagram of direction of IMF in plane of ecliptic from IMP-1 data, illustrating quasi-stationary structure of sectors. Arrows indicate direction of interplanetary magnetic field. From (160) via (1205).

have shown that in the plane of the earth's ecliptic the IMF has the shape of an Archimedes spiral and a varying angular direction to the earth–sun line. For several days the direction of the IMF remains either at an angle of ~135°, corresponding to the direction away from the sun (positive sector), or at an angle of ~315°, corresponding to the direction toward the sun (negative sector, Fig. 7). The interplanetary sectors, in which the IMF retains its main polarity rotate along with the sun. The change of polarity in them is rapid, and hence there is a distinct boundary between the sectors (see Appendix). The polarity of the IMF sectors varies regularly with a period of 26⅞ days; in the first half of the period the IMF is directed toward the sun, and in the second half it is directed away from the sun

(160, 161). The author of the papers cited has expressed the view that the core of the sun (or its inner layer) rotates with the above period and has a magnetic field directed sometimes toward the sun and sometimes away from the sun.

Investigations have shown that the variations of the IMF are related to the variations of the magnetic field at the earth's surface and the nature of geomagnetic activity (160, 161, 326, 327, 1149, 1150). In particular, it has been shown that the IMF has a particularly strong effect on the variation of the vertical component of the GMF at high latitudes (326, 327, 1149, 1150). It was found initially that if the sectoral structure of the IMF has a direction away from the sun, the variations of the vertical component of the IMF in both hemispheres are directed away from the sun, and when the IMF is directed toward the sun the variations of the Z component are directed toward the earth. At lower latitudes ($\sim 78°$) the effect of the IMF is manifested in a change in magnitude of the horizontal component of the IMF: When the IMF is directed away from the sun the field shows an increase, and when it is oppositely directed the field shows a decrease.

Subsequent investigations revealed that the ground variations of the IMF in the pole regions are closely related not so much to the polarity of the IMF (directed toward or away from the sun) as to the direction of the azimuthal IMF component Y_{SE}—the south–west component directed perpendicularly to the earth–sun direction (160, 161). From the information available at present the ground variations of the vertical component of the IMF at high latitudes can be used to determine very accurately and reliably the direction and strength of the azimuthal Y_{SE} component of the IMF and thus to determine the polarity of the IMF sector and the moment of intersection, to within a few hours, of the sector boundary in the IMF by the earth. These indices are important, because in the period of intersection of the sector boundary by the earth and the day after the intersection, various geophysical processes are greatly enhanced (1210), including geomagnetic activity. Consequently, there are changes in biological objects (19, 615). Other parameters of the IMF, particularly the increase in the north–south component Z_{SE}, are closely connected with the appearance of magnetospheric substorms affecting the whole planet. Observations in space and ground determinations of the index *AE*, which characterizes polar magnetic disturbances, provide a means of predicting the occurrence of magnetospheric substorms in the vicinity of the earth. The Japanese scientist A. Nishida has established that a change in orientation of the north–south component of the IMF at the earth's surface leads to another type of geomagnetic disturbance (DR-2), which occurs in phase over the whole planet from the pole to the equator with a period of about 1 h (1014).

Thus, the cited material shows how closely the earth is linked with the space surrounding it and how sensitively it reacts to changes in the parameters of the interplanetary magnetic field.

It is apparent from the above account that the GMF has a very complex structure and properties, and the factors responsible for its formation are also complex. This, of course, affects the relations between the GMF and the vital activity of living organisms on the earth, which we discuss in the following chapters.

CHAPTER 2

Role of Geomagnetic Field in Vital Activity of Organisms on Earth

Before considering the question of the biological significance of the GMF we will point out that it is a controversial question. Although this subject has an ancient history, there are still no grounds for accepting completely the geomagnetic theory of development of the earth's biosphere. During the last ten years, however, scientific and technical progress has immeasurably widened the horizons of human knowledge and research, with the result that the important role of the GMF in the life of the biosphere has become apparent.

The development of geophysics and astrophysics (including solar physics) has led to an understanding of the importance of solar activity for various processes occurring on the earth, particularly for the biosphere (117, 153–155, 301, 588).

The term "solar activity" usually means the internal processes occurring on the sun, and their external manifestations (spots, faculae, protuberances, chromospheric flares, etc.), which are accompanied by the emission of colossal energy and the release of matter in the form of particles and fields (620). Investigations of recent years have shown that one of the main pathways by which the sun affects the earth is the GMF.* A close relationship has been found between the sun's magnetic field, its environment, and the earth's magnetic field. This relationship is mediated by redistribution of the magnetic lines of force of the solar wind and the earth's magnetosphere (160, 161, 326, 327, 824, 1048).

The effect of solar activity on geophysical processes on the earth is reflected in a close correlation between magnetic activity and natural processes. This correlation has been revealed by investigations carried out in different areas of

*The indirect effect of solar activity via atmospheric circulation, and the direct effect of solar processes, cannot be excluded [B.R.].

atmospheric and terrestrial physics. For instance, it has been found that changes in the atmospheric pressure on the ground (356, 357), occurrence of droughts (442, 443), the barometric circulation regime (496), the drift rate in the ionospheric *F* layer (925), the temperature of the upper atmosphere (904, 905), outbreaks of cold on the earth (619), the formation of fronts and cyclones (150), and other processes (507, 508), are closely correlated with variation of the GMF. These investigations have reinforced the view of an important role of the GMF in the dynamics of climatometerological processes and phenomena on the earth.

On the other hand, biophysicists who have studied the effects of natural electromagnetic fields on living organisms and the concomitant changes in physiological and biochemical processes have also found a relation between the investigated effects and solar activity. The results of scientific research on the effect of solar activity on living organisms have been summed up in papers by various specialists in the fields of medicine, biology, and agriculture (see the following collections: *Solar Activity and Life,* 1967; *The Sun, Electricity, and Life,* 1969; *Effect of Solar Activity on the Earth's Atmosphere and Biosphere,* 1971) (118).

Investigations have shown that the variations of the functional–dynamic parameters of living organisms are not a random scatter around some mean value, but are orderly changes in biological characteristics under the influence of several factors, including heliogeophysical factors. The extensive and comprehensive investigations of heliobiologists are closely connected with biorhythmology, the science of rhythmic processes in living organisms (671, 869, 877, 1133, 1151). By making continuous observations of the rhythm of very diverse processes in a constant external environment over a long period, researchers (446, 746, 751, 752) have found that natural magnetic fields have a considerable effect on biological processes. Such investigations (see the following chapters) have produced data indicating a biological effect of the GMF. Thus, quite independently developing processes of scientific cognition—geophysical and biophysical—indicate an effect of the GMF on the biosphere and have led to the development of new disciplines: *biogeophysics* (308, 451, 644, 942), which is concerned with the effect of all geophysical factors on living organisms and the biosphere as a whole, and *geomagnetobiology* as one of its branches.

Globality and Universality of Synchronous Course of Some Biological Processes

The GMF, like the gravitational field, is an all-pervading and all-embracing physical factor. Hence it must, by virtue of its properties, have an effect on processes occurring on the earth and the space surrounding it.

The effect of these factors on the biosphere can be gauged from the following interesting feature. Investigators (90, 92, 123, 124, 132, 516, 751, 767, 809, 810) have compared different biological, physicochemical, chemical, and geophysical processes over many years and have found that they are synphasic and synchronous. Synchronism is revealed in the distinct simultaneous variation of biological phenomena, for instance, numbers of animals or insects, recurrence and spread of epidemics or epizootics, mass migrations of animals and insects out of season, variation of the cell composition of the blood in humans.

In addition, distinct synphasic variations of biological processes over the extent of a year (seasonal rhythms) and many years (when indices of the processes as a whole for a year are considered) have been observed.

The reported rhythmicity can be distinctly manifested in some years and distorted in others, since all processes are affected not only by geomagnetism and gravity, but also by the most diverse factors, whose mode of action and points of application are unknown (618).

Examples of a synphasic course are shown in Figs. 8 and 9 (123, 128), which compare the results of work of three different investigators (515, 751, 1049), who studied phenomena and processes of different kind in 1956: the composition of human blood, plant respiration, and the course of a physicochemical reaction. As Fig. 8 shows, the annual courses of the three different processes at geographical localities thousands of kilometers apart are almost completely synphasic. The processes investigated in different conditions were the leukocyte count in healthy people on holiday in sanatoria in Sochi, the respiration of potato plants in a completely constant environment in a thermobarostat in Evanston (the respiration was measured by automatic respirometers), and the hydrolysis of bismuth chloride in ordinary laboratory conditions in Florence.

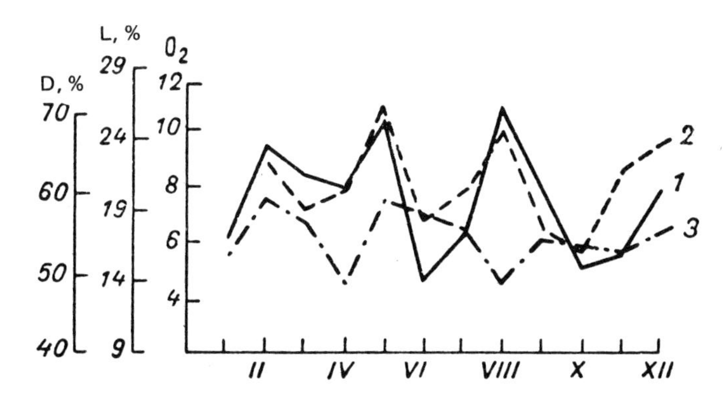

Fig. 8. Synphasic variation of different processes in 1956 (123). (1) Level of leukopenia, L (Sochi, USSR), (2) dynamics of oxygen uptake by potato plants, O_2 (Evanston, Illinois), (3) hydrolysis of bismuth chloride, D (Florence, Italy).

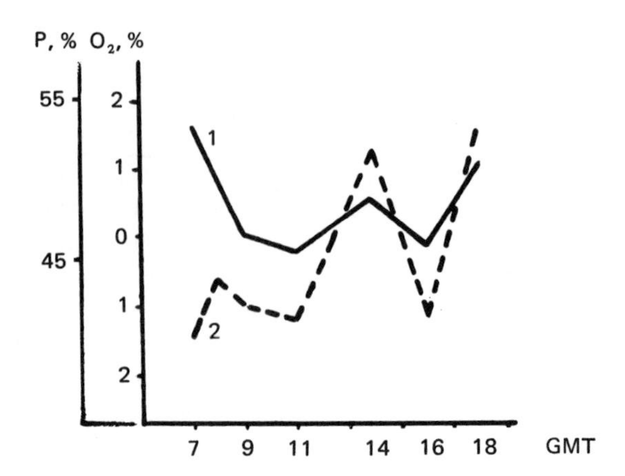

Fig. 9. Diurnal rhythm of hydrolysis of bismuth chloride, test P (1) for 1967–1968 (Florence, Italy), and of oxygen uptake by potato plants (2) (mean value for 1956–1966) (Evanston, Illinois). From (128).

Many similar examples of global synchronism and synphasic variation of natural processes on earth could be cited. For instance, a regular pattern has been found in the recurrence of extreme values of various physicoclimatic processes (323) and in the periodicity of the increment of some types of conifers (320, 469). It has also been reported (751) that in different geographical localities on the earth the annual variations of the phases may coincide, e.g., the ability of algae to reduce nitrate, the effect of weak gamma rays on planarians, the variation of the metabolic rate of beans, and the growth of grass. Cases of global disturbance of the course of biological and physicochemical reactions have also been observed (132), however. A special center has been set up under the Smithsonian Institution to investigate short-term phenomena occurring synchronously on the earth (769). The globality of the synchronous course of different and identical processes has previously been reported in heliobiology (767, 999).

A global pattern of synphasic variation of such phenomena as birth-rate of people in different countries or the dynamics of onset of some diseases, such as measles in children, has been discovered (999).

Other investigators (481) have shown that hypertensive crises in people can occur synchronously in different geographic localities (Fig. 10). It should be noted that the localities themselves are in different climatic zones and that such synchronism would be better revealed if medical data were grouped according to the solar calendar, based on the 27-day period of revolution of the sun around its axis.

All these facts of synphasic variation of biological processes occurring at

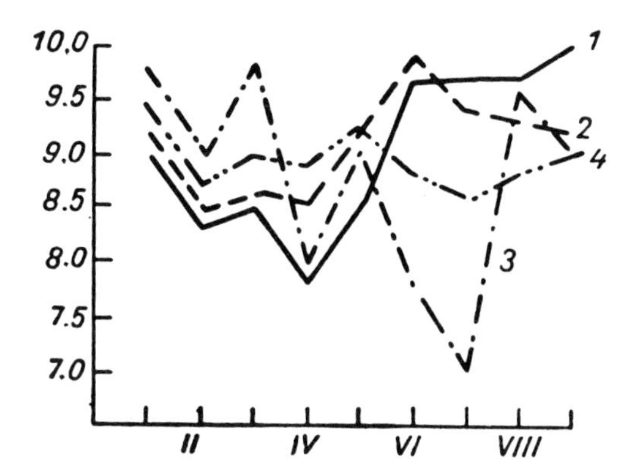

Fig. 10. Synchronous incidence of hypertensive crises in people living in different towns, 1963 (481). (1) Alma-Ata, (2) Stavropol, (3) Moscow, (4) Leningrad. Ordinate, number of crises in relative units; abscissa, days of solar calendar, grouped in threes.

different points on the globe, exceptional in themselves, are simply explained if one assumes that the main synchronizing factor is the action of the GMF. We will later give examples confirming the postulated role of the GMF in the synchronism of various processes. Here, however, we will cite the data of only two investigations, the author of each of which analyzed biomedical indices for different towns in one country.

For instance, Koval'chuk (275) found, from an analysis of many years' data on erythrocyte and hemoglobin content, that in different towns of the USSR (Kirovsk, Petrozavodsk, Moscow, Ternopol, Uzhgorod) the nature of their variation was similar and was correlated with the dynamics of global variation of geomagnetic activity. The author of another investigation (1079–1081, 1083) discovered a similarity in the dynamics of such diverse indices as birth rate, death rate, industrial injuries, and road accidents, and they were all closely correlated with geomagnetic activity and ionospheric disturbances during the same period. Thus, the information presented above indicates the globality and universality of the course of various phenomena observed at different (and widely separated) points on the globe and in different towns in one country.

The synchronism of various processes can be detected on an even smaller spatial scale. For instance, an interesting feature discovered in the analysis of biological research conducted on small areas is the great similarity in the periodicity of the variation of processes in different objects during the same time. These facts often appear inexplicable to the investigators themselves. We give several examples.

In 1966 an investigation of the excretion of biologically active substances by tree root systems (allelopathy) and the accumulation of these substances in the tree litter in plantations of white acacia, oak, and pine of different ages revealed a similarity of the allelopathic process. The author (332), taking into account the species of trees that he had investigated, wrote:

> "An analysis of the obtained data shows that the variation of the allelopathic process in all the investigated plantations is *of the same type*" [italics added]. These data were obtained on the territory of a single forestry in Voronezh Region. Other authors have found a corresponding similarity in plants growing in the same field. For instance, a monotypical diurnal rhythm of respiration of beet (not only of the whole plant, but also of individual leaves) in the field on each specific day of investigation, despite the widely varying external conditions, was observed (303). The authors (303) write: "The experiments revealed an extremely irregular course of respiration of sugar beet leaves. . . . It should be noted that the irregular course of respiration in the two investigated leaves *coincided* fairly well, which suggests that the leaf respiration curves are controlled by *common factors*" [italics added].

It was subsequently found that the diurnal periodicity of respiration is very similar not only in leaves of one plant, but even in the most diverse objects (potato, carrot, beans, mealworm, mouse; see Fig. 11) (752). This similarity in the rhythm of the functions of different organisms indicates the effect of a single acting factor.

An investigation of processes (biochemical and physicochemical) occurring in such distinct formations as soil also reveals synchronous variation of all the indices. As an example we indicate the data of Baranova's work, which gives the exact dates of the soil investigation. Baranova found that the annual (seasonal) course of activity of catalase, ammonium nitrogen, and mobile forms of iron was very *similar* in different soils when they were investigated at the same times. This was the case even though the hydrothermic regime in these soils, as the author reports (49), was different: ". . . the water and temperature regime throughout the growing season *differed* in the different soils" [italics added].

Thus, a comparative analysis reveals an extraordinary synchronism and synphasic variation of very diverse manifestations of vital activity in the biosphere. This applies not only to biochemical reactions in plants and animals and physicochemical processes in soil, but also to several diseases of man. The period of manifestation of this feature may be a few hours to a year or more. This raises the logical question: How does one account for the synchronism of different reactions and processes on global and local scales? What are the reasons for it if the usual external factors (light, temperature, humidity, and pressure), as is apparent from the cited works, are not the main cause of the similar periodicity of the phenomena and processes?

We certainly cannot attribute these coincidences of the course of such different phenomena and processes purely to chance. In fact, the abundance of coincidences in the rhythm of the biological processes occurring in living organisms, and of physicochemical reactions investigated in laboratory conditions or in natural complex media, like soils, and also the synphasic nature of the time

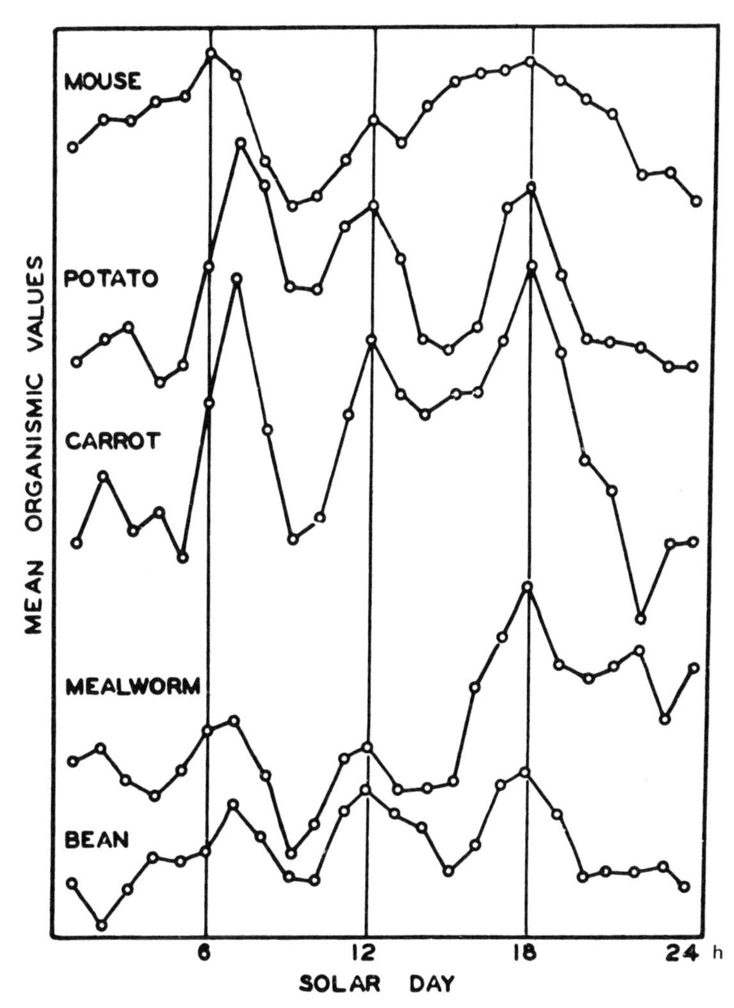

Fig. 11. Diurnal rhythm of respiratory activity in various organisms—mean values of activity in relative units (752).

periods suggests that there are *common factors* and laws operating. Confirmation of the nonaccidental nature of these phenomena is provided by the scale of their occurrence (from global, where they are reported at geographic points separated from one another by thousands of kilometers, to distances of a few centimeters—on the same plant). All these phenomena are affected by a sole factor, and there is a common reason for their periodicity, rhythmicity, and cylicity.

This raises the question: What factor (or group of factors) can be responsible for the synchronism of the reactions and processes on the scale of the whole planet, if not meterological indices? We can make the following inference: These features in the course of processes in different objects and different groups of effects can be explained if one accepts that the GMF has a direct effect on the biosphere. We will show that the question of the effect and role of gravitation in these processes will have to be considered separately.

A number of observations definitely indicate in favor of the view that the GMF is the active factor. This is indicated, first, by the universality of the discovered correlations, which include water molecules and entire organisms. Many researchers have also confirmed experimentally that both natural and weak artificial magnetic fields can act on water molecules (68, 253, 260, 426, 1049), and on animal organisms (162, 180, 243, 676, 678). Additional evidence of the close correlation between biological processes and the GMF is provided by the sudden changes in the cyclic variations of processes in living nature, synchronous with magnetic storms, and also the distinct annual course, similar to the course of geomagnetic activity, and the 27-day recurrence of natural phenomena, coinciding with the 27-day recurrence of magnetic disturbances (51, 1191).

One must beware, however, of an oversimplified view of the biological effect of the GMF. The fact is that the effect of this factor on living systems with a different level of organization is produced and mediated by complex ecological and climatometeorological chains and links, since this relation is based on the complex laws operating in nature. This is why the effect of the GMF in many cases is blurred or concealed by a series of intermediate links of a purely biological nature, and by the stochastic development of processes in biogeocenoses. Hence, in addition to synchronous and synphasic manifestations the course of some processes can be opposite, or have maximum or minimum values that do not coincide with the corresponding values of geomagnetic processes. In addition, according to recent data, a very important factor may be the functional dissymmetry of biological objects, which imparts so much diversity to the responses of living organisms to the GMF and other factors (125, 126, 135). Thus, the GMF acts against a background of complex development of various biogeocenotic processes and very specific relations between them. The steplike hierarchy in the system of energy and matter circulation and the special transformations of organic matter in the earth's biosphere also have their effect on the biological action of the GMF.

It should be noted that in addition to the indicated conditions and processes occurring in ecological systems there are also very important physical factors, such as the nature of ionization, the electric field of the atmosphere and earth (47, 345, 366, 458, 672, 906, 907, 930, 931, 932, 1023, 1027, 1084, 1178), and the electromagnetic field of biological objects themselves (189–194, 445, 446, 564, 731). One should also stress the role of such factors as temperature and humidity of the air and light, which further complicate the interrelations of processes in the biosphere and the GMF. Their influence and role can sometimes be decisive in the development of biological phenomena over a long period of time. For instance, during the winter in central latitudes or in the monsoon period in low latitudes it is almost impossible to detect the effect of the GMF on living nature in natural conditions. In such cases either special methods have to be used, or particular groups of living organisms that lend themselves to investigation at these periods have to be studied. For instance, in winter, when the trees are in a special state of anaerobiosis and their functional–dynamic characteristics are practically constant, soil microorganisms can show sharp changes in activity even at negative temperatures. In particular, special biometers (463) have shown a change in the dynamics of soil bacteria, fungi, algae, and the content of total, ammonia, and nitrate nitrogen in winter conditions due to solar activity (463).

However, despite the factual nature of all the data on the synchronism and synphasic variation of natural processes, this correspondence in time is merely indicative of the common nature of the acting causes or factors, and is not incontrovertible evidence of the influence of the GMF. The influence of the GMF on biological objects requires further evidence and direct confirmation. We discuss this material next.

Direct and Indirect Evidence of Effect of GMF on Biological Objects

Before discussing in detail the relation between biological processes and the GMF we shall give some factual data indicating that this effect really exists. This statement of the problem is quite valid, since the strength of the GMF does not exceed 0.5 Oe on the average, and the variations of the elements of the GMF on the horizontal vector even in the period of very strong magnetic storms reach 600 γ at the most, whereas the intensity of artificial electromagnetic fields present in the urban environment greatly exceeds the natural background (447).

Incidentally, in industrial and domestic conditions people can be exposed to fields that exceed the GMF by a factor of tens or even hundreds (646). We need only mention that a hairdryer's fan produces a magnetic field of up to 24 G (950). A television mast, for instance, with a transmission radius of 100 km and a carrier frequency of 200 MHz, emits into space a flux of 2×10^{-17} W $\cdot$ m^{-2} $\cdot$ Hz^{-1}. People living in the town or its vicinity, within the radius of the transmitting television center, are subjected for six hours a day to a strong flux of electromagnetic radiation, such as the sun produces rarely, and then only for a few minutes during solar flares (152). This, of course, raises the perplexing question: Why do we consider the biological effects of natural electromagnetic fields and regard their influence as decisive, if the effect of artificial electromagnetic fields is greater? It would appear that the radiation produced by industrial equipment and household appliances would be more effective than that produced by natural electromagnetic fields. In this connection it has even been suggested that the artificially produced electromagnetic background may affect the development of living organisms (400, 445–447, 1072).

Without underestimating the importance of artificial electromagnetic fields for living organisms we must point out that natural electromagnetic fields have a strong biological effect, and this can be attributed to the specific informational nature of their effect. This compels a careful investigation of the biological role of the GMF, and it is in this feature that we must seek the cause of the fundamental effect of natural magnetic fields (despite their small absolute strength) on the vital activity of living organisms and the development of the biosphere as a whole. Another feature of the GMF is the continuousness of its action and the ubiquitous nature of its effect over a long time (536).

The evidence that natural magnetic fields have an effect on living organisms can be classed as direct and indirect.

The direct evidence includes experiments whose results make it obvious that removal of the GMF greatly disturbs the functional state of organisms. These are primarily experiments involving the use of special shields that create a hypomagnetic environment, with a residual field of about 50–1000 γ, for the biological objects. This direct-evidence group includes experiments in which the GMF is compensated by various electromagnetic systems (Helmholtz coils, powerful permanent magnets, etc.).

The indirect evidence includes experiments that indicate a possible biological role of the GMF, but whose results do not allow definite conclusions, since the effect of other factors acting is not excluded. Such evidence includes observations of the effect on living objects of weak artificial magnetic fields commensurable in absolute intensity with natural fields. This group includes experiments with biological objects that orient themselves in space relative to the geomagnetic poles, and also investigations in space and under water.

Another form of indirect evidence—biomedical statistics—is of particular value at this stage of study of the biological role of the GMF. By simple comparison and calculation of the correlations between the course of some process and the simultaneous variation of the elements of the GMF and its activity in various frequency ranges one can infer the presence or absence of a correlation between them. This method, provided that events are accurately dated, can clearly reveal the effect of the GMF on the course of biological processes in the earth's biosphere (124, 127, 133, 136, 647, 648). As indirect evidence we can also include paleomagnetobiological research, i.e., the study of the changes in the fossil fauna in sedimentary rocks in relation to the variation of the GMF in the distant historical past (339, 664, 722, 725, 785, 786, 851, 878, 879, 884–886, 1102).

We shall consider in turn direct and indirect evidence indicating an important role of the GMF for living organisms on the earth.

Experiments Involving Shielding of Biological Objects from Effect of GMF

The principle of the method of shielding of any device from the effect of external electromagnetic fields has been known for a long time and has been used in radio engineering for the elimination of all kinds of interference (209, 798).

It should be noted that there have been a fair number of investigations on the protection of biological objects from the effect of electromagnetic fields. Experiments with solutions of chemical and biochemical substances, with shields of metal mesh, thin foil, copper and steel sheets, etc., are well known (408, 409, 539–541, 798, 1049, 1050, etc.). In view of the special problems confronting us we shall consider only investigations in which shielding of living organisms from the GMF has been studied, and where the shield employed was made of permalloy or mumetal, which reduce the GMF quite considerably. There have been few such investigations, which is probably due to some extent to the lack, until recently, of careful engineering physical calculations for the chambers required for work with biological objects. In addition, such experiments require considerable expenditure not only on the special shielding materials, but also on automated equipment, without which it is difficult to obtain accurate data. These difficulties have recently been overcome (40, 338, 527, 665, 705, 706, 731, 772, 1034), and experiments have been undertaken by several researchers (708, 775, 776, 1199–1209).

Investigations involving the shielding of biological objects are of interest in that they represent a direct attempt to completely eliminate, or at least greatly

Table 1. Some Properties of Materials with High Magnetic Permeability

Material	Composition (%), residual Fe, and impurities	Permeability: Initial	Permeability: Maximum	Saturation induction, $\times 10^{-4}$ T	Electrical resistance, $\times 10^{-8}$ μ m
Mild steel	0.2 C	120	2,000	21,200	10
Technical iron	0.2 (impurities)	150	5,000	21,500	10
Pure iron	0.5 (impurities)	10,000	200,000	21,500	10
68-Permalloy	68 Ni	1,200	250,000	13,000	20
78-Permalloy	78.5 Ni	8,000	100,000	10,800	16
Supermalloy	5 Mo, 79 Ni	100,000	1,000,000	7,900	60
Mumetal	5 Cu, 2 Cr, 77 Ni	20,000	100,000	6,500	62
Alperm	16 Al	3,000	55,000	8,000	140

reduce, the effect of the GMF. At present there are five methods of obtaining a space with a hypomagnetic environment (676, 776): (1) superposition of fields—alteration of the GMF vectors by means of a bar magnet; (2) astatization—reduction of the GMF to zero by an appropriate arrangement of magnets; (3) shielding by materials with very high magnetic permeability (mumetal, permalloy); (4) field compensation by means of Helmholtz coils; (5) combined shielding by means of mumetal and active electrical compensation.

The quality of shielding devices became much better when materials and alloys with high magnetic permeability (mumetal, permalloy, etc.) came into use. Since these materials are fully described in special textbooks (73), we give in Table 1 only a short list of materials that are frequently used in shielding experiments and information about some of their properties.

It should be noted that a shield of these alloys does not absorb the earth's magnetic lines of force, but merely concentrates them and diverts them from the shielded object in a direction of less resistance. Direct shielding entails the construction of a chamber with walls whose thickness is calculated to reduce the strength of the GMF. The effectiveness of the shield is proportional to its thickness, but a wall thickness of 1 mm is usually sufficient to produce a hypomagnetic environment with a residual strength of $50 \pm 20\ \gamma$. The chamber is constructed from sheets of one of the alloys listed in Table 1. An extensive review of experiments with biological objects in low magnetic fields has been published (776). An examination of the data in this review indicates that in most cases a short exposure to hypomagnetic conditions did not produce any changes in whole organisms or in cells cultivated in artificial media *in vitro*. In a number of cases, however, even a brief exposure to a hypomagnetic environment led to significant changes in biological objects. We discuss next the changes observed in biological objects in such cases.

Cells in Vitro. There have been investigations of biological processes in shielded spaces (SS) formed by thin-walled steel cylinders in which the GMF was reduced by a factor of 10^3 (181, 408, 409, 540, 541).

The experiments showed that a short stay of the biological objects in the shielded space led to appreciable changes in the ESR, enzymic reactions, and division processes. Tissue culture cells showed a reduction in the number of mitoses (by 35%) and cell area (by 12%) in comparison with an unshielded control (540).

A study of the blood of 923 tubercular patients showed that the ESR in SS conditions was reduced, depending on the blood group; the reduction was 14% for group I, 20.5% for II, 27.9% for III, and 40.7% for IV, in comparison with an unshielded control. The catalase activity of donor blood kept in a SS for 72 h was also reduced. It was also found that in a SS the ESR of patients with a mild form of the disease was reduced (to 79.2%) in most cases, whereas in the case of patients with severe tuberculosis the ESR in the SS was increased (in

69.3% cases) (181). It has been reported also that in a SS there is an appreciable reduction of the blood ESR of patients with bacterial infections (acute dysentery) (20 ± 1.33 mm/h as against a control rate of 29 ± 2.05 mm/h, $p < 0.001$), and a smaller reduction in viral infections (17 ± 1.9 mm/h as against a control rate of 24 ± 2.4 mm/h, $p < 0.002$) (541). On the other hand, experiments with cultures of mouse and Chinese hamster cells gave results that showed that mammalian cells outside the organism react in the same way to an increase and a great reduction (to $100\,\gamma$) of the GMF (548). It was reported in this paper that permalloy shielding and artificial magnetic fields of high strength (1–3 kOe) led to similar changes—an increase in the size of the cell nuclei and a sharp reduction of the mitotic index.

Culture of Isolated Cells. Experiments have been conducted with various cell strains. Diploid and polyploid cells of man, Chinese hamster, chick embryo, and others were investigated. The investigated cell cultures (HeLa, WJ-38, RB, etc.) were contained in three mumetal cylinders 1.25 mm thick, 60 cm long, and 20 cm in diameter. The cylinders, oriented with their long axis in the east–west direction, fitted one inside the other and were separated by an air gap of 2.5 mm (858).

The magnetic field in the shielded space was $50 \pm 20\,\gamma$. The hypomagnetic environment produced no changes in either cell morphology or in the number of cells in comparison with a control situated in the GMF.

Plants. The response of 24 species of plants to hypomagnetic conditions was experimentally investigated. A considerable reduction of the GMF either accelerated the growth of plants (cucumber, radish), or inhibited it (barley, corn). After two weeks in hypomagnetic conditions the seeds of many plants formed more roots and growth buds (825, 1118). Shielding of germinating conifer seeds prolonged the dormant period, reduced the germination of the seeds, oxygen uptake, and the dry matter content by 30% on the average (455) and retarded the growth of mustard (826).

In a reduced GMF (steel shields, $\mathbf{H}_T \approx 0.5 \times 10^{-3}$ Oe) the radicles of two- to three-day-old barley, vetch, pea, and millet seedlings grown in the dark were significantly shorter than in the control, and the seedlings showed various disturbances of growth (305). Respiration of shielded barley seedlings, on the other hand, was 70–100% higher than the control level (305), and the activity of some enzymes after 72 h was twice as high (catalase) or half as high (polyphenol oxidase, peroxidase) as in unshielded control seedlings (304). A disturbance of the diurnal rhythm of optical activity of the leaves of kidney bean, cucumber, barley, and geranium plants shielded from the GMF has also been reported (500). Onion seeds enclosed for a long time in a permalloy shield ($\mathbf{H}_T$ was reduced by a factor of 10^6) showed enhanced germination, and the number of dividing cells in their radicles was higher: The mitotic index was 3.99%, as against 2.90% in the control ($p \leqslant 0.05$), and the duration of the mitotic cycle was altered (512). It is of interest to note that right- and left-handed forms of seeds differed from one another in their division rhythm and after shielding these differences were more pronounced, whereas the level and spectrum of aberrations in the experiment and control and between the right- and left-handed forms showed no significant differences.

Subjection of plants to an environment with a residual geomagnetic field of $100 \pm 50\ \gamma$ for a short time did not alter their functional–biochemical indices (1181), and even an eight-day period in hypomagnetic conditions ($30\ \gamma$) did not lead to any disturbance of their development (994).

Plants exposed for a long time to a hypomagnetic environment showed histological disturbances (95): Cell differentiation in the primary core of the central cylinder of the root, formation of the annular vessels of the xylem, and initiation of the side radicles in the pericycle were inhibited; the primary core became thicker and was covered with peculiar tumors.

It should be noted that no significant morphological changes were found in the multiplication of lower fungi *Aspergillus* and *Penicillium* by conidia (vegetative form) after two years in an ultraweak magnetic field (to 10^{-4} Oe) (95).

Microorganisms. Distinct changes have been discovered in shielded microorganisms (21, 414, 415, 530). Cultivation of staphylococci (*S. aureus*) in a hypomagnetic environment (0.05 Oe) led to a fifteenfold reduction of the colony count at all dilutions and to reduction of the size of the colonies in comparison with a control in ordinary GMF conditions (696).

Azotobacter cells after three to four months in near-zero fields (10^{-4} Oe or less) showed changes in shape and size. They were six to eight times larger than cells grown in ordinary geomagnetic conditions (95). It is of interest that in hypomagnetic conditions uncharacteristic "filamentous" and "streptococcoid" forms appeared among the *Azotobacter* cells.

An effect of permalloy shielding of the GMF on various bacterial cultures has also been reported (15, 21, 85, 414, 415, 530).

Multiplication of *E. coli* (strain SA-23) in a permalloy shield was more rapid than in a control outside the shield ($\beta > 0.99$), and strains K-12 and G became incapable of fermenting maltose. *Staphylococcus, Salmonella, Shigella sonnei, Escherichia,* and *Klebsiella* developed antibiotic resistance much faster (after three to five subcultures) than an unshielded control (21).

In experiments of other investigators, permalloy shielding (for 46 days) of *Escherichia* altered their fermenting ability (but slightly), mainly in regard to sugars, and also their cultural characters—10% of the colonies were R form (85). It was reported that the shape, structure, and size of the paracolon bacillus were unaffected by cultivation in shielded conditions.

Permalloy shielding affected the conjugation of *E. coli* (412). The frequency of R transmission in *E. coli* up to subculture 53 varied and did not differ greatly from the control, but in subcultures 56–60 the frequency of R transmission became three to four times greater, and then remained constant (equal to the control) for 18 subcultures, but after subculture 80 and later it was two to ten times greater. This investigation and others (21) indicate that shielding of the GMF affects the biological properties of recipient cells. A study of the variation of the biological properties of the diphtheria bacillus (a toxigenic strain—gravis type, and nontoxigenic strain—mitis type) cultivated in a permalloy shield ($\mathbf{H}_T < 10^{-5}$ G) showed changes in the cultural and biochemical characters of both strains (530). Beginning at the fifth subculture the nutrient bouillon became turbid, a coarse gray film untypical of the strain was formed, and matte black slimy gelatinous colonies reminiscent of *Klebsiella* appeared. From subculture 25 the experimental strains of diphtheria bacilli lost the ability to ferment maltose, although they did not differ from the control in their ability to split other carbohydrates, or in urease and cystinase activity (530).

At the same time, a standard culture of *S. aureus* 209P, subcultured daily for 40 days in permalloy shielding (10^{-5} Oe), showed changes in several specific characters after subculture 18 (415). In the experimental *Staphylococcus* culture the number of cells that had lost their hemolytic properties increased steadily, the typical pigmentation of the microbes was altered, fibrinolytic and coagulant ability was lost, and resistance to penicillin was enhanced (415), but catalase activity was unaffected (414). Permalloy shielding led to considerable changes in the enzymatic activity of the bacteria, e.g., catalase activity steadily decreased in *Serratia marcescens* and *Alcaligenes faecalis*, while in other bacteria it decreased in the first to sixth subcultures (paracolon bacillus, *Salmonella typhimurium, Bacillus subtilis*), or increased (*Sarcina, Klebsiella*) and varied up to the sixth subculture. Thus, it is apparent that bacterial cultures react in the most diverse manner when they are shielded by permalloy from the GMF for a long time. They exhibit phenotypic and genotypic changes, accompanied by considerable changes in metabolism and physiology.

Insects. Termites are capable of orienting themselves in their nests in an east–west direction, and this ability is retained in laboratory conditions (685). However, if the insects are put in an iron box, where the geomagnetic field is greatly reduced, the compass orientation of their body is not manifested.

The effect of shielding from the GMF on the development, fecundity, and life span of *Drosophila* has also been investigated (31, 573). A reduction of the GMF by a factor of 10 led to a significant increase (31) in the mortality of *D. virilis* in a shielded chamber in comparison with a dark control and to a significant reduction in life-span in comparison with a control under natural light ($p >$ 0.999). Observations of 12 generations of flies (*D. melanogaster*) kept in a shielded space ($\mathbf{H}_T \approx 0.5 \times 10^{-3}$ Oe) showed that the progeny per parental pair in the first seven generations was significantly lower ($p < 0.05$) than in the control, and there was a subsequent increase in the number of individuals per parental pair and some increase in their rate of development (573).

Higher Animals. Shielding of animals for long periods leads to irreversible changes in their body (760, 874). The experiments were performed at the request of the National Aeronautics and Space Administration (NASA). The test animals (white mice of the Swiss-Webster strain) were put into mumetal cylinders; the windows and ends of the cylinders were covered with a close-meshed wire grid made of nonmagnetic stainless steel. The control animals were put into similar aluminum cylinders.

The adult population of animals in each cylinder consisted of eight mice. Each cylinder initially contained one family (group I—one male and three females). The time spent by the parental animals in the shielded conditions varied from 4 to 12 months.

The first generation (F_1) of mice was then divided into two halves on the day that they were removed from the mother (day 21). One-half of the litter was put back into the mumetal cylinders (group II), and the other half (group III) was put into aluminum cylinders as a control.

Females of the parental group (group I) were mated continuously with males of the same group. Initially, despite the premature mating and frequent pregnancies, the females bore fairly large and quite normal progeny. By the fourth generation (F_4) reproduction ceased. But even earlier, in the second generation (F_2), there were more miscarriages and cannibalism than in F_3 and F_4. A large number of the mice kept in shielded conditions became inactive and weak at an early age. Their behavior was unusual—they spent a long time lying on their backs. Approximately 14% of the adult mouse population showed progressive loss of hair, which began at the head and moved down to the middle of the back. Many of the animals died by the sixth month.

The organs and skin of the parental mice (group I) were subjected to careful histological analysis. Epithelial connective-tissue tumors were found in different parts of the body. The skin in the depilated mice was greatly altered, there was proliferation, hyperplasia of the horny layer, plugging of the hair follicles, and hyperplastic squamous epithelium. The liver tissue of all the shielded mice showed a pronounced change in the nuclei, which had greatly enlarged nucleoli. Hemosiderin was found in the Kupffer cells. The kidneys of animals kept in the shielded space were also changed. They showed multilocularity, the presence of cysts, which compressed the cortical parenchyma, and other abnormalities. In the mice, especially those that had died suddenly, the bladder, filled with urine and an unidentified white deposit, had a hyperplastic mucous membrane with partitions and polyps.

Thus, experiments with animals clearly show profound changes in their body as a result of total shielding from the GMF.

Man. Experiments with people kept in a shielded room in which the residual magnetism was less than 50 γ (708) were of short duration. Two of the subjects were kept for five days in a completely shielded room and for three days

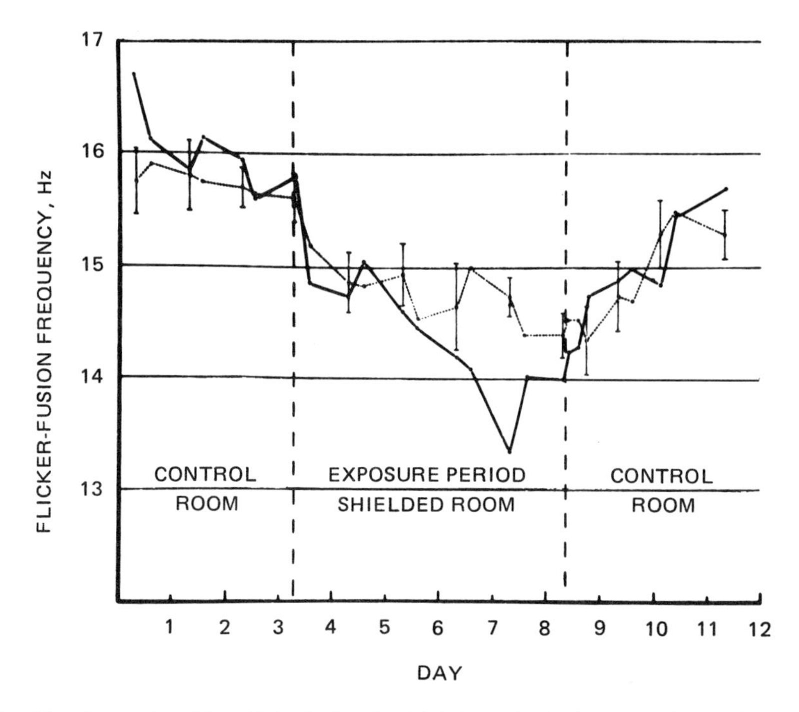

Fig. 12. Change in critical flicker-fusion threshold (cycles/sec) of two people kept in a room shielded from the GMF (760).

before and after the experiment in an unshielded room. During their stay in the shielded room their critical flicker-fusion threshold was altered, but after their transfer to an unshielded room with ordinary geomagnetic conditions it became normal again (Fig. 12). In people kept in a shielded underground bunker, where the GMF was reduced by a factor of 100, the period of circadian rhythms* was increased (1199, 1203–1208). On the average it was 25.65 ± 1.024 h, whereas in the unshielded space it was 25.00 ± 0.55 h; this difference is statistically significant ($p < 0.01$). Shielding of the subjects also led to desynchronization of the rhythm, i.e., a disturbance of its regular sequence.

Thus, although the experiments of different investigators are contradictory, one can conclude that living organisms are not subject to any visible physiological changes in a hypomagnetic environment. In such conditions man, however, shows a rapid change in the response of the central nervous system, which can be

*Organisms are characterized by periodically recurring variations of physiological processes—biological rhythms with cycles of varying length. When an organism is subjected to constant (aperiodic) conditions the period of the "circadian" rhythms differs from 24 h (20–28 h). In normal conditions 24-h rhythms predominate [B.R.].

detected if the critical flicker-fusion threshold is used as a test, and a change in the rhythm of some functional processes.

Prolonged total shielding of biological objects (animals) greatly alters their physiological and biochemical properties. There is atypical growth of cells and tissues, changes in morphology and functioning of internal organs, and premature death. Microorganisms in hypomagnetic conditions produce mutant cell forms.

The above-mentioned investigations give clear evidence that the GMF significantly affects the state of living organisms and hence is an important environmental factor (140, 141, 243, 400, 738, 739, 748, 788, 865, 932, 1027). Some indirect confirmation of this is provided by studies of submariners and cosmonauts.

Observations on People under Water and in Space

Of particular interest are the data indicating a change in the functional indices of submariners and cosmonauts, who, owing to the specific nature of their work, have to spend long periods in shielded locations and in hypomagnetic conditions. We consider first the information on the state of the physiological functions of members of submarine crews.

Submarine Conditions. As is known, the hull of a modern submarine consists of a layer of especially strong steel. The long isolation from the external environment and the autonomy of the submarine are almost equivalent to complete shielding. These circumstances provide an opportunity of assessing the role of the GMF in the vital activity of the crew. Despite the life support systems, which eliminate as far as possible any deviations from the norm and allow the crew to stay for a long time under water, considerable disturbances of the functional indices of personnel have been found. For instance, the basal metabolism is lowered (532), the total leukocyte count in the peripheral blood is reduced and digestive and myogenic leukocytosis is suppressed (310), the diurnal periodicity of various functions is disturbed, and premorbid states (490) and stomach diseases (383) occur. Although these authors merely established the occurrence of functional disturbances and did not attribute them to shielding, one can postulate that these disturbances are due to the hypomagnetic environment and the change in the natural background of electromagnetic frequencies.*

Space Conditions. Experiments in space are of great importance for revealing the role of the GMF. The most important feature of these experiments is that living objects are deprived simultaneously of two important factors of "earth" existence—gravitation and the GMF (702).

*One can hardly postulate that these changes are due solely to geomagnetism. Hypodynamia, uncongenial meterological conditions, absence of natural light, etc. must have some effect [B.R.].

It has been suggested that the GMF should also affect aviators in high speed flight, especially in an equatorial direction (367), and in space flights (992).

The response to weightlessness occurs immediately after escape from the earth, whereas the hypomagnetism of the environment begins to operate only at a great distance, approximately ten times the earth's radius, i.e., 60,000 km. In this case the cosmonaut's body is subjected to the effect of the magnetic field of space, the electromagnetic field of the spacecraft itself, and the solar wind, which carries magnetic fields "frozen into" the plasma.

The bodies of cosmonauts in flight show deviations in comparison with terrestrial conditions. Changes in metabolic reactions, particularly in calcium metabolism, have been reported (59, 856, 945). Other effects are reduction of the erythrocyte count (1108), changes in circadian rhythms (728, 872, 1103), and disturbance of sleep (1012).

A comparison of the reactions of human beings in submarine and deep-space conditions indicates a definite similarity between these reactions. Hence, one can postulate that the great reduction of the GMF is the main factor responsible for the similar changes in people in such very different environmental conditions.

Our analysis suggests that to the well-known space-flight factors—weightlessness, hypodynamia, restriction of living space, vibrations, psychological stress on the human and animal organism—we must add the effect of the continuously varying GMF, particularly in long flights (510).

The geomagnetic variations on artificial earth satellites (AES) are due both to the variable and constant (main) GMF. Variations of the first type are of a temporary nature and are due to changes in the current systems in the ionosphere, within the magnetosphere and at its boundary during magnetic storms. Such variations are manifested at the earth's surface and throughout circumterrestrial space, and their intensity depends on the distance from the sources.

Variations of the second type occur all the time, but only on board the AES, and are due to the motion of the satellite in the GMF. These variations are due mainly to the nonuniform latitudinal distribution of the main geomagnetic field and to the diminution of the field with increase in height.

Recently published papers indicate that variations of the variable geomagnetic field are of biological significance in space flights (131, 702, 760). So far, however, due significance has not been ascribed to variations of the main geomagnetic field resulting from the motion of the AES in its orbit, although these changes in the field are approximately two orders of magnitude greater than the natural geomagnetic variations. Calculations of the main laws of variation of the GMF on board the satellite due to motion in the main GMF have been made, and the effect of the orbital parameters on such field variations has been indicated (510). For a circular orbit with period $T_0 = 90$ min (altitude $\sim$ 280 km) the scalar magnitude of the GMF intensity along the AES orbit varies sinusoidally with period $T_0/2$ and variable amplitude, and the amplitude increases with increase in

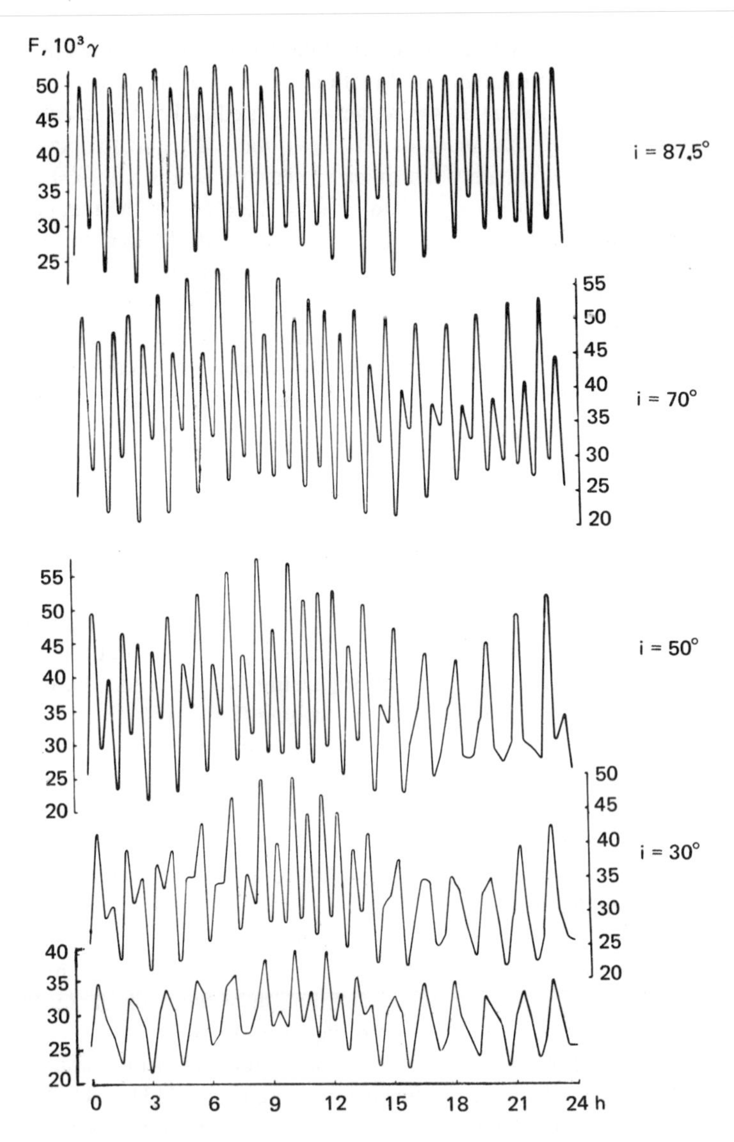

Fig. 13. Variation of geomagnetic field modulus F on artificial earth satellite in circular orbit at height $h = 300$ km for several angles (i) of inclination of orbit to plane of equator. From (510).

the angle of inclination i of the satellite orbit from 20×10^3 to $85 \times 10^3\gamma$ (Fig. 13). The lower value of each curve is limited by the value of the field at the geomagnetic equator for the given flight altitude, while the upper envelopes of the extreme values are limited by the values of the field along the parallel $\varphi = i$ at this altitude.

Owing to the angle of inclination between the geographic and geomagnetic axes, equal to 11.5°, and the change in gradient of the GMF with latitude, the upper envelopes also vary sinusoidally with an amplitude that depends on the angle of inclination of the orbit.

With increase in altitude of flight of the satellite above the earth, the period T_0 increases and the variation amplitude decreases. In this case the above-mentioned variation of the field for a circular orbit still occurs, but the field variations themselves will be rarer and have a smaller amplitude, whereas in the case of an elliptic orbit the field variations become quasi-sinusoidal. For small AES altitudes, however, the field variations differ considerably from those calculated with due regard to the higher harmonics of the Gaussian series, owing to the presence of pronounced global anomalies of the GMF.

Calculation of the local energetic parameters of the GMF (energy density W and power flux density S) for the geomagnetic variations occurring on board a satellite gives the following vales: $W = 9 \times 10^{-3}$ erg/cm^3, and $S = 7 \times 10^3$ erg/sec cm^2. If they are compared with the corresponding parameters for different classes of magnetic storms (359), where the minimal values are $W = 3 \times 10^{-6}$ erg/cm^3, and $S = 10^{-4}$ erg/sec/cm^2, it is apparent that the two local energetic parameters of the GMF on the AES are seven to nine orders greater, to say nothing of the threshold of sensitivity of the human organism ($W^* = 10^{-12}$ erg/cm^3; $S^* = 10^{-9}$ erg/sec/cm^2). Hence, interaction of the human organism with the varying field on the AES orbit is quite possible, and these features of the interaction must be taken into account in simulation of the geoelectromagnetic complex in a long flight by a man on an AES (136).

It is much more difficult, however, to distinguish a specific effect of a hypomagnetic environment in space, since many important environmental indices (gravitation, ionic composition of air, natural electromagnetic fields in different frequency ranges) are different from those on earth.

Hence, despite the clear evidence of a possible effect of hypomagnetic conditions in submarines and spacecraft, we cannot attribute certain identical changes in functional activity of people entirely to disturbance of geomagnetic conditions, but these changes can be regarded as indirect evidence of a possible biological role of the GMF.

Experiments Involving Compensation of GMF

In addition to experiments in which biological objects have been directly and completely shielded from the GMF, there have been experiments involving compensation of the GMF. Compensation of the GMF is usually effected by a system of coils (usually two or three pairs) lying in mutually perpendicular planes. Through these coils, on which calibrated copper wire is wound, a direct current is passed. The current is calculated so that the magnetic field produced by induction compensates the GMF. Helmholtz coils in various modifications are the devices most commonly used.

Compensation experiments differ radically from experiments involving total shielding, although the aim of both methods is to create a hypomagnetic environment. The difference lies in the fact that in compensation experiments the natural electromagnetic complex in its entire frequency range is unaltered, whereas complete shielding eliminates this complex. Hence, in compensation experiments short-period variations of the GMF, atmospheric electricity, and other kinds of natural electromagnetic fields can act on the organism. At the same time, the physical parameters of the environment (e.g., ionic composition of the air), and the usual circulation of the air are hardly altered at all. In addition, in compensation experiments the usual environmental factors affecting animals, such as light and humidity, are quite unaffected, and the mobility of the animal, the presence of the usual diverse stimuli, etc., are not greatly affected.

Thus, although complete shielding and compensation remove the effect of the GMF, the experimental conditions in the two cases are not identical and each method has its advantages and disadvantages. It should be noted that complete shielding is preferable, since in experiments with Helmholtz coils only the constant GMF is compensated, and its long-period components, which contribute to the diurnal variations S_q are not affected. In addition, if Helmholtz coils do not have an automatic servo system and are calculated for compensation of only one particular average level of the main field in the particular locality, then in long experiments and particularly on highly disturbed days the object may be affected by the aperiodic storm-time variation D_{st}. Hence, in experiments involving compensation of the GMF it is essential always to employ a special electronic system that follows very precisely the variation of the level of the GMF and its constituent elements (708, 994).

The investigations showed that living organisms react distinctly to the change in the constant GMF due to its compensation.

Plants. Barley seedlings kept for a short time in Barenbek coils, where the GMF was compensated to $\pm 90\,\gamma$, showed disturbances in the diurnal rhythm of excretion of organic substances by the roots (119, 121) in comparison with control plants.

Birds. It should be noted, first, that there have been no investigations in which birds have been subjected to complete isolation from the GMF or to a compensated GMF.

There have been investigations, however, in which the GMF was completely compensated by a pair of Helmholtz coils, while an artificial magnetic field with different parameters was created by another pair of coils. The results of experiments with birds subjected to such conditions are controversial, despite the refined technique of the experiments, which have been repeated many times and have been subjected to careful statistical analysis (829, 916, 984, 985, 986, 1215, 1216, 1219, 1220).

Merkel *et al.* (985, 986) investigated the selection of migration direction by birds. Experiments with robins (*Erithacus rubecula*) were conducted in an apparatus consisting of two pairs of Helmholtz coils: One pair, set in a north–south direction, compensated the GMF (0.41 G), while a second pair, perpendicular to the first, produced a magnetic field with an artificial "north" in the W (0.47 G) or ESE (0.38 G) direction. The control birds were kept in a timber house 30 m from the Helmholtz coils.

In the control experiments the birds mainly selected the NE direction, corresponding at the investigated time (March and May) to the spring direction of migration. When the GMF was compensated and the artificial magnetic field was created, however, the activity vector of the birds was directed toward the artificial "north" (see Fig. 68).

Investigations of the role of magnetic fields in the elaboration of a conditioned reflex in buntings (*Passerina cyanea*) gave negative results (829). The cage containing the birds was placed in a system of modified Helmholtz coils—two parallel coils separated by a distance of 1 m.

The bird was contained in a cage with a wire floor to which a stimulating electric pulse of 0.1-sec duration was applied, causing the bird to jump. This was accompanied by alteration of the direction of the artificial magnetic field for 1 sec or the sounding of a buzzer.

Thus, a conditioned reflex was elaborated in response to a change in magnetic field or to a sound, and the response of the bird (jumping) was automatically recorded.

In the first series of experiments the coils were oriented perpendicular to the axis of the horizontal component of the GMF, while the created artificial magnet field had a direction opposite to that of the normal field.

The resultant horizontal component of the magnetic field was equal in intensity, but altered completely (through 180°) or partially (90°) in direction relative to the natural field. The experiments showed that the birds learned to avoid the electric shock accompanied by the auditory stimulus, but did not respond to the short-term alteration of direction of the magnetic field (829).

Fishes. Experiments with young eels in a special maze showed that compensation of the GMF by Helmholtz coils led to equal selection of directions of movement in the maze, i.e., to the disappearance of GMF taxis (171, 173, 174). By altering the strength of the horizontal component of the magnetic field from 0 to $5 \times 10^4 \gamma$ and completely compensating the GMF in the maze, Khodorkovskii and Gleizer (233, 235) found that the mean directions selected by the fish were altered in comparison with the distribution in the natural field.

Mammals. White mice and rats subjected to a compensated magnetic field

(Helmholtz coils) showed changes in the leukocyte system of the peripheral blood. In the animals the total number of leukocytes was increased (especially neutrophils, $p < 0.01$), the total number of lymphocytes was reduced (409), the activity of phosphatase and transaminase in macrophages was reduced, and so on (776).

Insects. In a compensated GMF, insects selected a different orientation in space (687, 688, and others). In the case of bees, compensation had no effect on the nature of the dance, which indicates the direction of food, the site of landing at the food, and so on (840, 1086). Carefully conducted experiments, however, revealed that in a compensated magnetic field indications of the direction of food by the dance of the foraging bees were more accurate (951).

Since the dances are performed by the bees in a vertical plane, an effect of the GMF on orientation in the gravitational field and on the mechanism of gravireception becomes highly probable (951, 1196).

Man. A ten-day experiment on people showed that compensation of the GMF, like complete shielding, led to distinct changes in the central nervous system (701, 703, 707). This experiment was carried out in the following way. A system of large electromagnets was formed by three modified Helmholtz coils with an edge of 8.5 m perpendicular to one another. The whole system was connected to an electric clock and magnetometer for determination of the GMF. At the center of the space the strength of the GMF in compensated conditions was almost zero, and at a distance of 2.5 m from the center it was not more than 100 γ. In the space occupied by the subjects the strength was 50 γ with a gradient of 30 γ/m. The subjects were six men 17–19 years old. Four occupied the chamber with the compensated GMF, and the other two provided a control—they were in a chamber of similar design in which the GMF was not compensated.

Before the start of the experiment the subjects were kept for five days in a chamber in ordinary conditions and were then transferred to the test chamber, where the GMF was compensated for ten days. When compensation of the GMF was stopped, the subjects were kept in the same chamber in normal GMF conditions for another five days. The assessment of the effect of changes in the GMF was based on various physiological tests—weight, body temperature, respiration rate, arterial pressure, blood analyses, changes in the electrocardiogram and electroencephalogram, psychophysiological tests, and several other indices (about 30 altogether). During the experiment, i.e., during the ten days in which the GMF was compensated, none of the test results or physiological indices for the persons subjected to hypomagnetic conditions differed from those for persons in geomagnetic conditions. However, as in total-shielding experiments, the critical flicker-fusion threshold, an important functional characteristic related to the response of the central nervous system, was greatly reduced (Fig. 14).

Thus, the data presented show that compensation of the GMF alters some

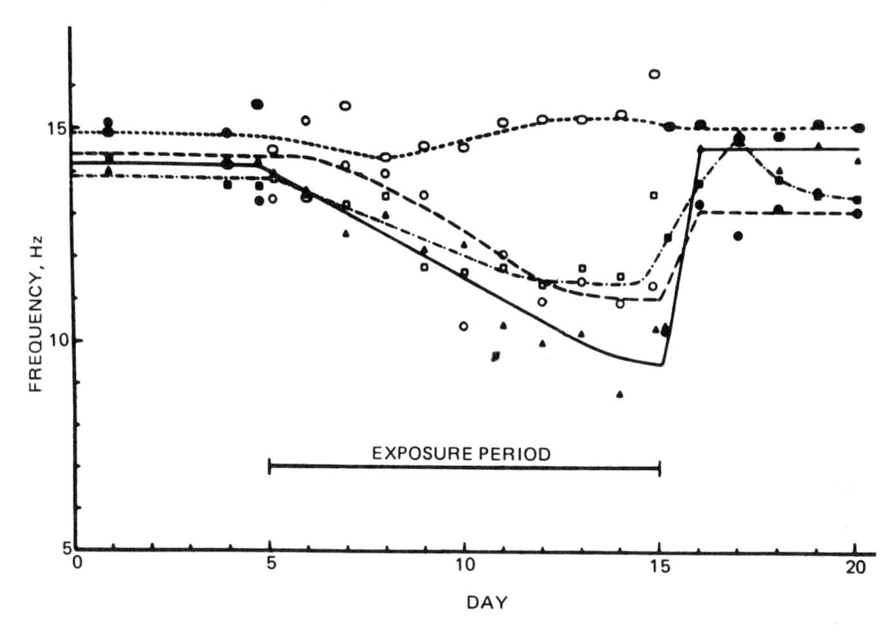

Fig. 14. Change in critical flicker-fusion threshold (Hz) of four people in a compensated GMF (760).

functional characteristics of various organisms and indicates an important role of the GMF.

Biological Effect of Artificial Weak Magnetic Fields

An examination of the material presented suggests that artificial electromagnetic and magnetic fields similar to natural fields in amplitude and frequency range should also have an effect on biological objects. Experiments have shown that weak artificial electromagnetic and magnetic fields do have such an effect (65, 66, 84, 96, 222, 258, 259, 340, 363, 364, 393, 405, 424, 549, 560, 627, 640, 641, 654, 711, 764, 775, 846, 860, 930, 932, 933, 941, 944, 956–960, 967, 981, 1027, 1028, 1038, 1188, 1189, 1202–1208, 1219, 1222, 1228). It should be noted that some investigations failed to reveal any biological effect due to weak electromagnetic fields (827, 873, 900, 901, 961, 1095, 1144, 1228). Here we apparently encounter the problem of irreproducibility of the experiments, which is also due to the natural electromagnetic environment (20, 132, 260, 261, 767, 1049). It is obvious that the effect of weak artificial magnetic fields will be manifested in particular conditions of the electromagnetic environment, and the level of magnetic disturbance will be of great importance.

Protozoa, Annelids, Microorganisms. Brown, his colleagues, and his followers have fully confirmed the biological effect of weak artificial magnetic fields (735–739, 742–746, 750, 752, 754, 855, 1030). They showed that any organism can distinguish the intensity of a magnetic field and can sense the direction of the magnetic lines of force passing through its body.

In particular, it was discovered that an artificial magnetic field of 1.5 Oe affects the choice of direction of motion of free-moving paramecia, snails, and flatworms (736, 740). Magnetic fields with a strength close to that of the GMF were always most effective (747).

An artificial magnetic field of 0.05 Oe could alter the periodicity of the geographic orientation of flatworms. A 180° change in the horizontal vector of the magnetic field caused a responsive 180° change in phase of the monthly rhythm of geographic (northern) orientation of flatworms (Fig. 15).

Volvox (V. aureus, V. cenobia) also showed a definite ability to distinguish the direction of the geomagnetic lines of force and sense a change in total field strength (1030, 1031).

It has also been shown that cultivation of bacteria of the *Salmonella, Staphylococcus,* and other types in an alternating magnetic field simulating the range of short-period variations of geomagnetic field of the *pc*1 type ($f = 0.6$ Hz; $H = 1\ \gamma$) leads to an appreciable reduction of their multiplication rate (17, 20, 628). At the same time, in electromagnetic fields of frequency 0.1, 0.5, and 1 Hz and strength 0.3–0.4 V/m the multiplication rate of the bacteria and the number of colonies were increased (623).

The effect of a variable magnetic field simulating geomagnetic pulsations of the *pc*2 type ($f = 0.1$ Hz, $H = 0.5\ \gamma$) on various strains of *Escherichia* and anthrax bacilli was greater than that of the *pc*1 type (18). The response of bacteria to prolonged exposure (30 days) to an ultraweak magnetic field was different: In the case of enterobacteria growth and multiplication were suppressed and the rate of gas production was reduced ($\beta > 0.99$), while in anthrax bacilli reproduction was stimulated ($\beta > 0.95$). The experiments of these authors on the effect of weak electromagnetic fields on microorganisms showed changes in metabolism, antibiotic resistance, and morphological and cultural characters (15, 16, 18, 21, 22).

Plants. Laboratory and field experiments have confirmed that a magnetic field has a biological effect on plants. A solenoid constructed from five rings of thick copper wire and powered by a 1.3-V mercury storage battery produced a magnetic field of 280 γ.

When dandelions were placed in such a solenoid the flowers showed a delay in opening and closing, and after a long exposure they wilted and died (1021).

It was shown in laboratory conditions that weak magnetic fields (0.05–3 Oe)

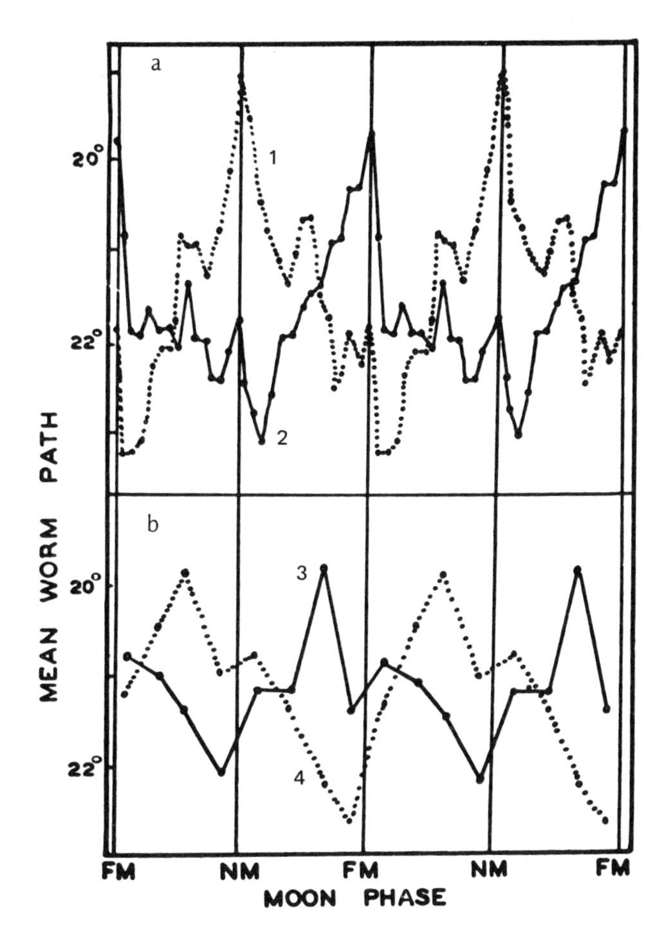

Fig. 15. Change in direction (degrees) of movement of flatworms, initially oriented northward (0°), in natural conditions (a) and in the presence of an artificial magnetic field of 0.05 Oe (b) in relation to lunar phases (752). (1,4) Exit corridor of apparatus oriented northward; (2) alteration of initial orientation of corridor through 180°; (3) superposition of artificial magnetic field of 0.05 Oe with vector opposed to the GMF. FM, full moon; NM, new moon.

affect growth and morphogenetic processes in plants (522–526, 571, 572, 933, 1006, 1168). When a magnetic field of 0.05 Oe acted on steeped seeds for two days the seedlings developed more rapidly, the differentiation of the stem metamers in the apical meristem of the growth point began earlier, and the formation of the lateral and adventitious roots by the plants was stimulated.

Birds. Experiments with artificial magnetic fields (0.14–3.46 Oe) showed that birds were highly sensitive to them and that their migrational orientative ability depended on the total strength of the magnetic field (915–917, 1215–

1219, 1221), and also on its polarity (258, 259, 321, 322). The birds could select their normal migration direction only in the GMF, and in artificially increased (0.73–0.95 G) or reduced (0.14, 0.30 G) fields their movements were random and showed no distinct orientation (1215).

Fishes. A high sensitivity of fish to low-strength magnetic fields was discovered in experiments in a specially designed maze, where the frequency of selection of three or six compass directions by the fish was determined (231). In laboratory conditions (constant temperature and light intensity) young European eels showed the ability to perceive a constant magnetic field of only a few tens of gammas and to distinguish the polarity of the magnet used (234). The fish could move in a directed manner along the lines of force, in the direction of increasing field, even when the gradient between the points of the path was only 52 γ (235), and showed a change in motor activity when the field inductance was reduced from 0.3 T to zero (594). Much earlier (952, 953), experiments on weakly electric fish showed their ability to respond with a characteristic motor reaction to a magnetic field of 0.01 Oe produced by a moving permanent magnet (the test fish was the Nile fish), to alter the frequency of the electric pulses (*Gnathonemus*), and to orient themselves in a magnetic field (*Gymnotus*). The ability of some species to react specifically to weak magnetic fields and orient themselves in them led to a critical analysis of the effect of the GMF on orientation and navigation, and their possible mechanisms (449, 450).

It has been suggested that the electroreceptive system in a fish's body might be responsible for orientation in the GMF (67, 712, 961, 1007). Experiments with rays showed that the electrosensory apparatus of the ampullae of Lorenzini exhibits an exceptionally high sensitivity not only to weak electric fields and currents (66, 450, 800), but also to weak magnetic fields (67). Behavioral experiments carried out on eels show that the absolute threshold of sensitivity of young eels to the gradient of a nonuniform magnetic field was less than 300 γ /cm (233). A high sensitivity to weak electric and magnetic fields was found in eels when reduction of the heart rate was used as a test (961). These experiments showed that the tested individuals could be divided into three groups: in some the changes in the test indices were greater, and in others less, than in the control. The third group showed no response to a weak electric and magnetic field. This work provides more good evidence in favor of our hypothesis of the existence of various types of functional dissymmetry in animals (135).

Mammals. The most comprehensive experiments have been carried out with magnetic and electromagnetic fields in the same frequency range as the short-period fluctuations of the GMF (365, 626, 628, 643). The experimental animals were placed between the plates of a $1 \times 1 \times 1$ m capacitor (642) on which a sinusoidal voltage of 0.5–1.0 V with frequencies of 2 and 8 Hz was applied. A single exposure lasted 3 h and was repeated up to ten times at daily intervals. A single exposure of rabbits led to reduction of the heart rate, while five to ten exposures even led to ventricular extrasystolia. In addition, the number of formed elements of the blood was increased. There were increases in the number of leukocytes and segmented neutrophils, changes in the hemoglobin content, etc.

The peripheral blood neutrophils of rabbits subjected to a magnetic field of low frequency and very low strength ($f = 8$ Hz; $H = 0.02$-$2\ \gamma$) showed a reduction in enzyme activity (alkaline phosphatase, NAD oxidase) and glycogen content (557–559). The enzyme activity was very greatly reduced after exposure to a magnetic field of 2γ (by 72–78%, $p < 0.01$). These data are of interest, since a magnetic field of this strength is observed during very weak magnetic storms with a sudden or gradual commencement. Unfortunately, this experiment has not been repeated by other investigators, although it is of fundamental importance for an understanding of the biological effect of GMF pulsations. At the same time, subjection to a magnetic field of frequency $f = 0.2$ Hz and strength $H = 5\gamma$ for 3 h affected the nonspecific immunity of white mice, increasing the peroxidase activity of blood neutrophils and enhancing the ingestive power of the leukocytes (562). Male mice subjected for 56 h to horizontal or vertical magnetic fields varying slowly in time (320 sec) from 0 to 3 Oe with a sharp cutoff in hundredths of a second showed a reduction in mitotic activity of the bone-marrow cells and corneal epithelium cells (96, 98). The investigated magnetic fields had a distinct stimulating effect on the hemopoietic system: The number of cells of the leukocyte series in the peripheral blood of the animals was increased by 10–68%, depending on the season of the year, and the functional state of the liver was altered—the number of binucleate hepatocytes was increased by 20–25%, the glycogen content was reduced, and so on.

In one review of the biological effect of weak electromagnetic fields (627) the authors concluded that the fact of the ecological significance of natural electromagnetic fields can be regarded as reliably established. They express the view that weak electromagnetic fields act as a nonspecific stimulus. There is no doubt that the most sensitive system of the organism is the nervous system (243). At the same time, weak electromagnetic fields exhibit pronounced biological activity, affecting the cardiovascular system (46, 75, 614), the blood system and hemopoietic organs (65), the immune system (23, 499), animal behavior (1036, 1040, 1046), the endocrine glands (1029, 1044), physiological function (1039, 1042), and sexual differentiation (1028).

Subjection of rabbits to a magnetic field of $10{,}000\gamma$ for 1–1.5 months led to a great increase in the clotting power of the blood, as well as to reduced consumption of food and increased consumption of water. The weight of the animals was reduced, and cases of death were observed (659, 660).

Careful experiments on the effect of weak magnetic fields (less than 1 Oe) in the range of geomagnetic pulsations from 0.01 to 20 Hz on white rats indicate definite and very precise changes in many systems of the organism (542). Subjection for 12–60 h to magnetic fields (whose frequency characteristics simulated a magnetic storm) led to various responses in the animal organism. For instance, the adrenals and posterior lobe of the hypophysis showed responses of an adaptive nature in the form of an increase in activity and a return to normal in two and a half days. At the same time, the parenchymatous organs (liver, kidneys, etc.) and the brain showed progressive changes, culminating in necrobiosis and necrosis. A magnetic field with the indicated parameters caused statistically significant ($p < 0.05$) changes in the state of the clotting and anticlotting system of the blood of white rats—the fibrinogen level was increased by 62%, and the recalcification time was increased by 40% (362); the ESR of the blood of rats became 89% higher than the initial level after 12 h of continuous exposure. The ESR subsequently decreased and returned to normal after 60 h (35). The experiments showed changes in the uptake of neutral red by intact and regenerating rat liver (34). Animals shielded for 60 h in a magnetic field of 0.50–0.66 Oe with a frequency of 0.5–20 Hz showed a distinct (by more than 30%) reduction in the absorption of dye due to reduction of permeability of the cell membranes. Analyzing their work on the effect of weak magnetic fields in the extralow-frequency range (0.01–20 Hz) the authors concluded that such weak alternating magnetic fields are biologically active and affect the whole animal organism. The response of the biological object largely depends on the combination of amplitude, frequency, and vectorial characteristics of the field—the nonuniformity of the field, the nature of its topology, and the structure and degree of complexity of the frequency spectrum (361, 543, 637). The central nervous system of animals also reacts to low-frequency electromagnetic fields (549, 628). Animals subjected to low-frequency electromagnetic oscillations and also to a magnetic field (simulating the short-period variations of the *pc*1 type with a carrier frequency of 3 Hz, modulation period of 30 sec, and strength of 1 γ) showed a change in the functional state of neurons of the cerebral cortex (626). A short exposure (15–30 min) led to acceleration of the cortical rhythm (8–10 Hz) and an increase in the amplitude to 50–70 μV. The authors reported that the pathological changes in rabbits after exposure for three hours persisted for one to two days in the form of prolonged alterations of the main parameters of the biopotentials. There were also various other changes in the electrical activity of the brain.

Several investigators have reported that subjection to weak electromagnetic and magnetic fields leads to fixation of an assimilated rhythm by nerve cells of the cerebral cortex (637, 640, 846) and by the heart (1046). A weak sinusoidal magnetic field with $f = 4.5$ Hz and $H = 0.70$ and a pulsed field with a switching frequency of 0.5 Hz and $H_{max} = 3.4$ Oe altered the power spectrum of the biopotentials of deep structures of the brain (637). On exposure to the pulsed magnetic field the spectral density was increased by a factor of 4. The delta-rhythm frequency fell from 3.1 (normal level) to 1 Hz, while the theta rhythm increased accordingly from 5.9 to 9 Hz, and the alpha rhythm to 15 Hz. The spectrum returned to normal 1–2 min after removal of the field. It was reported that a similar subjection to low-strength electromagnetic fields causes various changes in the cardiac activity and functional state of the cerebral cortex of animals of different ages (640).

It should be noted that a biological role of weak magnetic and low-frequency electromagnetic fields is indicated in several more papers (74, 76, 613, 806, 866, 1052, 1078–1083, 1184).

Man. To determine the role of artificial and natural electromagnetic fields, scientists have investigated several physiological processes in people (341, 342, 1199–1201, 1203–1207). The experiments were conducted in an underground dwelling: in a room shielded from the GMF (the field strength was reduced by a factor of 100) and in a room with ordinary GMF conditions. The subjects in the shielded room were exposed to an artificial electromagnetic field of 25 mV/cm and frequency 10 Hz applied in mutually intersecting directions, for 1 sec. The subjects were unaware of the shielding of the room or of the devices for exciting the artificial electromagnetic fields of 10 Hz, or when it was applied or removed. The experiment lasted three to four weeks, during which the times of activity and rest and the body temperature of the subjects were measured. Some indices and the excretory functions of the kidneys and electrolytic composition of the urine were investigated.

In this experiment the period of circadian rhythms was reduced by 1.27 h ($p < 0.01$), and the accelerating effect of the field was highly significant ($p < 0.001$). Internal desynchronization was also observed, more often in the shielded room, and consisted in abnormal extension (up to 30–40 h) of the period of activity, whereas the period of the simultaneously monitored vegetative functions was normal (about 25–26 h). There was no strong phase relationship between the observed periodicities. When the artificial field was applied, the internal desynchronization of the subjects disappeared. In the unshielded room the period of activity was also increased (in five experiments), but it differed in that the period of activity of the subjects was exactly twice as long as the period of variation of the body temperature and showed a strong phase relation.

Since the experiments were carried out on the same group of people, and also in symmetrically situated rooms that differed only in shielding, we can conclude that weak electromagnetic fields, both artificial and natural, affect the circadian rhythms and some physiological functions of people and, hence, their general state. The author rightly points out (1199–1201) that the two fields prevent desynchronization, which is observed in the absence of a natural and artificial field, but this interchangeability does not mean that the 10-Hz field is the only component of the natural field that affects human beings.

There are investigations that confirm this conclusion and indicate the importance of electromagnetic fields in other ranges. We are referring here to low-

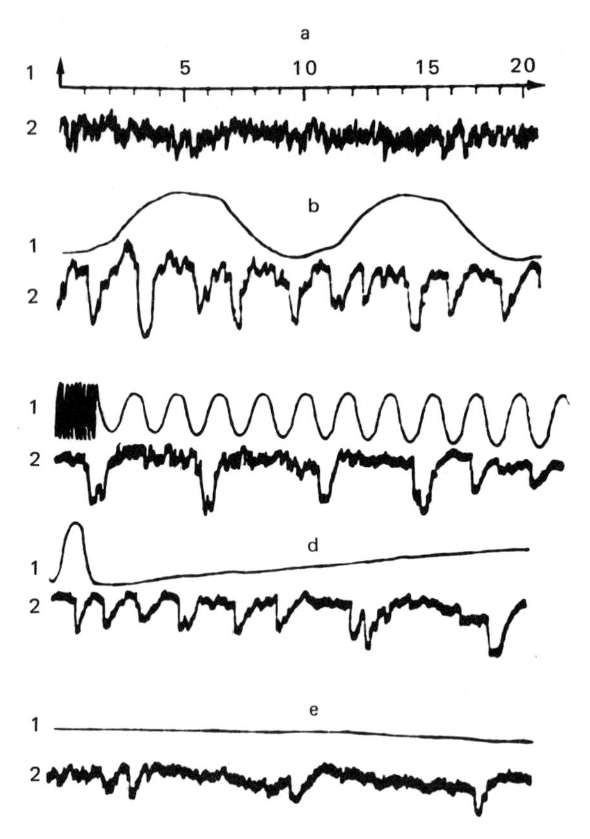

Fig. 16. Change in electroencephalogram (2) of man subjected to an artificial magnetic field of different frequencies (1) and strength $\Delta Z = 1000\,\gamma$ (341). (a) Before application of magnetic field. Frequency: (b) 0.1 Hz; (c) 5–0.5 Hz; (d) 0.5–0.01 Hz; (e) after exposure to magnetic field for 12 min and 1.5 min rest.

frequency natural electromagnetic fields in the frequency range 2–8 Hz, which affect the human reaction time to an optical signal (929), and to artificial magnetic fields of frequency 0.2 Hz and strength 5–11 G, which alter the reaction time of man (698, 839).

Two series of experiments (341, 342) were performed to find out if people could perceive weak magnetic fields with a strength of not more than 3–5% that of the GMF. A magnetic field of fixed frequency (0.01–10 Hz) was created by two-component (*H* and *Z*) Helmholtz coils of diameter 1.5 m. The subject was exposed in a sitting position to the action of the coils. In the first series the subjects were conditioned to fall asleep at the sound of an auditory signal of a particular frequency (for example, f_1 = 1000 Hz) and to wake up at the sound of an auditory signal of a different frequency (f_2 = 300 Hz). These signals reinforced the effect of a magnetic field of frequency 0.01–5 Hz and strength 1000–2000 γ. By this method three out of ten subjects developed a conditioned reflex to short-period variations of the magnetic field of this frequency and strength (the auditory signal was the unconditioned stimulus). In a rural locality where the level of interference was reduced to 10 γ people could perceive magnetic field variations of 200 γ (343).

A second series of experiments showed that even a brief subjection of man to a field of frequency 0.01–5 Hz and strength H = 1000 γ sharply altered the electroencephalogram (Fig. 16). The application of weak alternating magnetic fields increased the pulse rate of the subjects and brought on symptoms of illness (tiredness, headaches, shivering, etc.) and, as the records showed, greatly altered the electrical activity of the brain.

Another laboratory experiment showed that the heart rate in man was reduced by approximately 5% ($p < 0.02$) when an artificial magnetic field with H = 1 G, varying in frequency from 0 to 10 Hz, was applied to the subject's head (58).

The results of the above-mentioned experiments indicate that short-period variations of the GMF can directly affect the human central nervous system. This is of great importance, of course (698).

Carefully conducted experiments with weak artificial fields indicate a high sensitivity of living objects to these fields and thus the great effectiveness of the various frequency ranges encountered in natural conditions. All this provides additional evidence in support of the well-substantiated view that natural magnetic and electric fields can have a pronounced biological effect (445). That such an effect exists is clearly indicated by experiments on the spatial orientation of organisms.

Orientation of Biological Objects Relative to the Geomagnetic Poles

The evidence indicating an important biological role of the GMF includes another group of investigations that have been fairly numerous recently. These are experiments on the orientation of biological objects in the GMF.

These experiments are essentially as follows. The objects of investigation in the laboratory or field are oriented in a particular way relative to the geomagnetic poles or simply in a north–south or west–east direction. Since the earth's magnetic field lines are oriented along the magnetic meridian, their effect on the state and properties of living organisms and even nonliving systems is very pronounced.

Such experiments began a considerable time ago, but they were sometimes not related to the problem of the biological role of the GMF. The results of numerous experiments in various countries have now provided convincing evidence of the great importance of the GMF for the orientation of living organisms.

Orientation experiments can be classed as active or passive. Active experiments include investigations of the orientation of living organisms on the earth's surface, in water, and in air, e.g., the orientation of parts of plants in the field, the bodies of insects at rest and on landing, of birds in flight, of migrating fish, and so on.

In passive experiments the experimenter himself orients the biological objects relative to the geomagnetic poles and subsequently investigates the discovered changes. These two kinds of experiments, although they differ fundamentally from one another, contribute to the clarification of the problem posed and in this respect are of almost equal value.

The results of experiments of both kinds will be discussed more fully in the appropriate chapters on plants, insects, etc. Here we will refer to final results and studies that have an important bearing on the general problem of the possible biological role of the GMF. The orientation of objects relative to the geomagnetic poles and the great effect of this orientation have recently become the subject of research (52, 148, 149). Many fantastic theories have been advanced to account for this phenomenon. There have been cases where biological objects artificially oriented in mutually perpendicular compass directions have shown differences in their properties.

In natural conditions living organisms select particular compass directions for migration (736, 750). It is obvious that such migrations may have important value, since a genetic investigation of population polymorphism in natural conditions in 1943 established a difference in the number of *Drosophila* in different compass directions (801).

A later agronomic investigation also showed that in natural conditions the side roots, in beet, for instance, were arranged in a regularly uniform compass direction: the preferred direction was east–west (1117). This was not attributed at the time to the GMF. It was not until 1960 that experiments revealed that if plant seed embryos were directed toward the south geomagnetic pole the roots oriented themselves in a particular way and the root growth rate was altered. This effect was confirmed in experiments with artificial magnetic fields and was called magnetotropism (297, 673). It should be noted that the importance of orientation of an object relative to the magnetic poles had previously been pointed out and experimentally confirmed, but there had been no inference of magnetotropism (1074).

Many investigators in various countries have not only confirmed the phenomenon of magnetotropism, but have found features indicating that the manifestation of magnetotropic reactions is more complex than had previously been believed. Since plants have been more frequently investigated in this respect, we will discuss these features in the chapter devoted to plants.

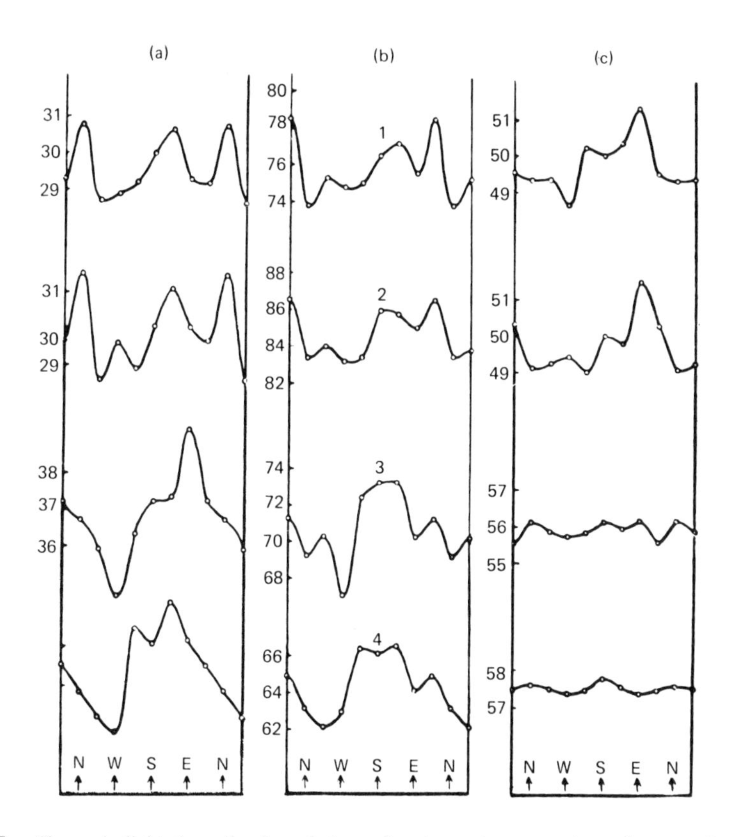

Fig. 17. Change in light absorption by solutions of various substances when tubes are oriented in different compass directions. (a) Kaolin; (b) erythrocytes; (c) protein; (1) May 26 to midday; (2) May 26 after midday; (3,4) May 27 before and after midday.

It should be noted that a geomagnetic orienting effect has been demonstrated on a very wide range of objects—from molecules of biological and inorganic substances to entire organisms, including man (52, 589, 590, 999).

How the earth's magnetic lines of force run in relation to the body is equally important for fish (63), insects (688, 951), plants (399), and other organisms (748–750).

In more than 20 years of research on the course of various physicochemical reactions of protein molecules, cell suspensions, and inorganic substances precisely oriented relative to the geomagnetic poles, Moriyama (999) found that any reaction, process, or investigated property of the same object depended on its compass orientation (Fig. 17). It was reported (52) that the sound obtained on sounding the human chest depends on the position of the body relative to the geomagnetic poles, as also does the functioning of the vascular system, the state of muscle tone, and the electrical properties of tissues, but the author did not confirm this with experimental data.

Czech investigators have recently studied the possible effect of orientation of man in the GMF on his sleep (935). A computer was used to analyze the bioelectric activity of the human brain in sleep stage IV in relation to the position of the human body relative to the direction of the geomagnetic meridian—along the plane of the meridian (position I) and at an angle of 45° to it (position II). It was found that in position II the bioelectric activity of the brain of the sleeper showed a reduction of approximately 13.3% in the number of delta and theta waves, accompanied by a 6.8% increase in their amplitudes, and an increase in the general activity of the brain regions, as indicated by the appearance of electromagnetic oscillations of higher frequencies.

The general conclusion that can be derived from this work is that the interaction of the bioelectric activity of the brain with natural short-period fluctuations (geomagnetic pulsations) depends on the position of the sleeper relative to the magnetic meridian. The state of bioelectric activity in a sleeping person lying at an angle of 45° to the magnetic meridian may indicate an involuntary transition to sleep stage III, which is less deep. Hence, we can postulate that the sleep and rest in this case will be less complete, particularly if this position is maintained for a long time.

From an analysis of these data we can conclude that geomagnetic orientation significantly affects the functional properties of biological objects. Individual objects, however, can show deviations and various forms of reactions, which can be attributed to many factors. It has now been established that this effect depends on whether the objects belong to a particular dissymmetric, "right-handed" or "left-handed," modification (125, 126, 379, 550, 552), on the phase of the solar cycle during these investigations (552), and also on the geoelectromagnetic features of the location of the experiments (94). The actual effect of the GMF, however, is quite distinct and raises no doubt. In conclusion we will discuss various indirect evidence that indicates a major biological role of the GMF. These data were not obtained in the course of laboratory research, but by an analytical examination and comparison of data.

Biomedical Statistical Analysis

In addition to experimental studies of the biological effect of the GMF there have been investigations in which a possible effect of the GMF has been revealed by a comparison of the changes in the GMF and some biomedical data in a specific period.

The investigations discussed here relate mainly to medicine, where direct experiments are not always possible, as, in particular, in the study of human pathological states and in the analysis of injuries (injuries and accidents in industry, road traffic accidents, etc.). These investigations, based on abundant factual data and the use of statistical analysis, have shown a correlation between the state of the GMF and particular medical indices.

Medical statistics are recognized as important evidence in the question of a correlation between the variations of natural electromagnetic fields and the state of man (54).

Theoretical calculations have shown that the energy density and power flux of the GMF during disturbances are usually several orders greater than man's threshold sensitivity (358). The threshold sensitivity of the human organism is 10^{-12} erg/cm^3 for energy density and 10^{-9} erg/(sec cm^2) for the power flux per cell (423). These values were obtained for the receptor cells of the visual and auditory organs, but they apply equally well to the perception of magnetic fields by cells. The values for the disturbed GMF are $10^{-4}-10^{-7}$ erg/(sec cm^2) for the power flux, and of the order of 10^{-7} erg/cm^3 for the energy density, which exceed the threshold sensitivity of the biological receptors by two to five orders (359, 360).

Medical statistics confirm these theoretical calculations and indicate that the human organism responds to disturbances of the GMF.

A number of authors have devised a statistical model for investigation of the effect of a complex of environmental factors (variation of vertical component of the geomagnetic field, temperature, pressure, humidity) on the dynamics of cardiovascular diseases (41, 42, 491, 492). The factual basis for the model consisted of the information relating to 80,000 calls for urgent medical help in Leningrad between January and September, 1969. The statistically treated data indicated that a leading factor affecting the incidence of cardiovascular diseases is the spontaneous variation of the GMF. Very intense and prolonged storms have the greatest effect.

The statistical methods used to detect the correlation between the GMF and biomedical processes (analysis of periodograms, the sliding-mean method, autocorrelation, cross correlation, and regression analysis) are well known and will not be discussed here.

Since medical statistical investigations will be referred to in subsequent chapters, we will mention only one as an example here. The research staff in the National Geophysical Institute (India) carried out a statistical study of road traffic accidents in relation to the degree of disturbance of the geomagnetic field in the locality (723). The accidents considered occurred in 1965–1969 in the adjacent towns of Hyderabad and Secunderabad, which have a population of several millions. Only official data recorded by the traffic police department were included in the analysis. The coefficient of correlation between the degree of disturbance of the GMF and road accidents was $r = +0.742$ at the 0.05 significance level. The reliability of the discovered correlation was confirmed by the fact that χ^2, calculated for the 5% level of significance with one degree of freedom (3.8), was very high (248). It was found that on 83% of the magnetically quiet days (*K*-index sum $\leqslant$ 15) there were no road accidents, whereas traffic accidents and catastrophes occurred on 81% of the highly disturbed days (*K*-index sum $\geqslant$ 25). On continuing their investigations, the Indian scientists found that during the sudden and exceptionally intense disturbance of the GMF on August 4–10, 1972, associated with the increase in solar activity in this period, the number of road accidents in these towns and the number of admissions of cardiac patients to hospital increased by 100%, and the general incidence of disease and mortality rate in this period were 10 and 33% higher, respectively, than in the preceding quiet period. The authors attributed the described effects of disturbance of the GMF to the increased activity of geomagnetic pulsations on the central and vegetative nervous systems (1142).

In our analysis of the biological consequences of the geomagnetic storm on August 4–10, 1972, we obtained distinct evidence of a correlation only in the case of admissions of infarct patients to the hospital of the Academy of Sciences of the USSR. The number of admissions was highly correlated ($r = +0.99$) with the mean amplitude of geomagnetic pulsations of the *pc*3 type (Fig. 18). An analysis of cases of infarct diseases attended to by the town first-aid station, however, failed to show any correlation with geomagnetic micropulsations.

Smith (1132) gives very interesting data indicating a sharp increase in the weight of beetle larvae [*Tribolium castaneum* (Hbst)] during this period of time, and associated with the unusually large geomagnetic disturbance.

It should be noted that this was not an accidental result, since a statistical analysis of other climatic factors—temperature, humidity, wind speed, pressure, precipitation, etc.—did not show such a high correlation between the factor and the investigated effect. Similar conclusions have been made for the course of cardiovascular diseases, and changes in the heart rhythm (210, 265, 266, 273, 481). Several statistical investigations of this kind have been made. In particular, a correlation between geomagnetic activity and epidemic processes has been shown on the basis of a vast amount of material (647) (Fig. 19).

For a long time there was no satisfactory theory that specified what links in the GMF–living organism chain were responsible for this correlation. Hence, it was suggested quite validly that the GMF did not have a direct effect, but acted via the bioelectric activity by induction of currents in the organism (60, 62, 63, 961, 1169).

Using a special method, however, magnetobiologists have found that living organisms have their own magnetic field (248, 319, 485, 731, 770), and also emit electromagnetic oscillations in the low-frequency range (189–191, 195, 501, 564, 867, 1076).

These facts open up the way for a deeper and broader investigation of the effect of natural magnetic and electromagnetic fields on living organisms, and for this purpose investigators can use certain specific techniques.

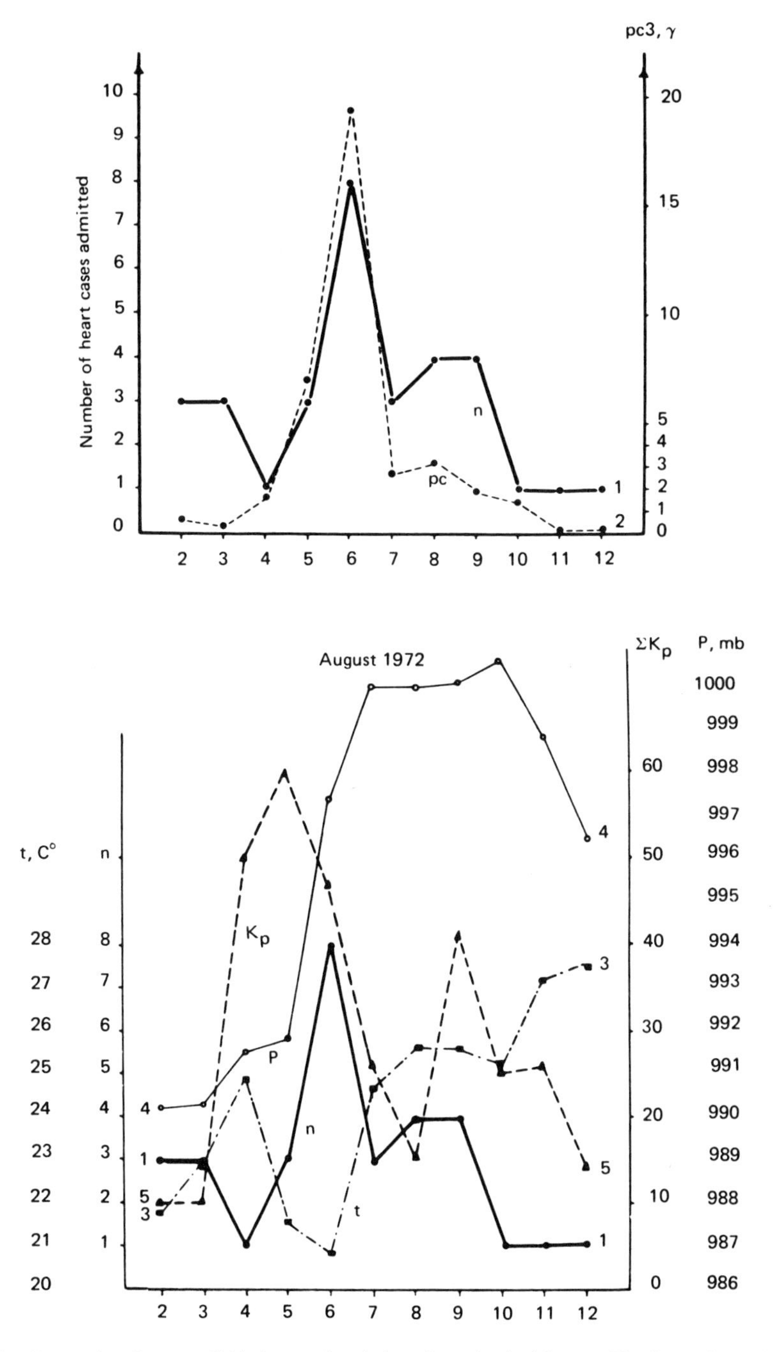

Fig. 18. Dynamics of myocardial infarct and variation of geophysical factors. The figure shows the admission of heart cases to hospital (1) and the variation of the amplitude of *pc*3 geomagnetic pulsations (2), temperature (3), atmospheric pressure (4), and degree of magnetic disturbance (5) in Moscow in August, 1972. From (141).

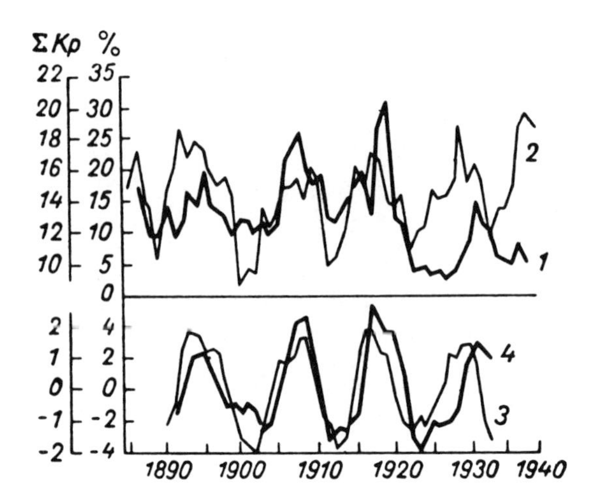

Fig. 19. Mortality from whooping cough in Moscow (1) and variation of magnetic disturbance ΣK_p (2). Variation of ΣK_p (3) and mortality (4) on removal of oscillations with a period of less than 5 years and more than 11 years. Curves 3 and 4 were obtained by subtaction of 11-year sliding means (647).

The *method of superposed epochs* is particularly widely used in geomagnetobiology and geophysics, since it also provides a means of assessing the dependence of a particular process on changes in the GMF (356, 443, 817, etc.).

The procedure is as follows. A day with a particular geomagnetic event (a disturbed geomagnetic day, the occurrence of the peak of short-period fluctuations of the GMF, etc.) is arbitrarily taken as the zero (reference) day. Biological or medical indices are then taken for the zero day, the day preceding it (−1, −2, −3, etc.) and the days following it (+1, +2, +3, etc.). Changes unaffected by geomagnetic disturbance will be randomly distributed, but if there is a relation the indices will be much higher on zero day or will be grouped around other days before or after the geomagnetic event. Using this method many investigators have shown a close correlation between geomagnetic disturbance and manifestation of its biological effect.

For instance, the use of the method of superposed epochs, where the chosen zero day was one on which the polarity changed, showed that patients with disturbed vegetative regulation function were very sensitive to a change in polarity of the sectors of the interplanetary magnetic field (IMF) (615, 616). The authors' analysis of 25 patients with vegetovascular dystonia, complicated by functional and organic damage to hypothalamic brain structures, revealed that the maximum number of vegetovascular paroxysms (160% of the mean level) coincided with the change in polarity of the IMF sectors to within ±1 day. The authors noted that out of a total of 45 recorded vegetovascular crises 71% occurred when the direction of the IMF was positive (away from the sun) and 29%

when it was negative (toward the sun). This indicates a distinct relation between the times of onset of crises and the position of the boundary between the IMF sectors. Discussing the possible causes of this relation, the authors suggest a possible role of the magnetospheric electric field, which is greatly altered when the earth crosses the boundary between the sectors (890, 1004). However, we can hardly come to any definite conclusion here, since it is known that sharp changes in very many geophysical processes affecting the biosphere occur when the polarity of the IMF sectors is altered (1210). We can also mention that a microbiological investigation based on the method of superposed epochs revealed a change in the vital activity of *E. coli* in relation to the polarity of the IMF (19). The authors reported that the suppression of vital activity was greatest on the second day after passage of the earth across the boundary between the IMF sectors.

In heliobiology and geomagnetobiology another method of treating data is well known—the method of the solar calendar.

The method of the 27-day solar calendar for treatment of experimental data has been known for a long time. This involves the compilation of a table with 27 or 28 columns corresponding to the days of revolution of the sun, beginning at zero heliographic longitude. The biological and medical indices are considered simultaneously for these days of the solar calendar (Carrington dates—see Appendix). If these indices depend in any way on geomagnetic disturbance, they will recur with a 27-day period, corresponding to the 27-day recurrence of geomagnetic storms. The recurrence of GMF disturbances is connected with the 11-year cycle of solar activity and is usually most pronounced in the years of decline and lowest level of this activity (28).

It is to this connection that various authors attribute the increase in the number of cases of various diseases in some years of decline of solar activity. This method has revealed that some neuropsychic (1068) and somatic (1020) diseases show a 27-day recurrence. A 27-day period has also been observed in the course of biological processes (740, 741) and physicochemical reactions (757).

The usual method of compiling the 27-day solar calendar has been modified by the introduction of relative numbers (from 1 to 27) instead of absolute values or anomalies of the investigated quantity (617). This excludes the effect of many-year fluctuations and the annual course of the process on the cyclicity, and allows an easy comparison of the variation of various heliogeophysical and biomedical indices and the calculation of rank correlations.

The *method of direct comparison* consists in comparing the variation of individual elements of the GMF with the variation of a particular medical or biological index over a particular period of time. This method has revealed a close relation between the variation of the GMF and the variation of the blood pressure and leukocyte count in man (669), the variation of the dark-adaptation threshold in man, which is an index of the functional state of the brain (89), and the variation of genetic, physiological, biochemical, and radiobiological processes in very diverse organisms (121–124, 809). The comparison was made for the range of short-period fluctuations of the earth's electromagnetic field. By comparing the time taken by man to react to an optical signal in periods when the strength of a field of frequency equal to that of the fundamental frequency of an ionospheric waveguide (8 Hz) was increased, and in periods where there were irregular fluctuations in the frequency range 2–6 Hz, König *et al.* (929) reached a definite conclusion. When the strength of the fundamental-frequency field was increased the reaction time was reduced by 20 msec, and in the presence of irregular fluctuations of 2–6 Hz it was increased by 15 msec.

Observations of the behavior of mice in the course of a day (989) also indicated the influence of various environmental factors. A special apparatus continuously recorded the activity of the animals for seven months. Multiple-regression treatment of the data showed that the activity of the animals in the course of a day was most affected by the following factors (in order of decreasing importance): maximum and minimum air temperature, geomagnetic activity, relative position of earth, moon, and sun, sunspot number, solar activity (Piccardi's chemical tests *P* and *F*), relative humidity, atmospheric pressure, and Piccardi's chemical test *D*. Field research confirms the effect of disturbance of the GMF on rodent activity (555).

An investigation of plant growth in constant environmental conditions and a careful statistical analysis of the results also revealed an effect of natural electromagnetic fields (375, 1185).

The use of this method revealed a correlation between the degree of disturbance of the GMF and the course of several diseases and mortality rate (818, 820–823). In particular, a correlation was found between the strength of the GMF and attacks of eclampsia, acute glaucoma, epilepsy (207, 208, 219, 675, 715, 716, 721, 1170), cardiovascular accidents (32, 112, 157, 164–166, 197–199, 316, 344, 369, 384, 432, 437, 440, 448, 477, 478, 528, 533–535, 545, 661), labor (196, 1152), and heart rate disorders (210, 213, 390, 391).

We will cite the results of the direct-comparison method later, when we discuss functional states in living organisms and diseases in man.

The *paleomagnetobiological method* occupies a special place in research, since it has features that link it to some extent with both the direct and indirect methods.

The physical basis of the method is the assumption that the direction of magnetism of ferromagnetic materials corresponds to that of the GMF acting at the time of formation of these materials. This method is used to determine the direction of the magnetic field acting at the time of formation of particular rocks in the remote historical past, since residual magnetism can persist for millions of years (218, 244–246, 418, 923, 964, 1008). By investigating the paleofauna and flora in these rock samples one can compare the changes in direction of the GMF

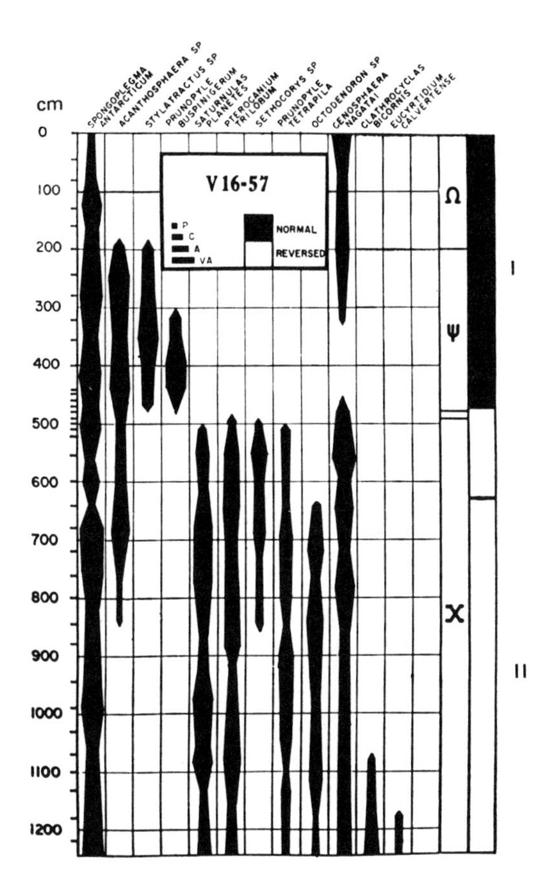

Fig. 20. Reversals of the GMF (A) from normal (I) to reverse (II) and occurrence of investigated organisms in different geological epochs (B). The height in centimeters of the sample of oceanic deposit is given on the left (1025).

and the species present at the time to which the specimens belong. Paleomagnetologists have found that the sign of the GMF has not been constant in different epochs, i.e., there have been polar reversals of the magnetic field (Fig. 20), which have been accompanied by pronounced changes in the species composition of the marine microfauna—radiolaria and foraminifera (254, 722, 725, 732, 785, 851, 878, 879, 884–887, 918, 1025, 1102, 1127, 1128).

Colloid Systems and Physicochemical Reactions

Another important group of investigations is of wide significance and indicates an effect of the GMF on reactions occurring in biological objects and inorganic systems (14, 216, 353–355, 403, 404, 420, 425, 426, 729, 999, 1049, 1153).

Living organisms are complex heterogeneous systems in which biocolloids and physicochemical reactions play the leading role. Hence, it is understandable that experiments with colloid systems indicating an effect of external physical factors are of interest. On the basis of 20 years of continuous research Piccardi showed that the rate of colloid chemical reactions depends on solar activity (420, 1049) and on position relative to the geomagnetic poles (999). Similar investigations in this region have revealed a correlation between colloid system reactions and such an important characteristic of solar activity as geomagnetic disturbances (729, 757). Thus, the conclusion made above regarding the global synchronizing effect of the GMF on biological and physicochemical reactions is valid for the colloid systems investigated by Piccardi. Figure 21, where data for the diurnal variations of the hydrolysis of bismuth chloride (P test) and the GMF are compared (128), shows that the diurnal course of the two processes is identical, i.e., that the properties of the colloid system depend on the state of the GMF. The main cause of this is the alteration by the GMF of the properties of water—the common component of these reactions in living and nonliving systems.

The range of geomagnetic pulsations (SPV) probably also has an effect on the course of colloid reactions. This is indicated by laboratory investigations of the precipitation of bismuth oxychloride by artificial magnetic fields of frequency 0.01–0.1 Hz and strength 50–1000 γ (426).

From their subsequent investigations the authors concluded that the nature of the changes in this physicochemical test due to weak artificial magnetic fields

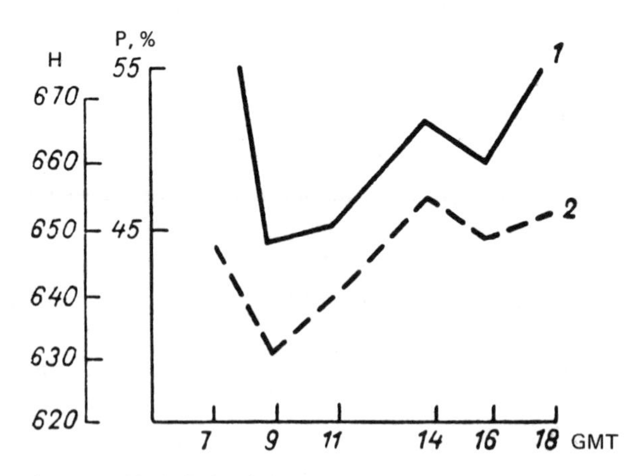

Fig. 21. Diurnal course of hydrolysis of bismuth chloride (P test) (1) and variation of horizontal component H of GMF (2) in August, 1967, in Florence. Geomagnetic data from Roburent Observatory, Italy (128).

in the infralow-frequency range and natural magnetic disturbance was the same and consisted in a significant increase in the reaction rate (403, 404, 426).

At the same time, using thorough permalloy shielding, which reduced the external magnetic fields by a factor of 10^6, the authors of the investigation found that the fluctuating nature of the precipitation of bismuth oxychloride in water, conducted in the usual conditions, was due to the artificial and natural magnetic fields in the environment. In a stabilized hypomagnetic space the precipitation of bismuth oxychloride showed hardly any fluctuations (425).

Interesting data on the role of the GMF in physicochemical reactions are also given in studies of the properties of tryptophan in solution (353–355).

These studies were based on the use of the indicator test, devised by Dickman and Westcott (1954), for L and DL tryptophan: Tryptophan alters the color of the red dye xanthydrol in a mixture of hydrochloric and acetic acids. Several thousand tests for the L and DL isomers of tryptophan were made with xanthydrol. In certain periods the observations were made hourly over a period of 24 h or even several days, and also synchronously in geographically separate localities (216). These investigations confirmed that the color changes depended on the geomagnetic environment at the time of the test ($r = +0.72 - 0.90$). The experiments showed a change in the properties of L and DL tryptophan during magnetic storms, and their correlation with lunar phases and solar activity. A long-term analysis in strictly controlled laboratory conditions also revealed that agglutination was reliably correlated with the K index of geomagnetic activity (424), and on magnetoactive days the expression of the agglutination reaction was significantly more pronounced than on magnetically quiet days. An artificial magnetic field with a frequency of 0.1 Hz and an amplitude of 500 γ had a significant effect on this immunological reaction (405). Yet no correlation was found between the planetary index A_p of geomagnetic disturbance and indices of human nonspecific immunity (597), such as complement titer, level of normal hemolysins, hemagglutinins, and total hemolytic activity of blood serum, serum lysozyme titer, etc., investigated in a group of presumably healthy people (200) in Tomsk and donors (1000) in Orenburg. The negative result obtained in this statistical investigation is due, in our opinion, to the fact that the values used in the analysis were not the local values of magnetic disturbance (K index) calculated from nearby geomagnetic observatories, as was done in (424), but planetary indices of disturbance, which in this case cannot be used, since they are too general and coarse. A similar error was made by other investigators who compared the physiological state of man and animals with the general planetary indices of GMF disturbance, instead of local indices. The methodological error of the experiments is obvious in this case. It is just as if the authors, instead of using the local air temperatures at the sites of investigation (Tomsk and Orenburg), had used the index of planetary disturbance and the temperature gradients for the planet earth as a whole. Hence, it is not surprising that for the monthly dynamics of lysozyme in Tomsk they found a significant negative correlation with the local temperature (−0.729) and absolute humidity (−0.677), and a positive correlation with the atmospheric pressure (0.663), and for the dynamics of complement titer correspondingly higher values (for the mean monthly temperatures in Tomsk −0.852, and for the daily temperatures in Orenburg −0.700), whereas the coefficient of correlation between the mean-monthly values of the lysozyme titer in Orenburg and the index A_p was −0.573. This example very clearly illustrates how careful researchers must be in performing statistical calculations and using indices of magnetic disturbance in the analysis of biological data.

An investigation of the effect of the GMF on the crystallization of a substance from solution showed that the GMF has an orienting effect on the nuclei of crystals of epsomite—magnesium sulfate heptahydrate (464). An orienting effect of the GMF was discovered for epsomite crystals originating and growing on the open surface of the solution, and also on the bottom of Petri dishes. The maximum of the crystal orientation distribution (Fig. 22) occurred at 80–90° relative to the meridian. This interval contained almost 30% of all the crystals, i.e., three times more than in the absence of the GMF, where the angular distribution was uniform.

The orienting effect of the GMF has been investigated also in the case of acicular crystals of morenosite—nickel sulfate heptahydrate (465, 466). The orientation distribution curves for the two substances—epsomite and morenosite—were similar, despite the difference in magnetic properties of these substances: The magnetic susceptibility of epsomite is $K_M = 0.5 \times 10^{-6}$, and that of morenosite is $K_M = +16.0 \times 10^{-6}$. This fact can probably be attributed to the anisotropy of the magnetic properties of the crystals: The direction of greatest magnetic permeability is perpendicular to the direction of greatest growth.

We investigated the orienting effect of the GMF on the crystallization of sodium nitrate, which has highly anisotropic magnetic properties, from aqueous solutions. We found that the orientation distribution for sodium nitrate crystals in the GMF was represented by a much more complex curve than those for epsomite and morenosite.

Synchronous experiments on epsomite crystallization were conducted at different geographical points in the USSR. The first synchronous experiment on epsomite crystallization was carried out December 10–13, 1973, in the following towns: Kirovsk, Leningrad, Moscow, Gorki, Belgorod, Voronezh, Uzhgorod, Tbilisi, Petropavlovsk-Kamchatskii, and Ashkhabad (466). Photographs of the crystals taken by the participants, with the direction of the magnetic meridian marked, and the collected crystals were sent to É. Rogacheva in Gorki for statistical treatment.

The overall distribution of the orientation of epsomite in the GMF, constructed from the obtained material, is shown in Fig. 23. The overall distribution

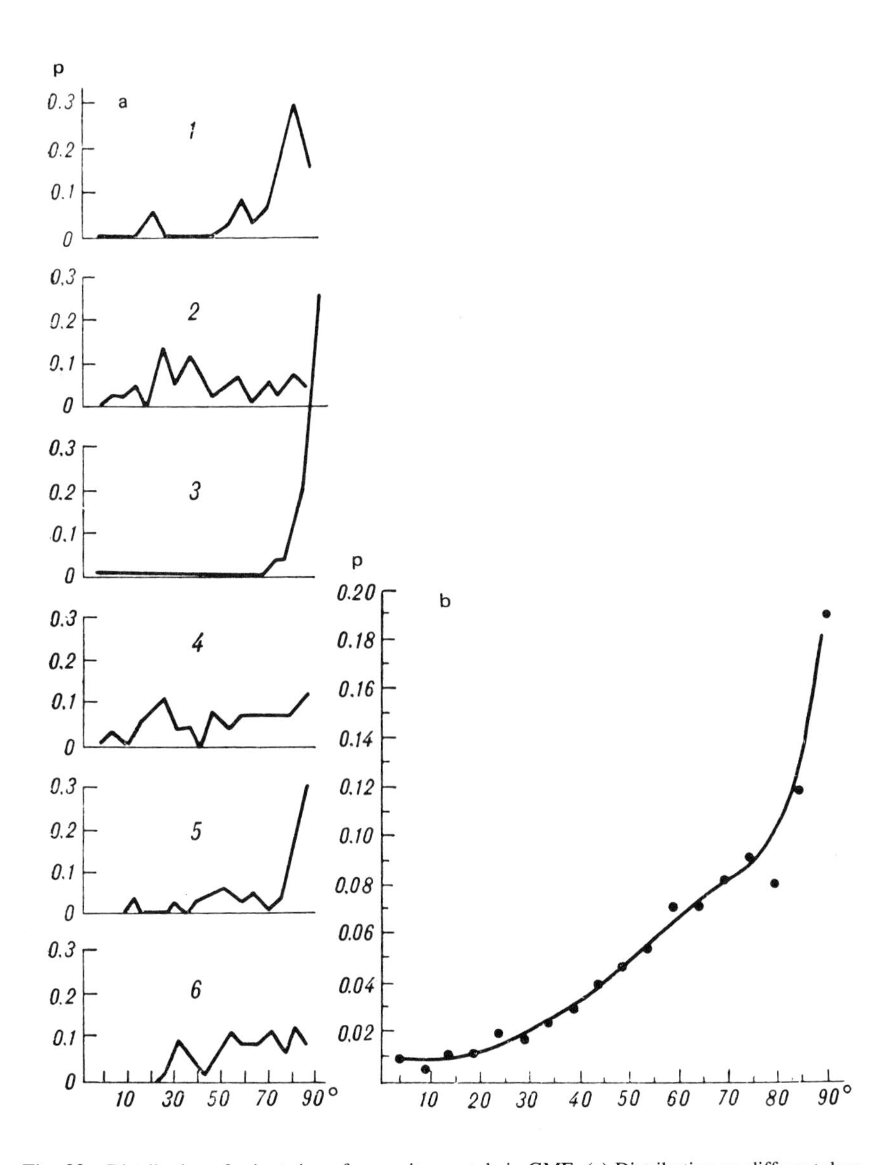

Fig. 22 Distribution of orientation of epsomite crystals in GMF. (a) Distribution on different days of experiments: (1–6) April 17–April 22, 1972, respectively, Gorki; (b) summary graph of crystal distribution. x axis, angles of orientation of crystals relative to magnetic meridian; y axis, relative frequency (464).

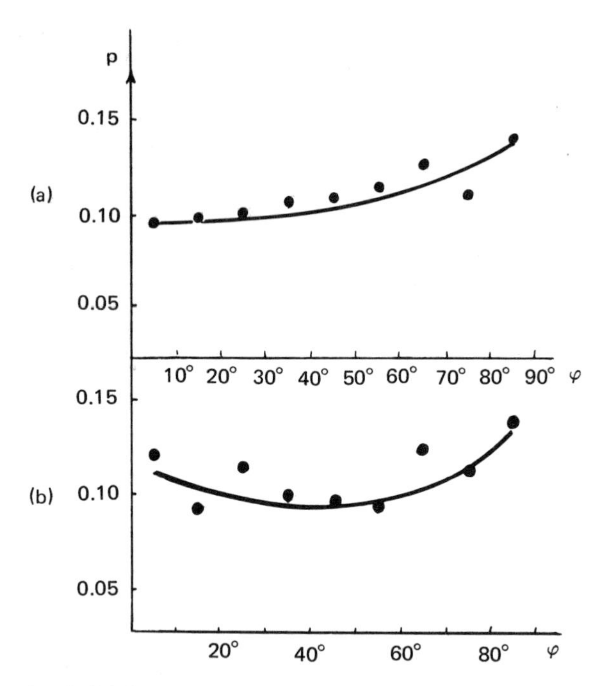

Fig. 23. Results of global synchronous experiments on crystallization of epsomite at different geographical points in the USSR. The first experiment was conducted December 10–13, 1973; the orientability coefficient β is the relative frequency of crystal orientations in the angle range 50–90°; for a uniform distribution, i.e., without any effect of the GMF, $\beta = 0.440$. (a) Experiment 1, total number of crystals $N = 1432$, mean value $\beta = 0.505 \pm 0.021$. (b) Experiment 2, $N = 1174$, $\beta = 0.455 \pm 0.034$. The second experiment was conducted March 11–16, 1974.

reveals a statistically significant tendency toward oriented growth of crystals in the GMF. In the second synchronous experiment, in the period March 2–16, 1974, the oriented growth of the crystals was less distinct (Fig. 23). As the data of our experiments over three years showed, the effect of the GMF on orientation is manifested differently at different times. At the same time, the correspondence of the data for simultaneous crystallization at several geographical points with the results of constant and prolonged observations at one of them (in Gorki, from 1973 through 1976) shows that in addition to the diverse local factors acting on nucleation and growth of crystals in solution there are common factors that lead to similar changes in crystal orientation. The GMF is probably one such factor. Experiments involving the shielding of epsomite solutions confirm this: In such conditions the crystals do not exhibit oriented growth (466).

The GMF can have a direct effect on the kinetics of chemical reactions. Indirect evidence of this is provided by an investigation in which the polyvariance of photocondensation of alkylphenanthrene hydrocarbons with maleic anhydride was investigated (337). This reaction can take three different directions and give rise to compounds with different properties. Being performed systematically for 3.5 years in identical conditions in a chemical laboratory, this reaction nevertheless gave a different yield of reaction products in different seasons of the year. In the search for reasons for the formation of compounds of such different nature, the authors investigated the effect of more than 12 possible factors: purity of the initial components, nature of the reaction mixture, concentration and proportion of reacting components, temperature, light intensity, etc. They found that none of these factors had *any* effect on the direction of the photochemical reaction. The authors indicated that the formation of these products does not depend on the conditions in which the reaction takes place, but is of a distinctly seasonal nature and depends on factors associated with solar activity.

We believe that, since the usual factors affecting the kinetics and yield of reaction products were constant and checked, the only factor left is the state of the GMF, which varies from month to month.

Direct evidence of this effect is the previously described reaction of L and DL tryptophan with xanthydrol (353–355) and an investigation of the role of geomagnetic variations on a chemical reaction—the rate of combustion of ethanol vapor in an oxygen and nitrogen atmosphere (589, 590). After numerous experiments the author found that the rate of combustion was reduced by an increase in strength of the GMF. From a careful analysis of the results the author concluded that the combustion process is characterized not so much by the relation to the total GMF vector (T), as by the relation to the individual H and Z components of the GMF.

Thus, numerous investigations by various methods and techniques indicate that the GMF plays an important role in the activity of living organisms and the course of physicochemical reactions on earth. This indicates the existence of a close link between the GMF and the biosphere both in the distant past and in the present. For a better picture of the whole diversity of this relation, we shall consider next some aspects of general and special geomagnetobiology.

CHAPTER 3

Questions of General Geomagnetobiology

In the preceding chapters we have given evidence for the possible role of the GMF in the vital activity of living organisms and have indicated the main methods that geomagnetobiologists use to detect and demonstrate the leading role of the GMF in biological processes occurring at different levels of organization of living matter. In the following chapters we shall consider the numerous and very diverse investigations carried out by scientists who have either specially studied the biological effect of the GMF, or have discovered its effect in an independent objective investigation of biological processes. Before turning to the account of these investigations however, we consider it essential to discuss the more general concepts of geomagnetobiology developed in our work, since only in this case can the reader acquire a complete picture of the biological action of the GMF from the complex and diverse, and at times apparently extraordinary and contradictory, mosaic of facts, observations, experiments, and hypotheses.

The fundamental concepts of the biological role of the GMF discussed in this chapter provide a basis for the formulation, in the future, of a geomagnetic theory of development of the biosphere. We formulate these ideas on the basis of a complete survey of the entire natural-scientific knowledge amassed by science to the present day and a consideration of the most advanced hypotheses on the structure and functioning of living matter. In undertaking such an extensive analysis we cannot, of course, within the confines of one book, cite many of the basic studies in the biology areas considered, and we must select from the number of fundamental works those which represent most completely the essence of the question being investigated, and which bring out most clearly and vividly those aspects which assist the understanding of our hypotheses on the decisive role of the GMF in the development of the biosphere. This chapter sets out to show the role of the GMF in such fundamental properties of living matter as rhythmicity, symmetry, evolutionary variation, and heredity, provides objective evidence of the effect of the GMF on these basic properties of living matter, and advances hypotheses on the basic mechanisms of this effect. We realize the difficult and major tasks that confront us, since our work presumes a completely new outlook on the theory of evolution, which is regarded as a paramount and universal biological theory (982). The problems that we discuss are the traditional problems of natural science, on which immense physical and mental effort has been expended by numerous investigators over almost two centuries, beginning with the theories of speciation and evolution of Lamarck, Darwin, and De Vries, and Mendel's discovery in 1866 of his famous laws of heredity, and terminating in the work of Morgan, Vavilov, Dobzhansky, and contemporary scientists in the field of molecular and quantum genetics.

These questions are of exceptional importance, since they impinge on the interests of scientists of all specialties, and on the interests of man not only from the standpoint of a purely scientific understanding of evolution itself, but also from a practical viewpoint. Increase and reduction of the yield of animals and plants, and the prognosis and control of sudden increases in numbers of crop pests and outbreaks of infectious diseases due to viruses and bacteria, are all problems in which the main and decisive factor is a change in genetic properties and the disturbance of the genetic homeostasis of organisms. It is also apparent that this problem is closely linked with population dynamics and with the population structure of the most diverse species of living organisms, i.e., with the micro- and macroevolutionary processes continuously occurring at various genetic levels. The most eminent geneticists and evolutionary theorists, however, in considering population dynamics and problems of ecological variation and speciation have devoted their attention mainly to the interrelations of organisms with one another in biocenoses, to natural selection, adaptation, and normal climatic conditions, and have regarded these as the only factors affecting evolution, and the course of evolution itself as a random, stochastic process (802, 803, 982).

Our investigations have shown, however, that evolution is based on rigorous physical laws of development of the genetic systems of species, and that the GMF is one of the main factors of this controlling mechanism. The evolution of living organisms—an apparently random, boundless, and wide ocean of changes and forms—now appears to us as a perfectly regular process governed by general laws of physical fields and symmetries, which impose their limitations, by virtue of the laws peculiar to them, on evolutionary processes, beginning with the deep bases of construction of genetic systems. The framework of evolution is simultaneously broad and narrow, and its potential is vast but, at the same time, limited by the general theory of systems (591), which indicates the limits of the large and broad variations that have taken place and will take place in the evolution of living organisms on the earth. Thus, we now have an opportunity of understanding the true physical laws of evolution and establishing it as an exact biological science.

Paleomagnetobiology

If we regard the effect of the GMF as decisive for the biosphere and reckon its history as many hundred millions of years, this influence should be traceable throughout the historical periods of evolutionary development of living organisms on the earth and should not be an exclusive prerogative of our time.

In fact, paleomagnetologists have reported on many occasions that paleomagnetic zones of normal and reverse polarity coincide with the times of appearance or disappearance of the most diverse species of marine and terrestrial organisms (444, 664, 785, 786). Further investigations have shown that thorough and extensive analysis reveals close correlations between the evolution of not only individual representatives of the fauna, e.g., marine protozoa (339, 537, 722, 882, 884, 885, 1102), but even between the whole evolution of the organic world, including man, and sudden geomagnetic changes—GMF reversals (225, 335, 554, 663, 785, 786).

A study of the periodicity of onset of sharp changes in the evolution of terrestrial and marine fauna and flora and geomagnetic reversals indicates a close relation between them (289, 664, 878, 884). This relation between the boundaries of biological zones and GMF reversals in also revealed by an analysis of the

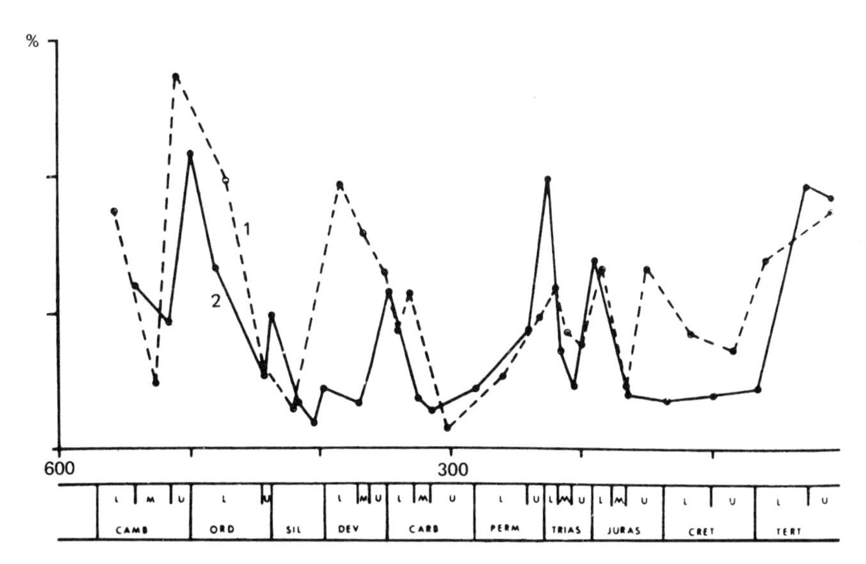

Fig. 24. Correlation between geomagnetic reversals and evolutionary changes in the organic world. (1) The index of frequency of reversals (964), (2) index of disappearance of living organisms (1127). *x* axis, time, millions of years; *y* axis, percentage of mixed measurements and percentage of disappearing families. From (785).

bottom muds in the Pacific (879), Antarctic (886), Atlantic, and Indian Oceans (722, 851). The response of organisms differing completely in organization and living conditions, from oceanic microfauna to terrestrial mammals, is similar and synchronous (664).

The major stages in the evolutionary development of the organic world have usually coincided with a change in magnetic polarity from direct to reverse. ''Biological catastrophes'' depend primarily on a change in polarity and strength of the GMF and, to a lesser extent (it would be more correct to say secondarily), on climatic changes, since they begin in relatively normal climatic conditions (554). Thus, there are grounds for regarding sharp changes in the GMF as a prime and direct factor in the evolution of living organisms (Fig. 24).

In fact, in the last 600 million years there have been five major intervals with specific GMF regimes:

1. The Cambrian–Wendian interval of reverse polarity (515–600 million years).
2. The Devonian–Cambrian interval of alternating polarity of the GMF (375–515 million years).
3. The Permian–Devonian interval of reverse polarity (230–375 million years).
4. The Mesozoic interval of normal polarity (70–230 million years).
5. The Cainozoic interval of alternating GMF (0–70 million years).

We can conclude from these data that changes in the GMF reveal a definite correlation with geological stages: reverse polarity of the GMF dominated in the Paleozoic, the GMF was normal in the Mesozoic, and an alternating field characterizes the Cainozoic (300). The authors of several studies report that the appearance of individual reversals is not characterized by strict cyclicity, and their distribution in time is probably due to the effect of a regular, but random, process. It has been reported, however, that the number of reversals and their frequency in a standard time interval, e.g., 10 million years, were greater in the Mesozoic and Cainozoic than in the Paleozoic. The most distinct paleomagnetic boundary in the Phanerozoic is the Permian–Triassic boundary (230–250 million years), when the number of reversals greatly increased and there was the only occurrence in the Phanerozoic of a change from an interval of reverse polarity to an interval of normal polarity (554).

A comparative analysis shows that these changes in the GMF were associated with major abrupt changes in the evolution of the organic world in the late and middle Cambrian (510–565 million years), the beginning of the Carbonaceous and end of the Devonian (300–400 million years), the end of the Triassic and Permian (200–245 million years), and the start of the Paleogene and end of the Cretaceous (35–75 million years).

If we consider the subsequent period, we see that in the scale of the

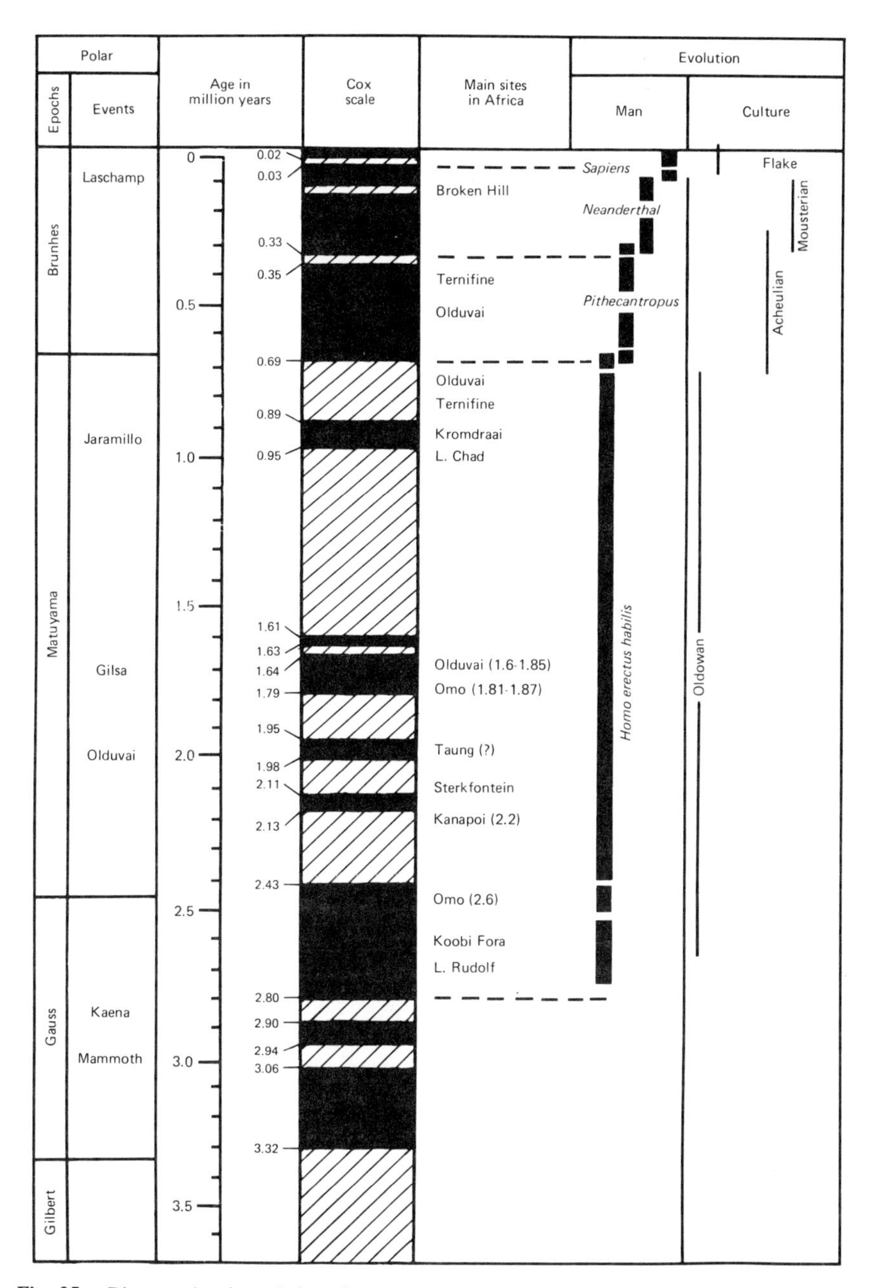

Fig. 25. Diagram showing relation of man and his culture with paleomagnetic epochs. In the Cox scale the dashed region indicates reverse polarity, and the black shading indicates normal polarity. From (336).

Neogene–Quaternary period in the earth's development, created on the basis of a radiological analysis and paleomagnetic data (783, 784), there were four distinct major epochs of different polarity: the Gilbert epoch (reverse polarity), 3.32–4.45 million years ago; the Gauss epoch (normal polarity), 2.43–3.32 million years ago; the Matsuyama epoch (reverse polarity), 0.69–2.43 million years ago; and the Brunhes epoch (normal polarity), from 0 to 0.69 million years ago, which is still continuing.

It should be noted here that the earliest finds of fossil hominids relate to the second half and end of the Gauss epoch, in which two events—the Mammoth (3.06–2.94 million years) and Kaena (2.94–2.8 million years)—were characterized by at least four changes in the polarity of the GMF. In this connection it has been suggested that a decisive factor in the appearance of hominids was the global reversal at the boundary of the Gauss and Matsuyama epochs (335, 336). Matyushin believes that the occurrence of types of fossil man in particular epochs and events suggests a correlation between GMF reversals and the appearance of particular mutations in man in subsequent time. For instance, the appearance of *Homo habilis* in the Matsuyama epoch and archanthropes, paleoanthropes, and *Homo sapiens* coincides with the reversals in the first and second halves of the Brunhes epoch (Fig. 25).

Questions of the evolution of living organisms have always excited the human mind and constitute a puzzling page in the history of the development of life on earth. Hence, it is understandable that the new approaches to this problem, which are now becoming apparent, should be received with great interest, but also with circumspection and criticism. The argument against the geomagnetic hypothesis of the evolution of living organisms on earth is that GMF reversals mark times of distinctive microcatastrophes in geological history and sudden changes in climatic processes, which affected the evolution of the biosphere (663, 878, 1011). It has been suggested that reversals were accompanied by a sharp increase in the intensity of ionizing radiation, which is a basic cause of the appearance of mutations (336, 1127, 1179). This argument, however, can be disposed of in the following way. First of all, calculations and analysis show that the radiation hypothesis is not valid in this case (725, 878, 1105, 1187), since even in the reversal period the fluxes of charged particles would be absorbed by the atmosphere and the effect of radiation would be insignificant, not exceeding the level at the present time, when the contribution of ionizing radiation to the natural mutation rate of man is estimated as 12%. In addition, the studies of paleomagnetobiologists show that evolutionary processes anticipated geological events and sharp changes in climate, i.e., preceded them (554), and hence the major evolutionary changes in the organic world depend primarily on the change in polarity and strength of the GMF, and are not mediated by changes in geological events, climate, or radiation regime.

In this connection it has been suggested that the GMF, depending on its intensity, during the periods of reverse or normal field, can accelerate or retard the individual development of animals (606) and this, in turn, can strengthen or weaken characters and physiological properties, or even give rise to new attributes. For instance, the duration of the breeding season of animals might be reduced so much that the animals remain underdeveloped and consequently become extinct due to inadequate reproduction, and the reverse might take place if acceleration occurs—the breeding season is prolonged and the number of animals increases. The originator of this hypothesis cites the following example in support of his view. The last wave of mass extinction of terrestrial animals occurred 12,000–10,000 years B.C. and coincided with the GMF reversal 12,400 years ago. At the same time, as the author notes, there are data indicating a reduction in the size of large animals and their underdevelopment at the end of the Pleistocene, which can be attributed to the retardation following the shortening of the breeding season of the animals. In our opinion, however, there may be more general mechanisms by which the GMF acts and leads to the extinction and appearance of animal and plant species. We attribute these mechanisms to the presence in organisms of various types of functional dissymmetry—left-handed, right-handed, and symmetric (L, D, S). These enantiomorphic types of organisms react differently to the north and south geomagnetic poles, which is probably due to their own biomagnetic fields. It is understandable that when a GMF reversal occurs and the polarity changes, living organisms that have a biomagnetic field of the same polarity as that produced by the reversal will be adversely affected and perish. The only organisms that will remain alive, develop, and produce progeny will be those whose own biomagnetic field corresponds in sign to the established polarity of the GMF, and symmetric (S) forms, since they are magnetically most labile and supplement the pool of L- or D-enantiomorphs (552). A direct test of our hypothesis might be direct measurement of the magnetic field of the fossilized remains of animal or plant species, but the strength of the biomagnetic field is exceedingly low, and the probability of magnetic reversal due to various physical processes is so high, that it would be impossible to make direct measurements of the magnetic polarity of paleoorganisms, unless, of course, some indirect techniques can be devised.

We can also mention in conclusion that paleontologists think that the discovery of a correlation between GMF reversals and sudden changes in the evolution of living organisms does not lead automatically to the discovery of cause-and-effect relations (290). These relations can be established only when the mechanisms of action of real physical factors, in this case the GMF, have been determined. This comment is valid, in fact, if we regard it as referring to the relation between GMF reversals and changes in fauna and flora in remote historical epochs, when it is simply impossible to recreate the actual picture of physical events. We must point out, however, that in our work we have shown the real nature of the effect of the GMF at present on all the genetic processes governing the course of evolution—gene drift, microevolutionary processes, and change in chromosomal apparatus—and have discovered the main laws of this effect (see the section Effect of the Geomagnetic Field on Genetic Homeostasis).

We can only hope that extensive and systematic research will in the future provide even more convincing evidence of the role of paleomagnetic effects in the evolutionary transformations of the organic world.

Archeomagnetobiology

The study of the earth's magnetic field as it existed in ancient epochs is assisted by the archeomagnetic method, which was introduced by the French researcher Thellier (70, 651, 1163). It consists in investigating the magnetization of objects found in archeological excavations—dishes, pots, bricks, remains of hearths, etc. These objects, constructed mainly from clay and sand, which contain many ferromagnetic compounds, acquired remanent magnetization during firing and provide evidence of the direction and strength of the GMF existing 2000–8000 years ago, since the objects can be dated by archeological methods (70).

Archeomagnetic data have provided a basis for exceptionally interesting investigations that reveal the role of the GMF in evolutionary changes in animals and man in ancient times (600–611). The direct-comparison method, described above, showed a direct relation between changes in the earth's magnetic moment and processes of acceleration and retardation in living organisms (131, 136, 598–612). Vasilik reported that a reduction in GMF intensity leads to an increase in body length, and an increase in the GMF intensity to a reduction of body length, i.e., there is an inverse relationship between the change in human body length and the change in GMF intensity ($r = -0.93$).

Human craniological series found in archeological excavations in the Ukraine were subjected to careful analysis (598, 606, 612). Calculations of the coefficients of correlation of several craniological features with the variation of the earth's magnetic moment over a period of 5000 years gave the following results: for the skull base length 0.09; for the height diameter $r = 0.67 \pm 0.14$; and for the face base length $r = -0.72 \pm 0.20$ ($p > 0.99$).

The data obtained from the study of craniological series of ancient inhabitants of the Ukraine and indicating an important role of the GMF in evolutionary anthropology were confirmed by independent investigations of other anthropometric indices of the population of Europe in ancient centuries (606). The study was based on generalized anthropological characteristics (1099), which were used to represent the craniological series in the form of an 11-term discriminant function (D_{11}) giving the position of the particular series of skulls between two extreme cases—broad-faced and narrow-faced. The results of the

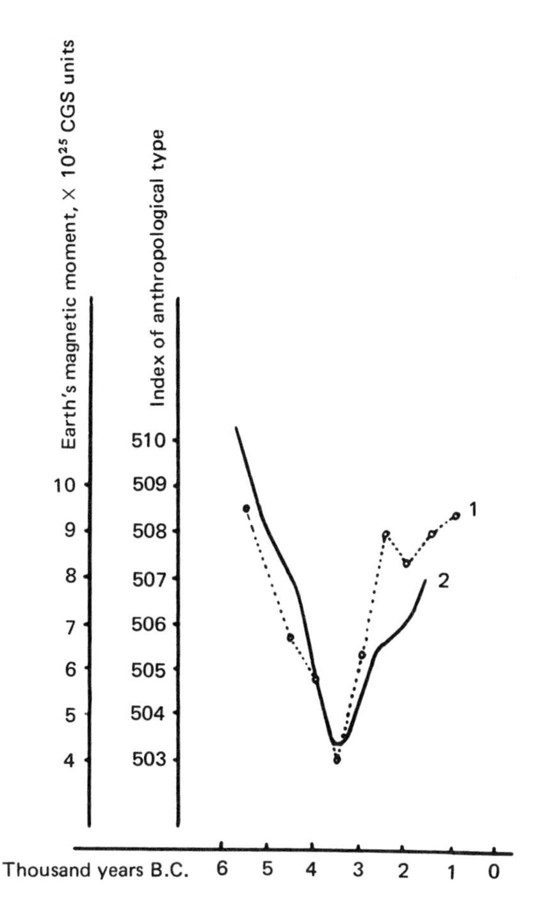

Fig. 26. Variation of earth's magnetic moment (1) and anthropological type of ancient population of Europe (2). From (606).

comparison showed that over a period of 4000 years the variation of the earth's magnetic moment and the anthropological variation of the population of Europe were almost completely synchronous. There was a gradual change in the population toward the narrow-faced type up to the middle of the fourth millenium followed, after an interval, by a similar gradual trend toward the broad-faced type (Fig. 26). These changes, of course, were accompanied by changes in other skeletal features.

The above-described variations of the physical type of the ancient population of Ukraine and Europe, correlated with the magnetic field intensity, covered periods that lasted thousand of years in the past. The conclusions drawn from the studies cited of the relation between the GMF and man's development are confirmed by an analysis of information now available (136, 603, 667). Similar data are obtained in a study of the variation of mammalian bone remains if one assumes that animal species are also subject to periods of acceleration and retardation due to changes in the GMF (602, 606, 610, 611). For instance, an analysis of the results of a study of the secular variation of the morphology of the Baltic elk showed a correlation with the variation of the GMF: the smaller the value of the earth's magnetic moment, the greater the size of the elk's bones, i.e., the inverse relationship between animal size and GMF intensity is again corroborated. Vasilik indicates, however, that the secular variations in size of the elk are also correlated with the variation of the mean annual and July temperatures. Hence in this case we cannot be sure of the predominant effect of the GMF.

Nevertheless, Vasilik's conclusion that the strength of the GMF is an important factor in the variation of animal body size is confirmed by examples taken from investigations of the intraspecific geographic variation of the fox at the present time (606). The author gives data illustrating the variation of the total skull length of the common fox in the USSR. The smallest skull lengths of this fox are found in two regions of the USSR—on the shore of the Caspian Sea and northwest of Lake Baikal, which in 1930–1935 and 1962 were centers of geomagnetic regions in which the strength of the GMF was highest. The effect of climate in this case is ruled out, since, as the author notes, foxes with skulls of the same size are found in regions with very different climatic conditions (e.g., the Volga River region and Yakutia). The author concludes that the GMF is the main factor affecting the rate of growth and development of foxes and is responsible for their size and distribution.

Thus, it is apparent from the above account that the GMF is one of the leading factors affecting the rate of growth and development of animals in the past and present, and is even more important than the known weather and climatic factors. To this general conclusion we must add one important comment. The highly important role in acceleration and retardation of living organisms is due not only and not so much to the total intensity of the GMF, as to even finer variations, namely, changes in *particular* elements of the GMF, which govern the dynamics of the growth indices of the flora and fauna in specific periods of time. We base this fundamental conclusion on the results of study of the effect of the GMF on the growth responses of animals and plants (124, 127) and on genetic homeostasis, which is the fundamental principle in the regulatory systems of living organisms (122, 138).

Changes in the total strength of the GMF undoubtedly affect living objects, but their effect can be traced only in an analysis of the *long-period* components of the variation of GMF intensity (hundreds or thousands of years). It is difficult at present to detect this effect over short periods of time, since it is masked by the effect of diurnal, seasonal, and annual variations of *particular* elements of the GMF. We shall discuss these aspects of the biological action of the GMF later, in the account of acceleration in man at the present time and in the examination of the effect of the GMF on genetic homeostasis.

Effect of Geomagnetic Field on Genetic Homeostasis

In previous chapters we briefly discussed questions relating to the effect of the GMF on the development of living organisms in remote historical epochs. Organic life on earth is a unified whole, an integral continuous process of appearance and disappearance of living forms.

There is no doubt that it owes this unity to the genetic systems present in any living organism. All the changes in the characteristics and functions of living creatures are directly or indirectly determined by the genetic apparatus and, primarily, by hereditary structures—the chromosomes. Genetic systems determine all kinds of biological phenomena—from physiological to evolutionary. In view of this we must show the decisive role of the GMF for genetic systems, since only then will the whole range of diverse effects of the GMF on living organisms be explicable and understandable.

This role can be revealed in two ways: first, by showing that the GMF is implicated in the creation of genetic systems and, second, by showing that the GMF continuously modifies these systems, interferes directly with their function, and has a decisive effect in comparison with other factors. In other words, we must show an effect of the GMF on genetic homeostasis—the dynamic equilibrium of the genetic system and the external environment, its flexible behavior—the stability of characters and properties and, at the same time, their variability. It is genetic homeostasis that underlies the stability, constancy, and conservation of characters and properties and the continuous changes, fluctuations, and lability, that we see in life, i.e., the appearance of numerous phenotypic and genotypic differences in organisms, which is covered by the general concept of polymorphism of populations (802, 803, 982). Thus genetic systems must, on the one hand, be flexible and respond appropriately to variations of the GMF and, on the other hand, be conservative and ensure the constancy of genetic characters, the precise and correct transmission of dominant genes, and the retention of recessive genes, i.e., the whole genetic complex that gives the organism advantages in the struggle for existence in biocenoses and in adaptation to the usual physical factors of the environment. This dynamic variation of the response of the genotype to the action of external environmental factors has been called developmental homeostasis (804).

This process is closely linked with mutational variation, and hence we must begin with its consideration. In our research in the Institute of General Genetics in 1966–1968 we first discovered the decisive role of the GMF for the mutational process at all its levels. The mutational process at the chromosome, gene, and genome level, with its variation of gene frequencies, chromosomal polymorphism, and sexual recombination, is the basis of evolution. Hence, it is apparent how important it is to demonstrate the role of the GMF in these processes.

To demonstrate this effect we used the above-described method of direct comparison. Our analysis showed complete synchronism of the periodic and cyclic fluctuations of the parameters of genetic processes and the variation of the GMF. The studies discussed below, taken from numerous genetic investigations and investigated by us, were selected on the following basis: first, they were carried out by geneticists as a result of major *continuous* research whose conclusions were validated by the high level of statistical significance of the results, and, second, the authors of the cited works *give the exact dates* (day, month, year) and *location* of the investigations, which is very important for our analysis. We will consider in turn all levels of genetic systems and the effect of the GMF on them.

We have already referred to the close relation between GMF reversals and evolutionary changes in living organisms. It should be borne in mind, however, that the sharp changes in numbers of species of living organisms revealed by paleontological strata were not, in fact, an instantaneous event or process. One must remember that each centimeter of an investigated paleomagnetic column represents events that lasted at least a million years and embodies a compressed, concentrated time of the past, and the events of those times. Those events, like the GMF reversals themselves, were not sudden, like an unexpected volcanic eruption or the incursion of a meteorite, but were spread over some time, and the GMF reversals themselves, as such, were merely the culminating phase of slow changes of the GMF in the direction of normal or reverse polarity prior to this time. Hence, during this transitional period, lasting approximately 5000–10,000 years, successive generations of living organisms were subjected to the action of a varying GMF and had to adapt themselves to the upward or downward trend of the total strength of the GMF and its individual elements.

The evolutionary process in principle involves mutability, i.e., a change in gene frequency in the population, and a change in its genetic structure due to fluctuations of the numbers of individuals in populations. Hence, the rate of evolution depends mainly on the frequency of mutations, particularly gene and chromosome mutations. Genetic investigations have revealed that the frequency of gene mutation and chromosome mutability in the same population are different at different times, but no explanation of this effect has been given or, more precisely, all such changes have been attributed to the action of unknown forces of natural selection (802, 982).

To reveal the role of the GMF in these processes, we carefully investigated studies of population polymorphism by genetic evolutionists. In particular, we investigated data on balanced polymorphism, which is one of the mechanisms of genetic population homeostasis. In one of these studies the investigators reported

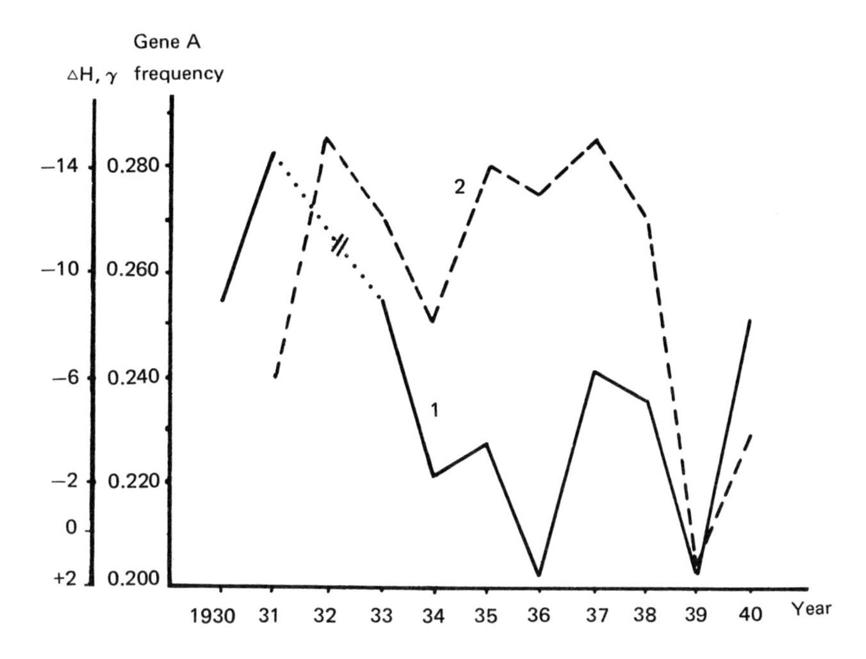

Fig. 27. Change in gene A frequency in *Adalia bipunctata* (1) from 1930 to 1940 in Germany and gradient of horizontal component (ΔH) of GMF (2) during the same period in Niemegk Observatory, Germany. From (136), where the data of Timofeev-Ressovskii and Svirezhev, published in *Problems of Cybernetics*, Vol. 18, Nauka, Moscow (1968), were analyzed.

fluctuations of genotype frequencies in polymorphic populations of *Adalia,* an insect whose wing-covers (elytra) have a red or black color, controlled by gene A. We showed (Fig. 27) that the annual fluctuations of gene A frequencies in *Adalia* were correlated with the GMF (136). We subsequently found (138) that enzymic polymorphism, which is connected with gene mutations, also depends on changes in the GMF (Fig. 28). In this case we cite examples of the effect of the GMF on two gene loci coding the synthesis of enzymic proteins in *Drosophila*. One of them—the enzymic protein phosphoglucomutase (1)—is sex-linked, and the other—malate dehydrogenase (2)—is autosomal. It is apparent from the figure that the changes in the frequency of appearance of particular alleles of some genes coding soluble proteins are synchronous with the changes of the GMF on quiet and disturbed days. Hence, from the cited data we must accept that the changes in allozymic alleles follow the cyclic changes in the GMF and are caused by its variations, or that the changes in gene loci and GMF depend on some other unknown factor of cosmic nature. Not only gene mutations, but also changes in chromosome structure, i.e., chromosome mutations, make a significant contribution to population variability and play an important role in the

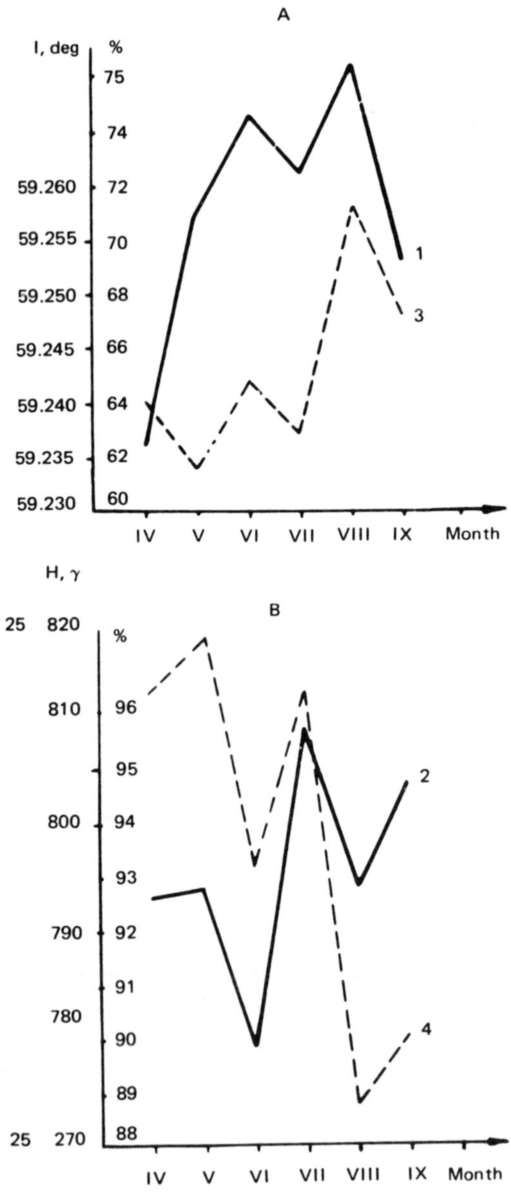

Fig. 28. Change in California in 1972 of allele frequency: (A) in sex-linked $p_{gm\text{-}1}$ locus (1) and (B) in autosomal locus Me-2 (2) in *D. persimilis* and variations of GMF elements during the same period: (3) dip (*I*); (4) horizontal component (*H*), disturbed days. Tucson Observatory. From (138), where data published by T. Dobzhansky and F. Ayala, *Proc. Nat. Acad. Sci. U.S.A.* **70**(3), 680–683 (1973), were analyzed.

Fig. 29. Seasonal changes in California, 1941–1942, in frequency of occurrence of inversions of different genes in third chromosome of *D. pseudoobscura:* Genes CH (1) and AR (2) and change in horizontal component of GMF (3) in same period. Tucson Observatory. From (138), where data by T. Dobzhansky, *Genetics* **28**(2), 166 (1943), were analyzed.

evolution of living organisms. The most thoroughly investigated chromosome mutations are structural changes like inversions and translocations, since they are most easily detected in a cytogenetic analysis. Chromosome inversions occurring in natural populations of the fruitflies *D. pseudoobscura* and *D. persimilis* have been best investigated. A chromosome inversion is the rotation of a portion within the chromosome through 180°, which leads to the appearance of a new type of chromosome with a different sequence of genes. Genetic investigations on the flies showed that the different gene sequence in the inverted chromosomes affects the organism, and in the same population of flies there were cyclic changes in the various types of inversions, the reasons for which were unknown. When we analyzed works of this kind we found (136, 138) that the seasonal variation of concentration of the sequence of genes ST and TL, CH and AR in the third chromosome of *Drosophila* salivary glands coincided with the variations of the GMF during a specific period of time in the locality where the observations were made (Figs. 29, 65). We found that the observed cyclic changes in the sequence of different genes depend on different elements of the GMF (Fig. 65), which indicates the highly specific nature of the relation between the deep structure of chromosomes and the GMF. Since chromosome polymorphism, examples of which we cite, is due to inversion of blocks of genes, we can conclude that these gene blocks are controlled by different elements of the GMF.

The correlated variation of genotype and phenotype within a population, associated with chromosome inversion, is ensured by sexual recombination of genes and chromosomes. This combinative variation is largely determined by the number of males and females in the population. Sexual reproduction ensures gene recombination and genotypic adaptation. Hence, the sex ratio of the population plays an important role in evolutionary changes in species. It is known that the separation of individuals into two different sexes depends on the genotype, i.e., the genetic apparatus is the primary link in this separation. In particular, the sex of *Drosophila* depends on the X chromosome, which carries many different genes governing the development of female characters, and on the autosomes, which contain the male genes (chromosome pairs II, III, IV).

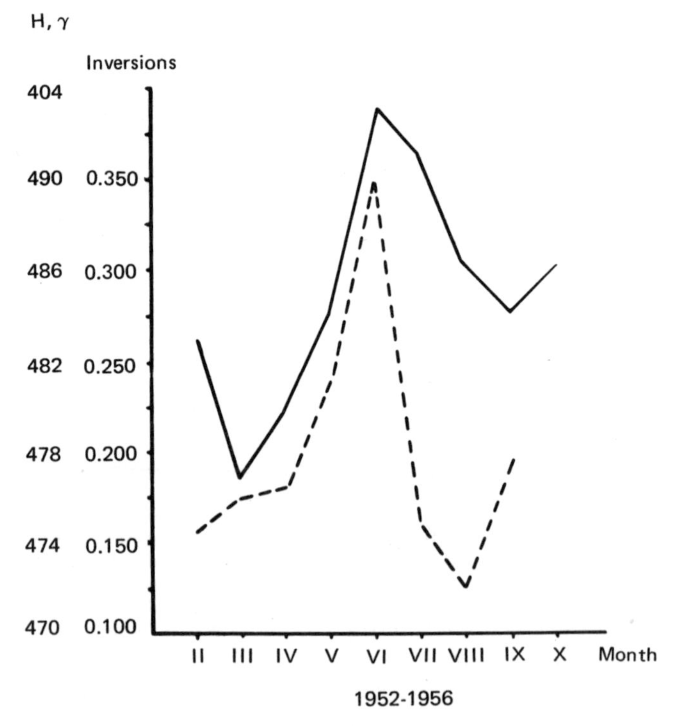

Fig. 30. Seasonal variation in frequency of inversions in X chromosome of *Drosophila* (1) at Pinon station in California in 1952–1956 and variation of horizontal component of GMF (2) in the same period. The inversion frequency was calculated for females and is given as a fraction. From (138), where the data by C. D. Epling *et al., Evolution* **2**(2), 225 (1957), were analyzed.

Our investigation showed that the GMF has a great effect on the appearance of inversions in the X chromosome of *Drosophila* (Fig. 30). This leads inevitably to a change in sexual dimorphism and the sex ratio of the progeny (138).

Our deduction that the genetic factors responsible for the sex ratio of the progeny depend on changes in the GMF is confirmed by rigorously controlled experiments in laboratory conditions. Such investigations, on diverse material, reveal that the GMF has a significant effect on the sex ratio of a population, i.e., the same result as we obtained from an analysis of the seasonal dynamics of *Drosophila* X chromosome inversions in natural conditions.

For instance, investigations on *Drosophila* larvae revealed an effect of orientation relative to the magnetic meridian on the numbers of males and females in the progeny (4). Other laboratory investigations on *Drosophila* in which the flies were subjected to a compensated GMF showed sharp changes in sex ratio due to the hypomagnetic conditions of their development (1155).

Our analysis of studies of regulation of the sex ratio in live-bearing aquarium fish (136) clearly indicated a regulatory role of the GMF in the sex ratio (Fig. 31). This, in the light of the above account, is not surprising, since a relation between the GMF and the genetic factors responsible for the sex ratio has been demonstrated.

We confirmed the universality of the relations with the GMF by an analysis of very diverse aspects of genetic homeostasis. A comparative analysis showed a

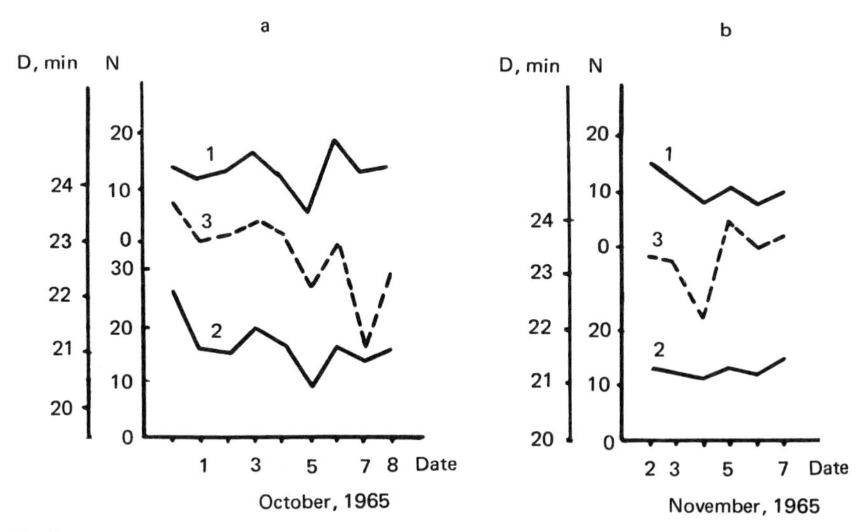

Fig. 31. Variation of numbers (*N*) of female (1) and male (2) guppies during laboratory experiments in October (a) and November (b), 1965, in Moscow and variation of declination (*D*) of GMF in the same periods (4). Krasnaya Pakhra Observatory, Moscow. From (138), where data by V. A. Geodakyan *et al.*, *Genetika* **9**(3), 15 (1967), were analyzed.

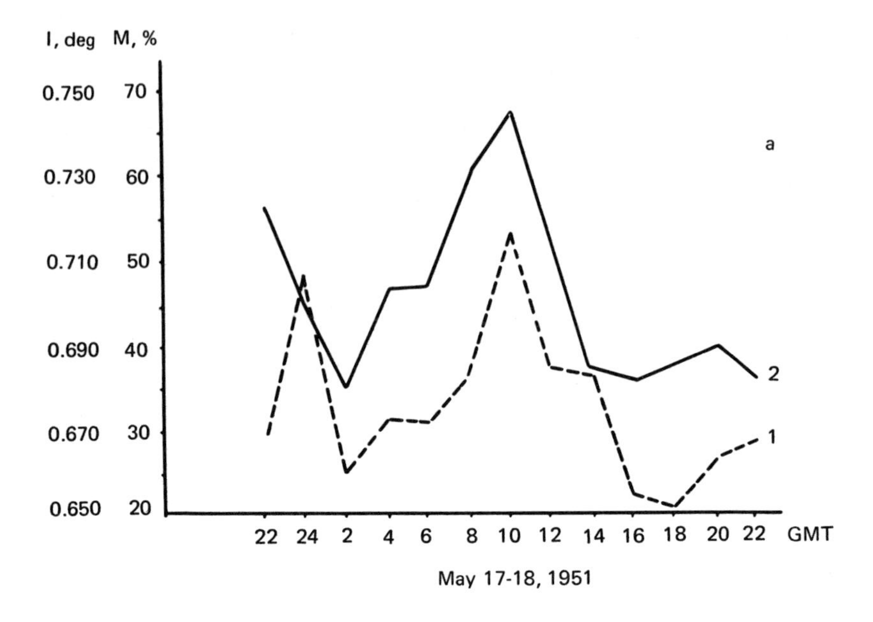

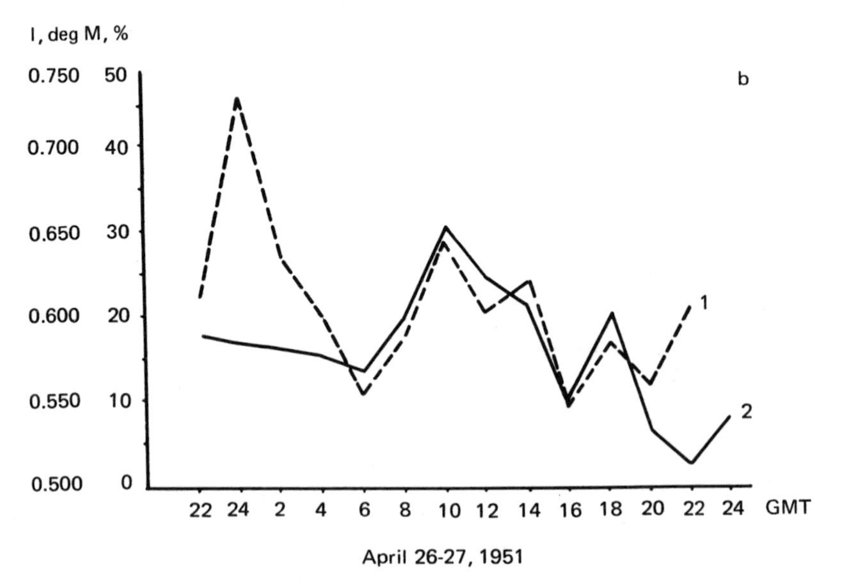

Fig. 32. Circadian rhythm of cell mitosis (M) in human skin carcinoma (1) on different days in 1951 in Helsinki and variation of elements of the GMF (2) on the same dates (*I*, inclination; *D*,

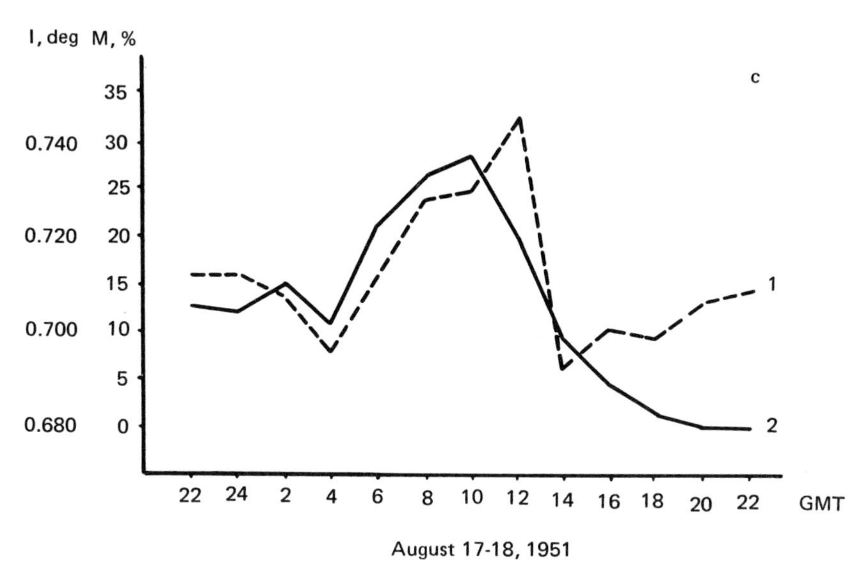

August 17-18, 1951

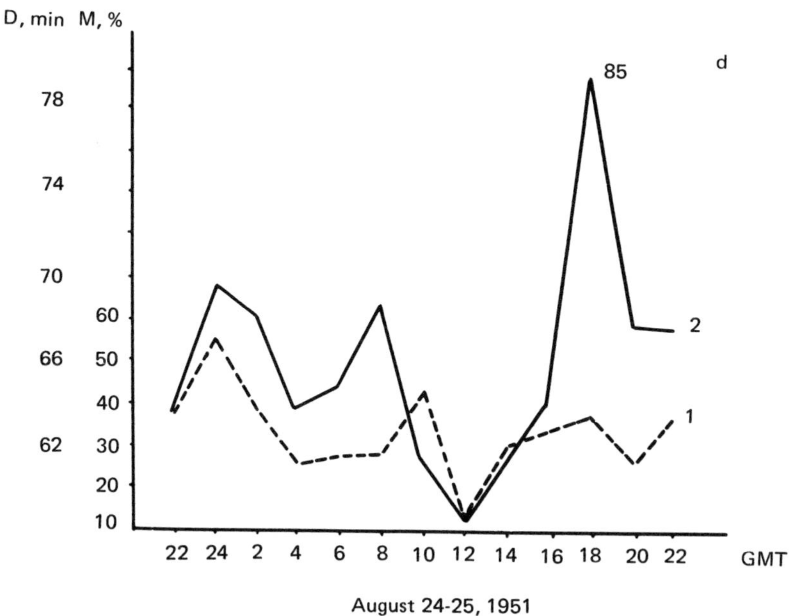

August 24-25, 1951

declination.). Voeikov Observatory, Leningrad. Greenwich mean time. From (136), where data by A. Voutilainen, *Acta Pathol. Microbiol. Scand.* **26,** Suppl. 91 (1953), were analyzed.

relation between genetic processes with different periods and the variation of the GMF. For instance, we found agreement between the curves (Figs. 32, 33) representing the variation of the GMF and a most important genetic index—the mitotic activity, i.e., the ability of the cells to divide, in an investigation of the circadian rhythm of this process in man (136).

It should be noted that the effect of the GMF on structural changes in chromosomes can be detected equally well in natural conditions (e.g., inversions in *Drosophila*) and in laboratory conditions in the case of induced mutagenesis. A good illustration of this is provided by the data that we obtained in an analysis of the change in frequency of chromosome aberrations in rat liver cells after the injection of dipine (Fig. 34). As Fig. 34 shows, the chromosome structural defects due to dipine vary in a cyclic manner similar to that of the GMF during the experiment. Moreover, our investigations showed that DNA synthesis in dividing cells, which is the main molecular mechanism of transmission of hereditary properties, also depends on the GMF. Convincing evidence of this is provided by the data for the incorporation of radioactive thymidine into plant cells (*Crepis capillaris*) during the mitotic cycle (Fig. 35).

Thus, the data presented above distinctly show that periodic and cyclic changes in the genetic apparatus are associated not only with natural selection, as geneticists believe at present, but also with variations in different elements of the GMF. Although the specific features and deep-lying mechanisms of the discovered universal relation of genetic processes with the GMF have not been completely clarified and investigated, all the data indicate that the GMF plays a decisive role in the mutagenic process.

We can cite further evidence in support of this view and suggest possible mechanisms of this relation with the GMF.

It is quite possible that, as well as acting directly on the hyperfine biomagnetic structure of molecular genetic systems, the GMF can also act on genetic processes through the membrane permeability mechanism, as occurs in physiological homeostasis. The GMF probably causes a general change in the electromagnetic conducting properties of membranes and the cell current systems associated with them, which leads to a change in the information entering the cell and to a change in genetic processes. It has been postulated that the frequency-modulated electric pulses associated with membranes are of decisive importance for the reception and conduction of information from the external environment to the internal nucleic- and amino-acid coding systems (1054, 1111).

On the other hand, our hypothesis of the decisive role of membrane permeability in genetic processes is confirmed by theoretical (1125) and experimental studies of the effect of various ions on gene activity in the giant chromosomes of dipterans (853, 948, 949) and on growth and differentiation of *Hydra* (965). It is understandable that the effect produced by ions depends greatly on their concentration, and the rate of entry to decisive genetic structures through the mem-

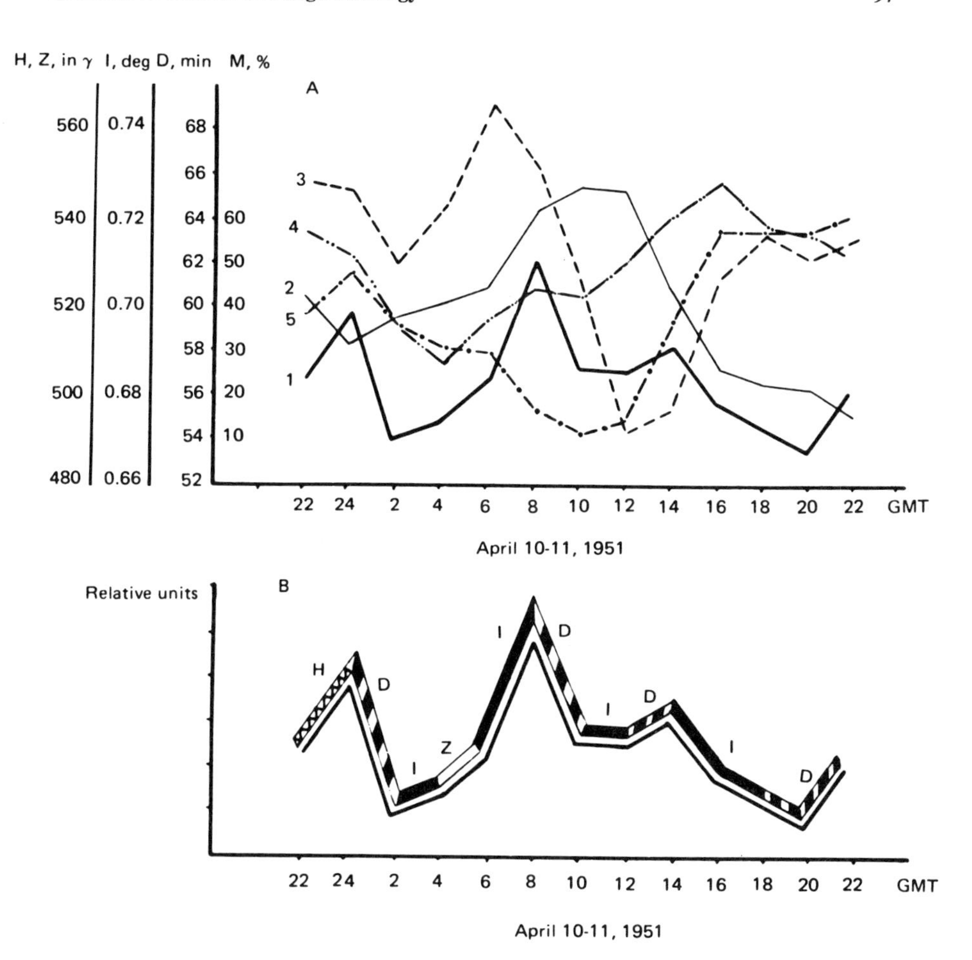

Fig. 33. (A) Circadian rhythm of mitosis (M) in human skin carcinoma cells (1) in Helsinki, April 10–11, 1951, and variation of GMF elements on these days: (*I*) dip (2), (*D*) declination (3), *Z*) vertical component (4), and (*H*) horizontal component (5). Other symbols as in Fig. 32. (B) Scheme showing possible effect of individual GMF elements on formation of circadian rhythm of cell mitoses on April 10–11, 1951. Each GMF element at the times indicated was connected with the rhythm of mitosis by a correlation coefficient $r = +0.99$.

branes, since this determines the induction or repression of the corresponding genes.

Thus the GMF, acting through the membrane permeability mechanism, can control the state of genetic systems and processes, and produce cyclic changes in them, although it is not impossible and, in fact, is highly probable, as we showed above and as we shall discuss later, that the GMF has a direct effect on the genetic code and on the gene as such.

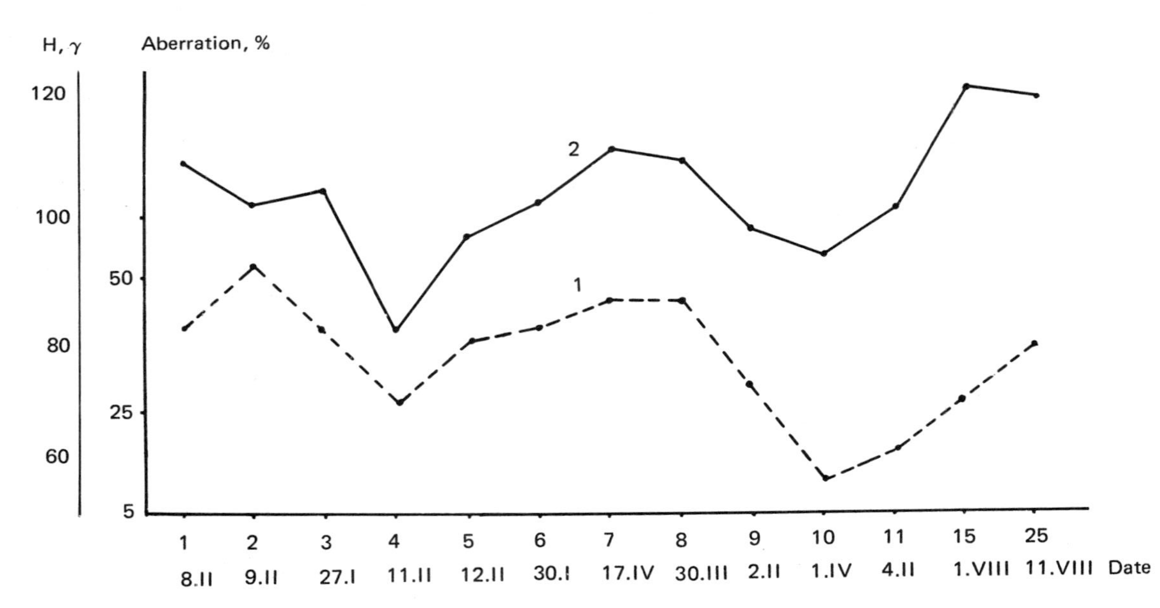

Fig. 34. Variation of frequency of chromosome aberrations in rat liver cells at different times after injection of dipine (1) and variation of horizontal component of GMF (2) during experiment in Moscow in 1968. *x* axis, days before hepatectomy from 1 to 25, with corresponding dates given personally by the authors of the investigation. The magnetic data for the same dates were obtained from Krasnaya Pakhra Observatory, Moscow. Analyzed in (138) from data of M. Movsisyan and A. P. Akif'ev, *Dokl. Akad. Nauk SSSR* **188**(3), 685 (1969).

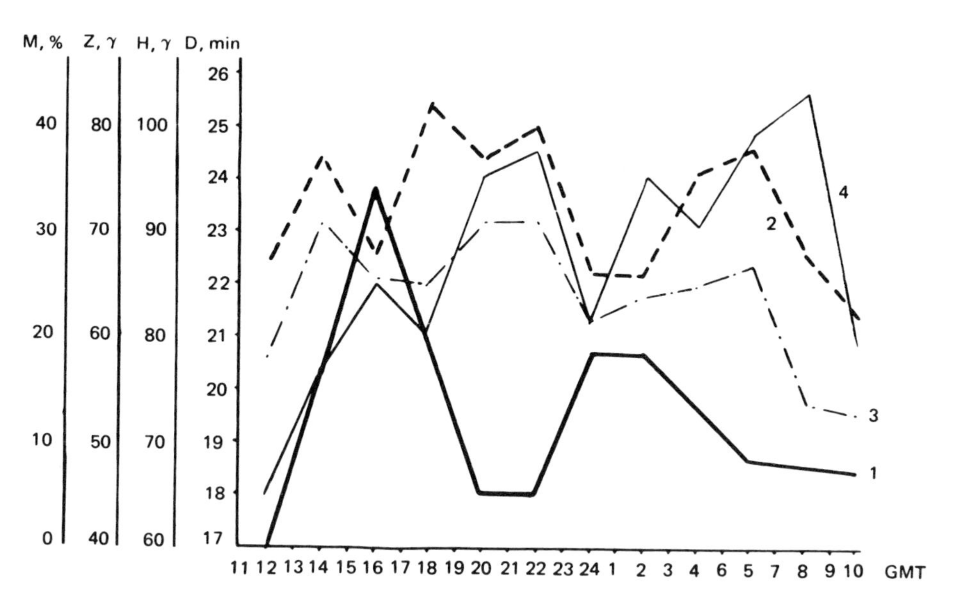

Fig. 35. Circadian rhythm of mitoses (M) labeled with radioactive thymidine in *Crepis* radicle cells (1) in the period March 17–18, 1966, in Moscow and variation of GMF elements during this period: (*H*) horizontal (2) and (*Z*) vertical (3) components, and (*D*) declination (4). Krasnaya Pakhra Observatory, Moscow. Greenwich mean time. Analyzed in (138) from data of Dr. G. Makedonov (Inst. of General Genetics, Moscow), personal communication.

Acceptance of the fact that the GMF is one of the main causes of cyclicity and periodicity in genetic systems of living organisms makes it a very important factor in the evolution of the organic world. Recognition of the leading role of the GMF in the evolution of the biosphere provides an explanation of some fundamental biological laws. For instance, Vavilov's law of homologous series, which reflects the objective genetic laws existing in nature and is expressed in the similarity of the cycles and a definite regularity of hereditary variation in phylogenetically close species, genera, and families, confirms our hypothesis of the universality of the effect of the GMF on genetic homeostasis and biological symmetry. The basis for this is the facts, discovered by us, indicating that the mutation process at all its levels involves a GMF effect of the same type on blocks of genes responsible for the genotypic and phenotypic features of close species, genera, and families.

In addition, our view on the role of the GMF in genetic processes provides an explanation of "spontaneous" mutagenesis. The reasons for mass mutations, leading to a sudden increase in their number in populations, and "systemic" mutations, which are responsible for the appearance of new species and genera, become understandable. These types of mutations are the result of sharp changes in the GMF.

From our standpoint we also have a good explanation of microevolutionary processes, which depend on gene mutations and recombinations. It is the connection between genetic systems and the GMF that accounts for the microevolutionary processes occurring at the present time: the sudden appearance of new types and the succession of types of viruses, the appearance of latent viruses in the animal organism, the appearance of new forms of bacteria, the synchronous appearance of mutations in different species of animals, insects, and plants, and also the different incidence of genetic diseases in man and animals in different years. These processes have been investigated (57, 647, 648) and now find a good logical explanation and complete experimental verification (15, 16, 18, 22, 412–415).

Underlying all the above-mentioned processes are molecular and submolecular changes in the genome of living organisms due to the GMF. These "geomagnetic mutations," as we might call them to distinguish them from mutations due to sexual recombination, are particularly well illustrated by bacteria and viruses, in view of their high rate of reproduction, rapid succession of generations, the possible stabilization of these mutations by natural selection, and the very wide range of variation of the parameters of the external environment.

On the basis of the view expressed above we can propose an explanation of the mechanism of carcinogenesis of viral etiology. Viral cells (e.g., the herpes virus), as is known, can exist in the organisms of higher animals for a long time without causing it any harm. A similar phenomenon is found in bacterial cells and is called lysogeny, where the unexpected death of cells is due to the activity of specific bacterial viruses (bacteriophages) contained in them.

As we see, in both cases the virus is in a latent repressed state for a long time and then becomes active, multiplies rapidly, and kills the host organism. In our view the normal mechanism of the genetic "cell–virus" relation is disrupted by sharp disturbances of the GMF, in addition to other factors. Owing to the change in the status quo oncogenic viruses escape from the control of the host cells and multiply rapidly, which leads to development of the pathological process. This hypothesis is confirmed by investigations showing that the level of spontaneous lysogeny, i.e., the death of bacteria, e.g., *E. coli* K-12, due to the multiplying bacteriophages, depends on solar activity and the GMF (96, 313). The GMF may have a specific enhanced effect on migrating genes (episomes); this is indicated by experiment at any rate (412).

Thus we can conclude that the GMF affects the deep-lying mechanisms of heredity—the formation and functioning of the gene and the ontogenetic mechanisms associated with it. The problem of future research will be to decipher these intimate relations between the GMF and the ultrastructure of the gene.

This relation may be found where we least expect it. For instance, it may be

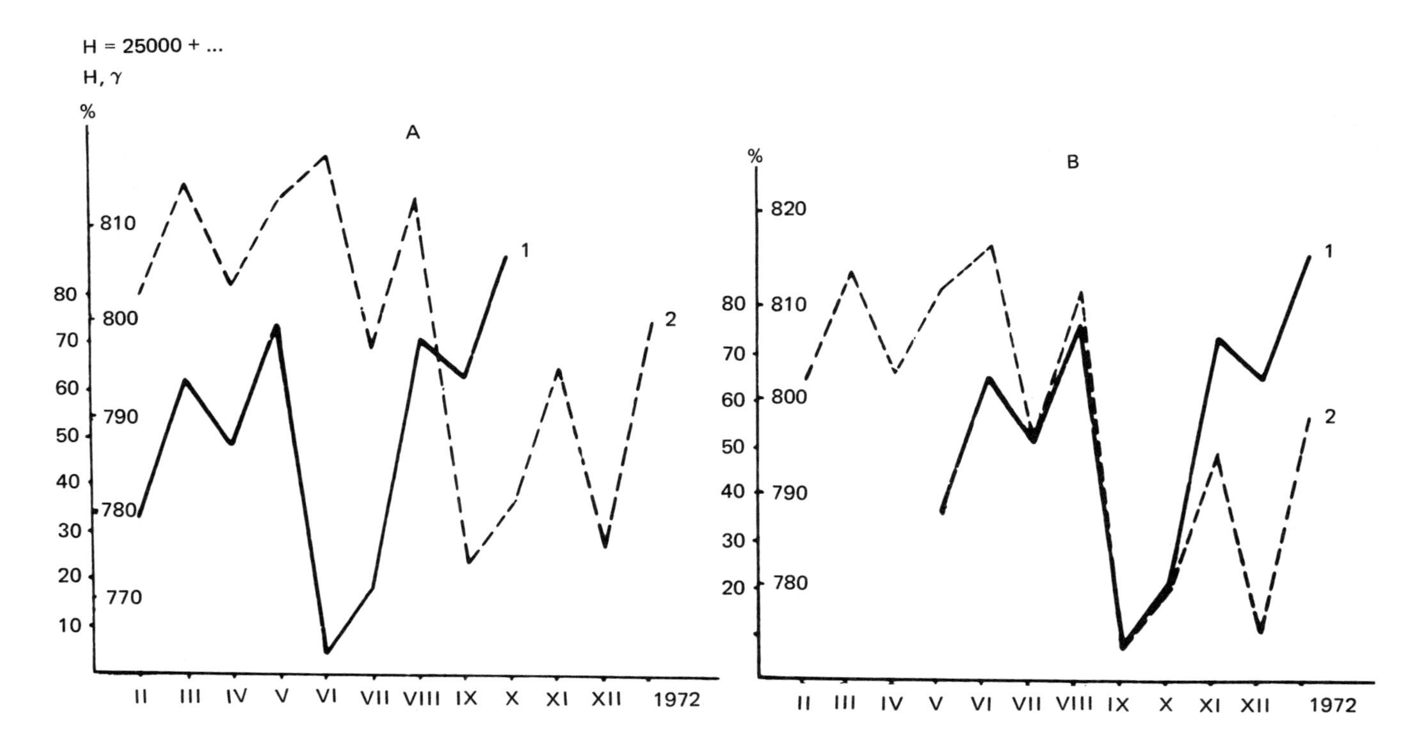

Fig. 36. Seasonal variation in California, 1972, of frequency of alleles in sex-linked locus Pgm-1 of *D. pseudoobscura* (1) and level of horizontal component of GMF (*H*) on disturbed days during the same period (2). The magnetic data were obtained from Tucson Observatory. (A) Actual course of investigated indices (1) and (2); (B) course of indices 1 and 2 when GMF variation is displaced two months. Analysis of data in T. Dobzhansky and F. Ayala, *Proc. Nat. Acad. Sci. U.S.A.* **70**(3), 680–683, Table 2 (1973).

associated with biosymmetry or the effect of the GMF on biological space and time, and it may be that an understanding of the facts, apparently simple at present, of a relation between genetic homeostasis and the GMF, will require an approach from the viewpoint of the topology of the biological microworld (682, 811, 813, 814). At any rate, such approaches have already been made. For instance, a correspondence between the general structure of the genetic code, the 2^6 binomial expansion, and the icosahedron has been established (638, 639). The author of these papers believes that pentameric symmetry is fundamental in the organization of living matter.

We would like to warn the reader against a simplified view of the relation between the GMF and biological processes, and, in particular, genetic systems. How complex this relation may be is illustrated by Fig. 36, which shows the seasonal variation of the frequency of one of two alleles coding the synthesis of an enzyme in *Drosophila* and the variation of the GMF in the same period. Two features stand out in this graph: First, the remarkable similarity between the variation of the gene frequency and that of the *H* component of the GMF (A), and second, if these curves are displaced by two months, we find (B) that the variation of the frequency of the alleles anticipates precisely the variations of the GMF (!). This indicates, in our view, that there is some other important cosmic factor that produces a response in living organisms much earlier than in the GMF. Such cases have already been described in the scientific literature (92, 767, etc.).

The Geomagnetic Field and Biological Rhythmicity

All biological and inorganic processes take place in time and space. This is why we have devoted here so much attention to studies of biological rhythmicity in which the exact date and location of the investigations are given. It is these circumstances that help us discover the very important role of the GMF for living and nonliving matter, and they form the basis of this entire volume. Their importance is very well expressed in (503):

> In particular, in the publication of experimental data it is essential to indicate the circumstances of the organization of the experiment in time, and also the calendar date [''and location'' should have been added] of its conduction. These dates not only increase the significance of the published data for reliable comparisons, but are also of independent scientific value: With such information previously obtained data can be used to determine the characteristics of the variation of specific reactions of an object in the course of time in a thorough diachronic investigation.

There are now many major reviews of biological rhythmicity (513, 746, 869, 870, 892, 893, 898, 1009). The science of biological rhythmicity—*chronobiology*—is developing very rapidly. Its rapid growth is due not only to the great scientific significance of its ideas, but also to the exceptional practical importance of this research for medicine, agriculture, and many other areas of natural science, including space biology (869–872).

One must not forget, however, that chronobiology is not an isolated science. Its vigorous growth and fruitful development reveal that it is a branch of a mightier tree—bioactochronics (814).

The essence of this science is that biological objects, by virtue of the unusual characteristics of living matter, interact with real time and the space surrounding them, and are not simply carried along in the fast-flowing river of time. If this is so, however, biological objects will interact continuously and closely with the surrounding fields, matter, and various forms of energy.

The first thing that is revealed by an analysis of biological objects in time is their fluctuating nature and the variability of biological characters in time and space. Chronobiology is concerned with the rhythmicity of the most diverse biological processes—cellular and populational, nervous and humoral, physiological and genetic. The duration of these periods varies from hundreds of years to microseconds. In all this we can see continual changes, oscillations, regular and irregular variations, cyclicity, and an endless sequence of phases and stages. It may appear at first glance that in this world of variations, this chaos of changes of all kinds, there is no unity and no single regulator or director. A careful analysis, however, convinces us that there are worldwide synchronizers, whose number and the significance that we ascribe to them depends on how far science has advanced in its analysis of the rhythmicity of natural processes and phenomena.

We must draw attention to one very important point that we mentioned at the start of this chapter. All biological rhythmic processes have been investigated by researchers at a *particular* real time and at a specific point on the earth. This is a very important circumstance and one that has hitherto been underestimated by investigators. If we turn our attention to this feature of biological rhythmicity we can discover its connection with one of these synchronizers—the GMF. This connection has been consistently demonstrated in all our studies, and numerous new confirmations of this relation are given in this book (Fig. 37).

A careful analysis of these studies reveals that variations of all the GMF elements may be involved in biological rhythmicity of any type—circadian, seasonal, annual, many-year, and also in any biological processes—physiological, biochemical, genetic, and biophysical.

Investigations in this region led us to the conclusion that the close relation discovered between biological rhythmicity and the GMF is based on one fundamental process—the close relationship between the permeability of biological membranes and the GMF.

To make our claims more convincing and better understood we have given a number of examples based on an analysis of investigations carried out by dif-

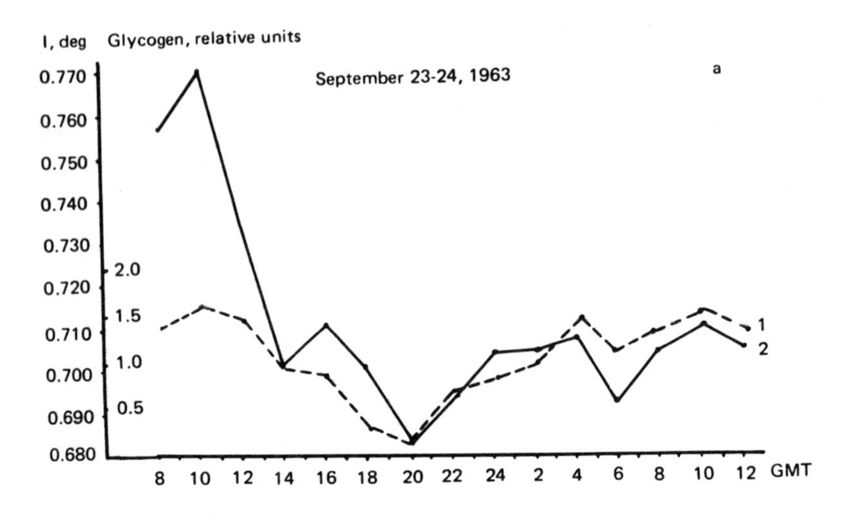

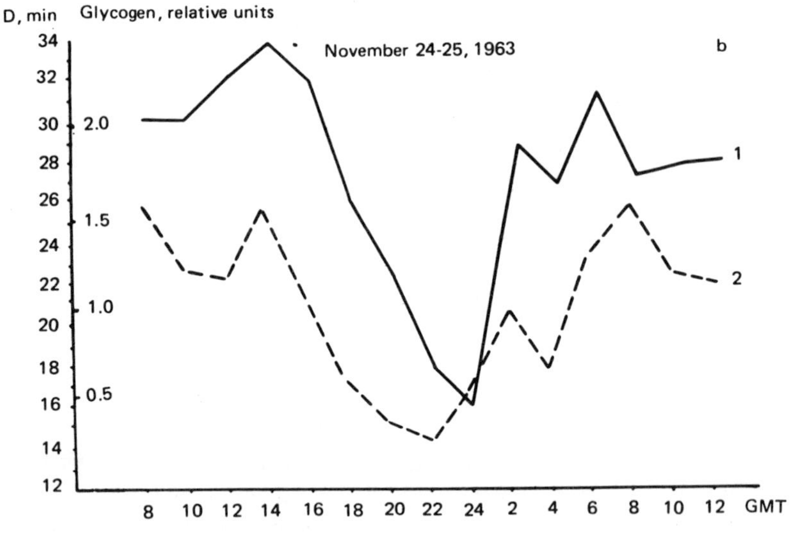

Fig. 37. Circadian rhythm of glycogen content of liver of female rats (1) and variation of GMF elements on different dates in 1963 (2). Analysis of work carried out by H. von Mayersbach in Nijmegen, Holland. The exact dates and site of the experiments were kindly communicated person-

ferent chronobiologists in different countries (Figs. 32–37). As these figures show, the circadian rhythm of any biological process is synphasic and synchronous with the variations of the GMF at the site of the experimental investigations. We must emphasize that the examples cited of coincident variation of biological processes and the GMF are based on studies carried out at *different* times by

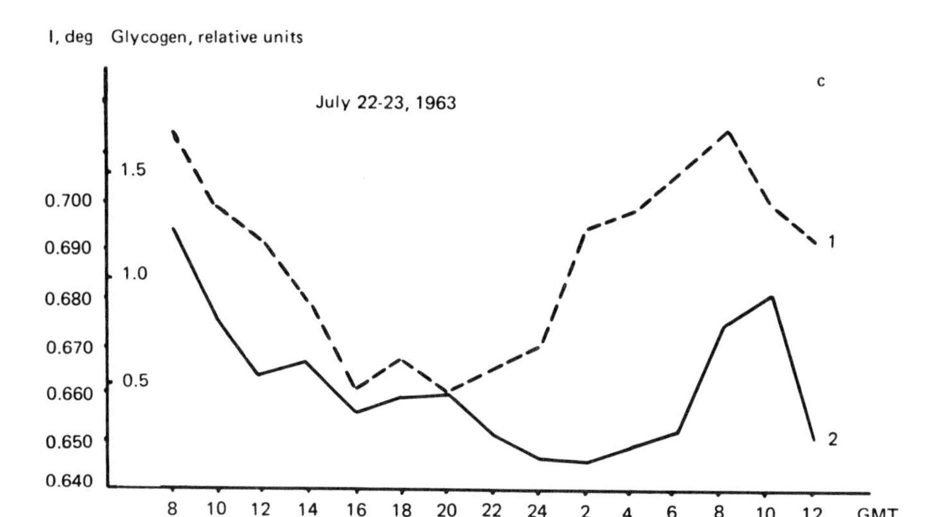

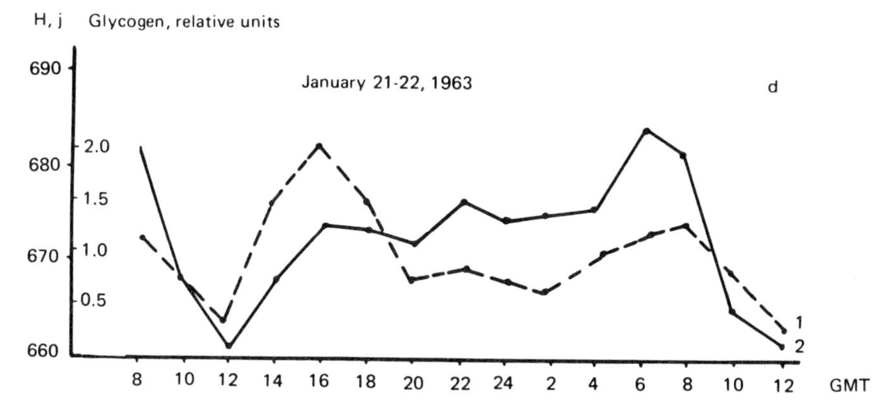

ally by Prof. H. von Mayersbach in 1970. Magnetic data taken from Witteveen Observatory, Holland. Greenwich mean time. From (809).

completely *different* investigators at *different* geographic points on the earth. There is nothing surprising in this from our viewpoint, since the relation that we discovered between the permeability of biological membranes and the GMF is a fundamental law of an all-embracing nature, acting universally. That we are able to show the correctness of our proposed ideas by such very diverse investigations is an additional and sound criterion of the validity of the discovered laws.

We must mention another important feature of the law that we have discovered: Biological rhythmicity is controlled by the change in diurnal variations of all the GMF elements. At each specific time of day the decisive factor can be any of the GMF components, its gradient, and its rate of change in time. This component is replaced by another GMF component with its own characteristic features, and so on, throughout the whole period of the biological process. Hence, the examples that we give show a correspondence between biological rhythmicity and variations of the H, Z, I, and D components of the GMF (Fig. 37). The analysis of the biorhythms would be more complete in a number of cases if we took into account the north and east components, since the GMF variations would then be described more accurately. Such a comparison, however, would require specially conducted biological research on biorhythmicity. But even when we take into account the four GMF components we can completely describe and account for the features of the biological rhythms (Figs. 33–38). As the figures show, the nature of the circadian biological rhythmicity corresponds exactly with the monthly course of diurnal variations of natural electromagnetic fields (Figs. 32, 37). In our work we usually cite examples of correspondence between the biological rhythm and the GMF component corresponding most completely with the biological process at any particular period of the day, and do not show the diurnal variation of the other GMF components because we do not want to overcomplicate the figure, and want to make it easier for the reader to analyze the figure and understand the basis of our approach to the problem of biorhythms. In principle we ought to provide a picture of the diurnal course of all the GMF elements for a complete analysis of the biorhythms (Fig. 33).

At the present moment we still do not know all the features of the relationship between biological objects and the GMF. Hence, we can postulate that the indicated variable effect of GMF components on biorhythms owes its origin to deep processes in the structure and functioning of biological membranes (e.g., to the phenomenon of biosuperconductivity). Confirmation of our thesis that the GMF acts on biological objects through permeability of the biomembranes can be found in many investigations: by doctors who have analyzed the variation of vascular permeability in relation to the GMF and have clearly reported this relation (164, 207, 208) and by botanists (755, 1151). This is particularly well illustrated by one of the works cited (755), where the author gives the results of a study of the circadian rhythm of gas exchange in dry onion seed, in which in the dormant stage there is no DNA replication, transcription of the DNA template to the RNA carrier, and no translation from informational RNA to protein. The author of the cited work correctly acknowledges all this, but comes to the conclusion that the energy cycle may be the main oscillator of the gas exchange rhythm in dry onion seeds. There is no doubt that energy exchange processes take place in a biological object so long as it is alive. These processes occur in the complex

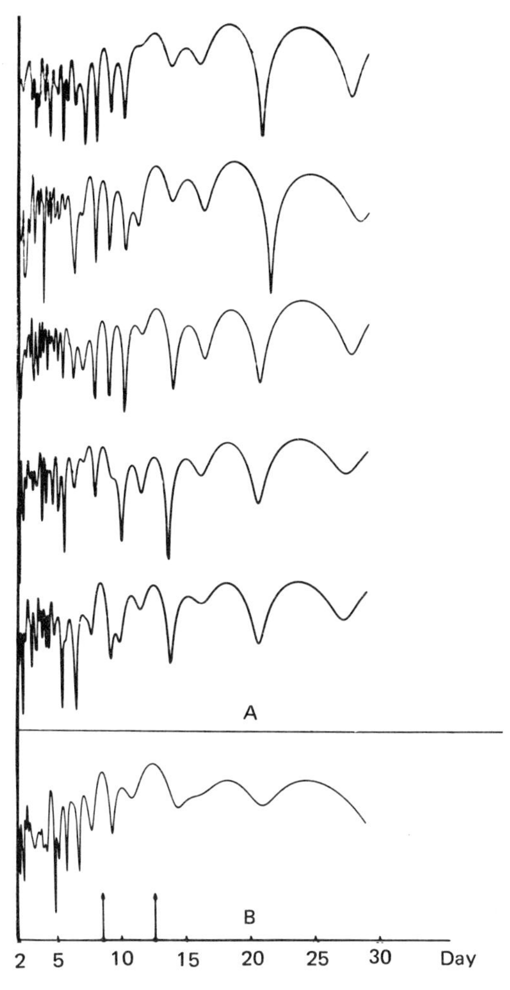

Fig. 38. Comparison of periodograms obtained from daily analyses of timber infested with five groups of termites (A) and magnetic disturbance (B) for 14-day periods of magnetic disturbance, selected and analyzed by computer from 82 days. From (692).

heterogeneous cell system with a vast mass of intracellular structures, distinguished by the presence of membranes, partitions, and walls, separated from one another by spaces. This is called compartmentalization in biology and is well exemplified by the principal energy structures—mitochondria and chloroplasts. Hence, energy processes themselves depend on the deep interrelation of the GMF with membranous structures and with the processes occurring directly in them, while the permeability, varying rhythmically, for substances of any nature, including gases, is merely a reflection of this relation between the GMF and the molecules and submolecular structures of membranes.

For instance, the changes in water molecules, which are contained in the ultramicroscopic pores and the membranes themselves and are responsible for their liquid-crystal structure (64), are of great importance. We showed (464–466) that the GMF had an orienting effect on crystals of various substances, and it is possible that a similar effect occurs in the liquid-crystal structure of biomembranes. It should be noted in general that water molecules, which are highly mobile and labile, can easily be altered by the GMF and have different spin and orbital moments or hygroscopic effects with different directions of rotation around the axis. The GMF, being a vector quantity, can have, along with gravitation, a continuously varying effect on the state of elementary particles in biological molecules and water molecules by, say, inducing a vortical electric field (96). Hence, when we cite examples of the synchronizing effect of sometimes one, and sometimes another, GMF component we are in fact, merely reporting the leading effect of *changes of the GMF vector* in space. The diurnal variation of the GMF vector in space, called the diurnal hodograph of the GMF, is of a complex nature, varies in shape, and shows continuous variation of the GMF in time and space (Fig. 3). A future problem of geomagnetobiology will be to determine the effect of the characteristic features of the diurnal course of the hodograph of the earth's electromagnetic field on biological rhythmicity. For instance, a study of the effect of the degree of polarization of the GMF hodograph on biological rhythms is of great interest, since polarization of the hodograph, like the GMF variations themselves, greatly depends on the geographical latitude of the point of observation. At the same time, the discovered features of the mechanism of the biological action of the GMF have enabled us to approach the question of normal responses of biological objects to the GMF and the specificity of biological rhythms from new standpoints. We have found that biological rhythms can be subdivided into three main groups: "left-handed," "right-handed," and symmetric (125, 126, 135). This division of biological rhythms reflects the specific responses of biological enantiomorphs to the action of any physical factors, including the GMF. This question will be discussed in detail in the chapter dealing with the effect of the GMF on the symmetric properties of biological objects.

Unfortunately, the leading role of the GMF in biological processes is ignored in current chronobiological research. The huge number of publications on circadian and seasonal rhythms deals with the effect of only such important exogenous factors as light, temperature, and nutrition, and ignores the underlying bases of biological rhythmicity, i.e., the role of the GMF and other physical factors—the synchronizers of biorhythms (Fig. 38).

If the laws that we have discovered are accepted as valid, then chronobiological research should include not only the usual study of circadian rhythmicity, but also thorough investigations of the *main causes* of circadian rhythms. As our analysis shows, these causes do not consist solely of the GMF, although it is the *main* synchronizer (*Zeitgeber*) setting the rhythm. The role of traditional factors affecting biorhythms, like light and temperature, must be considered from different standpoints. Contemporary science correctly attributes to these factors a very important influence on the physiology of living organisms, but their action is regarded solely from the viewpoint of the effect on the visual apparatus (animal) or pigment system (plant) in the case of light and the effect on the rate of metabolic reactions in the living organism in the case of temperature. Their effect, however, is much more diverse, and in particular, their role in biological rhythmicity is due to the ability of both these factors to induce in living organisms, as in semiconducting systems, complex dynamic changes in photocurrents and thermocurrents, which affect the current systems in membranes due to the action of the GMF and the manifestation of biorhythms as a whole. This action of photocurrents and thermocurrents produced in living organisms might account for the many aspects of the carefully devised and excellently conducted research on biological rhythmicity (871, 877, 892, 893, 898, 1009).

We should also mention the important role of cosmic factors in biological rhythmicity (118, 746, 1049). This is definitely indicated by the new kinds of radiations discovered by researchers: *Z* radiation (90, 766, 767), F radiation (1153), the X agent (999, 1000), and others (729). It is true that the effects they produce can largely be attributed to the action of natural magnetic and electric fields (128), but they definitely include completely new (undiscovered so far in physics (1098)), more universal forms of fields and emissions of physical factors, which can affect biological objects and also alter the GMF itself. There are cases where biological objects may respond to the action of these factors much earlier than to the GMF, since the scales of such systems as the GMF and biological objects are, of course, simply not comparable.

There is no doubt that an investigation of new aspects of biological rhythmicity will help to reveal these new and important principles.

For instance, in a series of exceptionally interesting and original investigations F. Brown and his colleagues (742–745) have shown that various kinds of living organisms—plants, worms, higher animals—can distinguish the rotation of a horizontal rotating magnetic vector close in magnitude to that of the GMF. From these studies the authors conclude that all organisms are distinctly oriented in geophysical space and in time and form part of a single integrated system in which magnetic fields play an important role. These experiments were conducted in different conditions and at different geographical points, which confirmed the great significance of the newly discovered factor. We think that these experiments can be interpreted as evidence, from the biological aspect, of the existence of a second gravitational field Ω, called the vortical gravitational field (761, 834), corresponding to the magnetic induction $\mathbf{B}$, which we will discuss in more detail later.

Hence, the conduction of synchronous experiments at different geographic points is of great importance for the study of the main synchronizing factors in biological rhythmicity, since it reveals the effect of such important global factors (144). A full program of such research, including all processes from physiocochemical to populational, has recently been proposed (139).

There are several distinctive features of such experiments: (1) At each geographical point at a particular time there is a specific heliogeophysical environment responsible for the local effect on the rhythms of living organisms in that particular area. (2) Heliogeophysical factors show, in addition to local diurnal variations, unitary global variations and hence such changes will also be reflected in the rhythms of the functional indices of biological objects or the characteristic dynamics of physicochemical processes, as was brilliantly demonstrated in the work of G. Piccardi (1049). (3) A point that is equally important and is often entirely overlooked by all researchers, synchronous experiments at different geographic points can reveal cases of simultaneous *irreproducibility* and failure of physicochemical and biological reactions (132). Table 2 briefly summarizes the results of our investigation of the irreproducibility of some biological processes. As the data in the table show, on June 5, 1968, investigators in different institutes reported the *total* failure of the processes that they investigated, whereas on June 12, 1968, the conditions were completely favorable for conducting the same experiments. We think that the logical sequence to all this must be the publication of a special international catalog of biological data that gives cases of complete failure of experiments conducted by research workers. The practical importance of such investigations and records of days of "complete irreproducibility and failure of experiments" is obvious, since this situation is equally common in laboratory and industrial processes—a subject that we will discuss in the chapter on the role of the GMF in working life.

Thus, worldwide synchronous experiments reveal an important role of local and global heliogeophysical factors and can thus lead to an elucidation of their role in biological rhythmicity.

As an example, we can cite the results of a synchronous investigation of the rhythm of excretion of organic substances by plant roots (SP test) at different geographical points (134, 144).

Synchronous investigations of the excretion of organic substances in identical laboratory conditions at several geographical points have shown that the rhythm could be very different and exhibited the following features. On the one hand, in some of the experiments there were cases where the circadian rhythm was similar at points a thousand kilometers apart (Moscow, Irkutsk, Florence) and, on the other hand, different for points situated relatively close to one another

Table 2. Results[a] of Comparison of Synchronous Failure of Various Processes in June, 1968, in Moscow and Kazan

Date	Institute[b] and process investigated							
	Institute of General Genetics	Institute of Geochemistry and Analytical Chemistry	Institute of Chemistry of Natural Compounds	Institute of Physiology	Central Institute for Higher Qualification of Doctors	Central Scientific-Research Institute of Antibiotics	All-Union Scientific-Research Institute of the Baking Industry	Kazan State University
	Effect of nitrogen mustard on viruses	Respiration of *Chlorella*	Photodynamic effect in nucleic acids Synthesis of isovaleric ester	Study of anaerobiosis in plant roots	Incorporation of ^{14}C-labeled sugar into animal blood	Fermentation for antibiotic production	Fermentative activity of yeasts	Light scattering by leaves
June 5	*	F	F F	F	F	F	F	F
June 6	*	F	F F	*	*	F	F	F
June 12	S	S	S S	S	S	S	S	S

[a]F, failure; S, success; *, not investigated.
[b]All institutes except Kazan State University are located in Moscow.

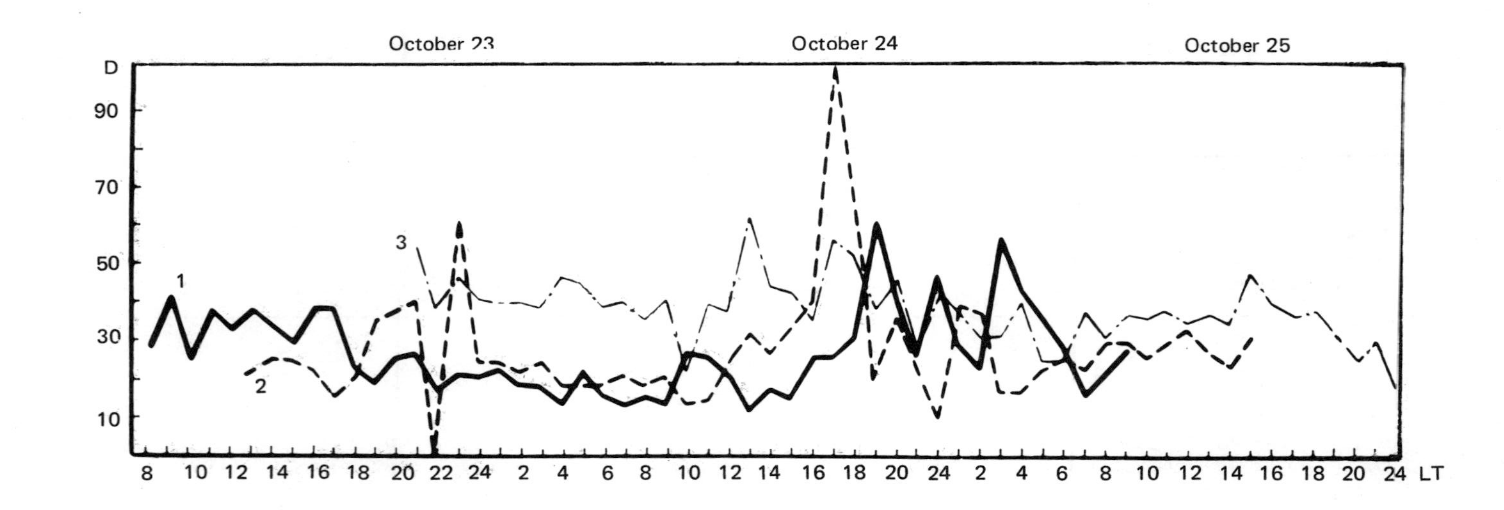
October 23
October 24
October 25
D
90
70
50
30
10
1
2
3
8 10 12 14 16 18 20 22 24 2 4 6 8 10 12 14 16 18 20 24 2 4 6 8 10 12 14 16 18 20 24 LT

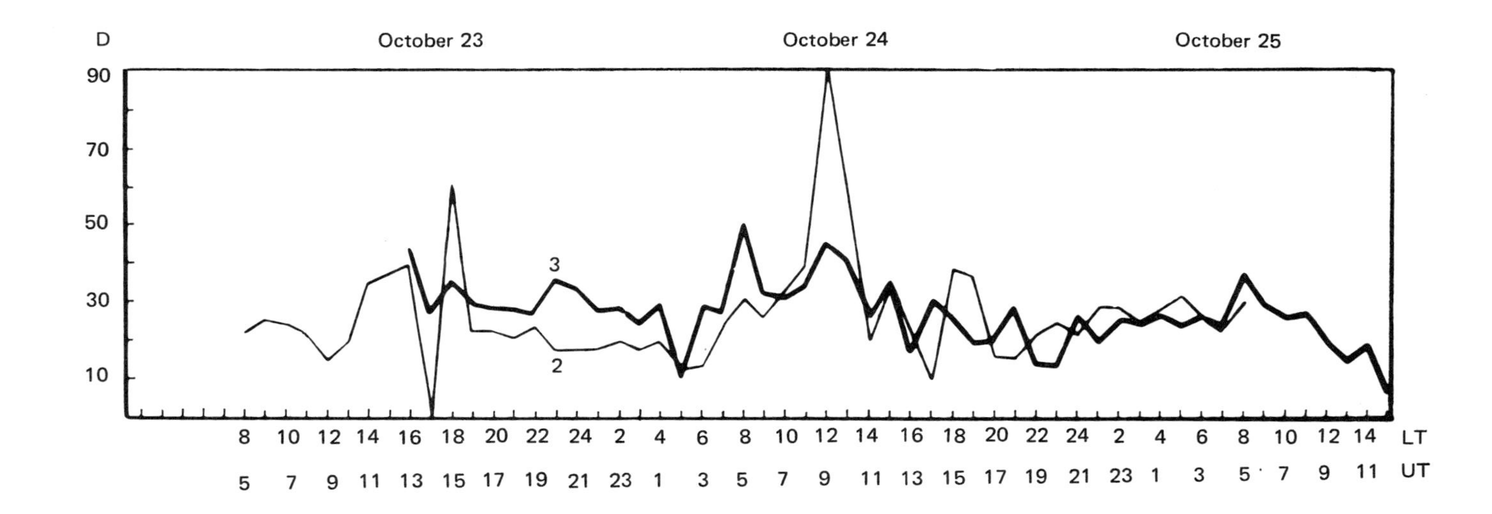

Fig. 39. Circadian rhythm of root excretions of four-day old barley seedlings in synchronous investigations in Irkutsk (1), Moscow (2), and Florence (3) on October 23–25, 1968. The experiments were conducted by Ada T. Platonova in Irkutsk, by Eleonora Francini Corti and Carlo Lenzi Grillini under the supervision of Prof. Piccardi in Florence, and by the author in Moscow. From (134).

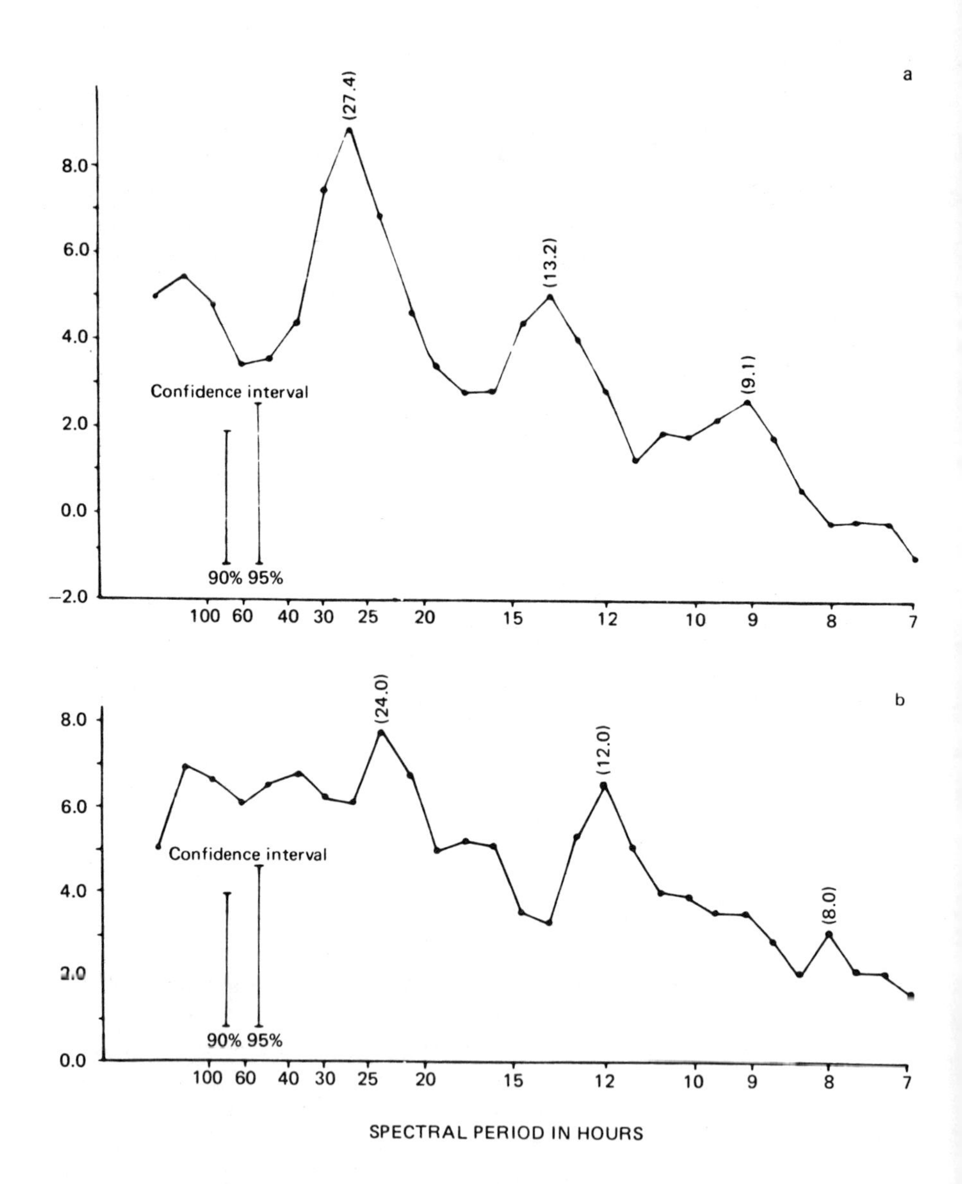

Fig. 40. Autospectra of "up-and-down" movements of *Phaseolus* leaves (a) and magnetic variations (b) in the period August 16–18, 1968. The figures above the peaks indicate that the factor investigated varied rhythmically with this period. From (1165).

(Moscow–Minsk, 700 km). Thus, in our investigations of the dynamics of excretion of organic substances by plant roots we noted that the excretory rhythm at each geographical point had its own individual features in addition to the general global character. This was manifested in the fact that the rhythmic activity of the plants had a general diurnal course and, in addition, each geographical point showed an additional maximum or minimum of excretion or phase shifts of the maxima or minima through 1–2 h. These differences were particularly clearly revealed by a continuous two-day record of the rhythms (Fig. 39). This conclusion is confirmed by an investigation of the relation between the leaf movements of *Phaseolus* plants grown in a biotron and variations of the GMF (1165). This was a thorough comparative investigation in which a time–frequency analysis, including an assessment of autospectra, cross spectra, and coherence for frequencies of 30 min to 200 h was carried out in accordance with a special program. The authors indicated that the investigation did not give rigorous proof that the GMF is the Zeitgeber for the rhythm of leaf motion. They reported, however, that a cross-correlation analysis for a period of 27 h showed coherence of the GMF fluctuations and the up-and-down movements of *Phaseolus* leaves (Fig. 40). In commenting on this study we would like to mention the following. First, it should be noted that autocorrelation analysis of the periods of rhythmic variations of the motion of the leaves and fluctuations of the GMF showed maxima at 27, 13.5, and 9 h for the plant rhythm, and 24, 12, and 8 h for the GMF fluctuations. The data cited show an interesting feature: The biological rhythms are respectively 3, 1.5, and 1 h longer than the GMF rhythms. Hence, we can conclude that the biorhythms lag behind the rhythmic fluctuations of the GMF by these times, and hence the situation should be analyzed in the same way as randomly correlated periodic processes, for which a special mathematical apparatus has been devised (629–636).

In addition, an analysis of this work shows pronounced differences in the response of plants to the GMF: In one plant, cross-correlation analysis revealed the presence of coherence for a 27-h period with a high degree of significance (0.770 as compared with 0.764 at the 2.5% level of significance), whereas in another plant this index was insignificant (0.616 as compared with 0.651 at the 10% level of significance). This difference of responses of plants is easily explained from the viewpoint of functional dissymmetry of biological objects: One plant was "left-handed" and reacted distinctly to the change in the GMF, while the other was "symmetric" and reacted weakly to the GMF. Differences in the responses of biological objects to the GMF are discussed in more detail in the next chapter. Incidentally, the data of the authors themselves confirm our conclusion: They reported that one bean plant showed synchronization of the rotational leaf movement with a magnetic storm that occurred during the investigation, whereas other plants did not show such synchronization.

The facts obtained confirm that the rhythmicity of functional activity is manifested in complex interaction between the homeostatic systems of the organism and geophysical (141, 451, 1177) and cosmic factors (130, 131, 137, 746). It is quite possible that living organisms are in critical states in relation to a whole series of heliogeophysical factors or particular parameters of them, and for some of them there are thresholds of action.

Investigations carried out over a period of many years have shown that the analysis of the rhythmicity of biospheric processes must take into account three important physical factors: the geomagnetic, gravitational, and cosmic fields. It should be noted here that the special features and complexities in the manifestation of the rhythmicity of biospheric processes that researchers encounter in practice can be attributed to special features of the manifestation of each of the above main physical factors and their transforming effect on the envelopes of the earth of most importance for the biosphere—the magnetosphere, ionosphere and atmosphere, lithosphere, and hydrosphere—and also to the complex interactions and influences of different cosmic factors on the radiation belts and powerful current systems of the earth.

In the rhythmicity of biospheric processes, by which we mean any manifestations of the activity of biological objects, we must distinguish two types of components: a *periodic* regular component due to unitary variations of the geomagnetic, gravitational, and cosmic fields, and an *aperiodic,* spontaneous, sporadic, but in principle cyclic component due to increased solar activity (flares, sunspot activity) and sudden changes in cosmic fields, including the interplanetary magnetic field and gravitational fields. If we are to understand the whole complexity of biological rhythmicity we will need to take into account that the changes induced by these factors are superposed on the "endogenic" biospheric rhythmicity associated with the internal interrelations and cyclic changes in biocenoses, i.e., large ecological systems.

While not underestimating the role of gravitation and cosmic fields we ascribe great importance to the GMF. All our generalizations, evidence, and facts indisputably indicate that the GMF is one of the main characteristics of the solar, planetary, and cosmic influence on the earth's biosphere.

The Geomagnetic Field and Morphological and Functional Biosymmetry

Along with gravitation the GMF is one of the main physical factors in the earth's biosphere. All objects on earth and in the universe have symmetric properties and, hence, it is important to find out how these fundamental properties of the world surrounding us are related to one another.

The symmetry of natural objects is a fundamental property permeating all living and nonliving matter (514, 843, 996, 1209). Yet it has been shown that symmetry is indissolubly connected with its violation—with dissymmetry. The introduction into symmetry of the concept of opposites, particularly the dissymmetric concepts of left-handed (L) and right-handed (D) has been very important for the understanding of the world around us (514) and of biological objects, in particular (583–586, 591).

A study of the problem of left- and right-handedness has led to the development of a new theory of dissymmetric factors and an associated theory of isomerism (585, 591). This work has led to the introduction of new concepts of fundamental importance for all natural science. The most important concept of dissymmetry theory is the recognition of the fact that all material objects are right-handed (D), left-handed (L), or intermediate (DL). Numerous investigations have established the *law of occurrence* of bioenantiomorphs: D and L biological objects—molecules, bacteria, plants, and animals—occur in nature in such a way that either $\Sigma D = \Sigma L$, or $\Sigma D < \Sigma L$, or $\Sigma D > \Sigma L$. A great step in the understanding of the nature of left- and right-handedness was the introduction of the concept of "dissymmetrifying" factors (diss-factors, for short), i.e., factors inducing left- and right-handedness of objects (591). This led to the establishment of the *law of properties* of bioenantiomorphs: On transition from a D to an L object some properties of the objects are altered in such a way that the properties of its L form cannot be derived from the properties of the D form by any symmetric or antisymmetric operation (591). The author of these theories gives a new definition of the dissymmetry of objects, which we reproduce completely in view of its universality and importance: "Objects are called dissymmetric if: (a) they are altered by mirror reflection in some respects to the exact opposite; (b) do not, consequently, coincide with their mirror images; and (c) exist in one, two, or more modifications." By the introduction of the concept of diss-factors the science of right- and left-handedness can be related to the general science of polymorphism and isomorphism, and it can be shown that any symmetry theory is a specific theory of isomorphism, which is important, as we shall show later, for an understanding of the action of the GMF on biological objects.

Investigators have previously stressed the theoretical and practical importance of the study of any properties of biological objects from the viewpoint of their dissymmetry (Fig. 41). For instance, one of the interesting and important features of D and L forms of plants is their different growth responses when the embryo is oriented toward the north or south magnetic pole (368, 369, 376–379, 512, 550–552).

On the basis of our own experimental studies and summarization of the results of other investigators, we have come to the important conclusion that the GMF is the main factor determining the functional and morphological properties of biological objects (119–141, 143, 809, 810) and hence is one of the fundamental causes of the symmetry and dissymmetry of biological objects (125, 126). A study of the differences in the functional properties of living organisms on this

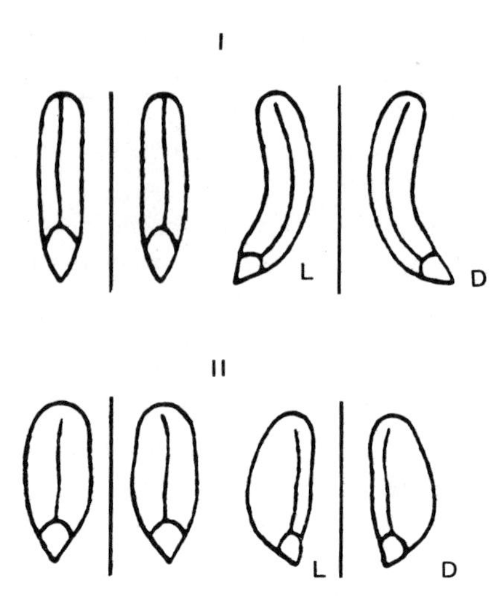

Fig. 41. Schematic diagram of left- (L) and right-handed (D) (based on position of endosperm) rye (I) and wheat (II) grains. From (552).

basis led us to introduce a completely new concept in biology—the *functional* symmetry and dissymmetry of biological objects (125, 135). The ideas that we have developed have been very fruitful for an understanding and explanation of the different responses exhibited by biological objects to the action of the GMF and any other environmental physical factors.

For a better understanding of our ideas we discuss in turn two interrelated aspects of this problem: the dissymmetry of biological rhythms and the dissymmetry of individual responses of biological objects. We concentrate on these aspects because the GMF, as was shown above, is one of the main *causes of rhythmicity* in the biosphere (136, 137). Hence, it is important to know what interrelations have been developed between biological objects and the continuously varying GMF, and what are the nature and the laws of this relation between two quantities varying continuously in time and space. We begin our discussion with a formalized description of phenomena.

Analytical Basis of Functional Symmetry and Dissymmetry

The particular individual response of a biological object to the action of the GMF or the instantaneous manifestation of functional properties of the organism (i.e., the single response) and the biological rhythms built up from them is a general and isomorphic property of all biological objects. However, the manifestation of these responses and rhythms in biological objects may be different. This difference in responses of biological objects to the GMF and other physical factors is so characteristic, and the differences in responses between organisms so specific, that they conform to the definitions and laws of biosymmetry (591) and hence can be assigned to a completely independent branch of knowledge—functional dissymmetry (135). If we follow the existing classifications, the kind of phenomena that we are discussing are those of *dynamic symmetry,* i.e., the symmetry of phenomena occurring in time (504).

To make our analysis clearer and more understandable and to allow progression from general definitions to more specific concepts, we examine some examples taken from the *published experimental investigations* of various authors. These examples will help us to arrive subsequently at an understanding of the role of the GMF in the functional–dynamic homeostasis of living organisms.

If we examine closely investigations of rhythmicity of any physiological or biochemical process in living organisms where the authors give *individual,* and not averaged, rhythm curves, we find that the rhythm of the biological processes reveals a general characteristic feature, irrespective of species, sex, or age-dependent characters of the investigated object. This characteristic feature of biological rhythmicity is the difference in the course of the investigated process in time, manifested in a particular inclination of the rhythm curves to the x and y

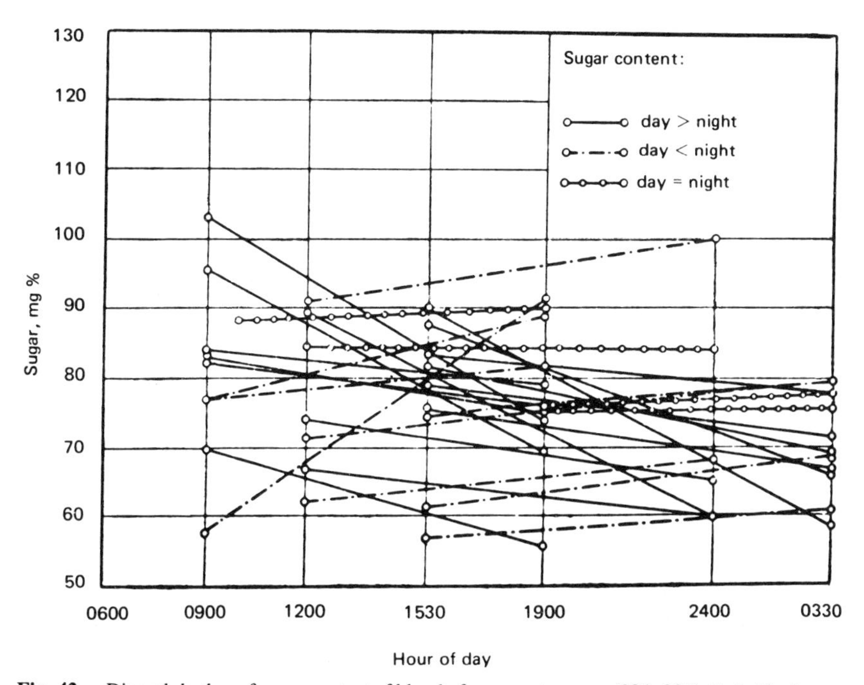

Fig. 42. Diurnal rhythm of sugar content of blood of pregnant women (284, 285). Individual curves represent cases in which sugar content was greater during the day than at night, lower during the day than at night, and equal during the day and night. From (135).

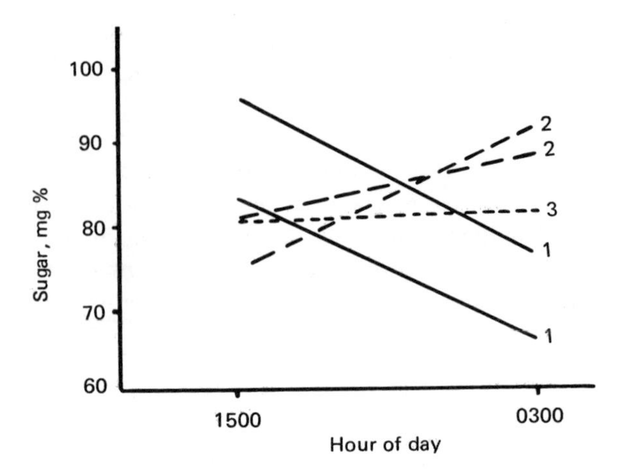

Fig. 43. Diurnal rhythm of sugar content of blood of nonpregnant women (284, 285). Individual curves represent cases in which sugar content was (1) greater during the day than at night, (2) lower during the day than at night, and (3) equal during the day and night. From (135).

(time and amplitude) axes (Figs. 42–44): from the initial point they can be directed upward, downward, or parallel to the time axis (x axis). After a second time point has been marked, the rhythm curves have a characteristic inclination to the left or right, or are parallel to the time axis. This behavior of the rhythm curve, specific for each object, indicates fundamental differences between organisms and their different response to the GMF. We will illustrate this by examples taken from investigations of the rhythm of functional processes in man and animals (Figs. 45–51).

Initially, to simplify the analysis, we shall reduce the whole set of points of the daily rhythm (24 h), representing the variation of functional properties during the 24 h (Figs. 42 and 46), to two points (Figs. 43–45) nominally representing the "day" and "night" response of the organism. In this case, complex diurnal rhythms will be represented by a system of straight lines with different inclinations to the x and y axes (Figs. 42–44).

From this set of straight lines ("simple" rhythms) we select extreme variants—the straight lines that reflect a change in the process or property, where: (1) the magnitude is greater by *day than at night;* since the straight line in this case is inclined to the right, we call it the "right-handed" or D form of the rhythm; (2) the magnitude is smaller by day than at night; the straight line is inclined to the left and hence this is the "left-handed" or L form of the rhythm; (3) we select another variant of the "simple" rhythm, where the *day and night values* of the change in the properties or characters are the same—this is the "symmetric" or S form of rhythm. The figures show that these straight lines, representing the D and L rhythms, are mirror images of one another, and hence

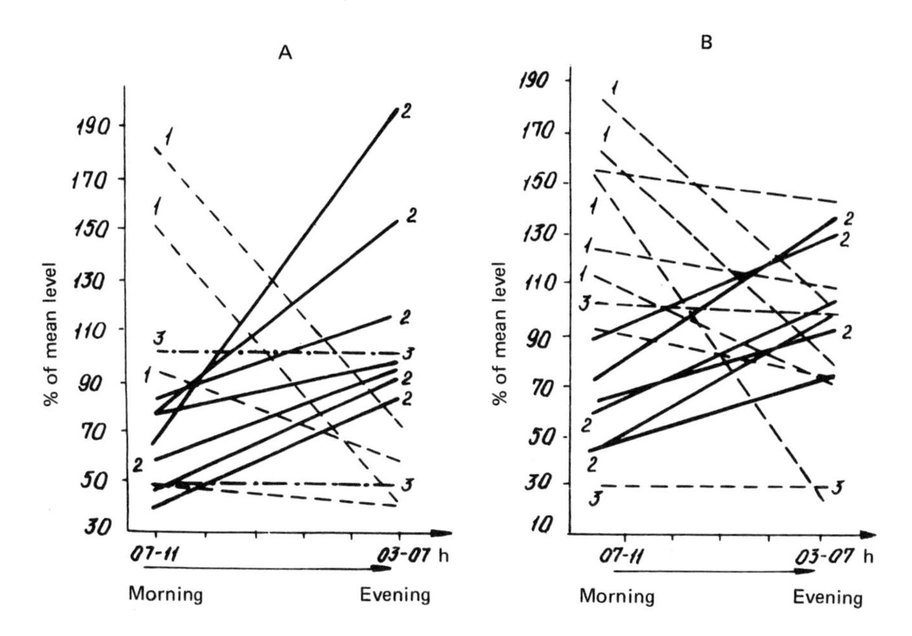

Fig. 44. Diurnal rhythm of excretion of sodium (A) and potassium (B) in patients with diencephalic pathology (187). Individual curves represent cases in which excretion was (1) greater during the day than at night, (2) lower during the day than at night, and (3) equal during the day and night. From (135).

we are justified in calling them "left-handed" and "right-handed" rhythms (Fig. 45). In fact, the transformation responsible for the mirror symmetry is essentially one dimensional: Reflection of the straight line can occur from any point O on it; this reflection takes a point P on the straight line to a point P', which is the same distance from O as P, but on the other side of it. This demonstrates that the simple rhythms illustrated in Figs. 42–44 are mirror images, since for an existing straight line we can always select a mirror-image rhythm line.

In the analysis of complete diurnal rhythms ("complex" rhythms) with a large number of time points, the resultant curves can be regarded as two dimensional figures in two-dimensional space. In this case the mirror-inverted forms of these figures cannot be made to coincide by any rotation, i.e., they can be regarded as D and L enantiomorphs of biological rhythms reflecting the functional dissymmetry of biological objects (Fig. 46). The changes in biological indices and properties, which we call functional diss-factors, and the degree of individual responses of biological objects may be very different, of course, which is manifested in the diversity of rhythms. In complex rhythm curves there is inexact mirror invertibility (reflectability), i.e., there are different degrees of dissymmetry, corresponding to the concept of magnitude of dissymmetry (ρ factor) in

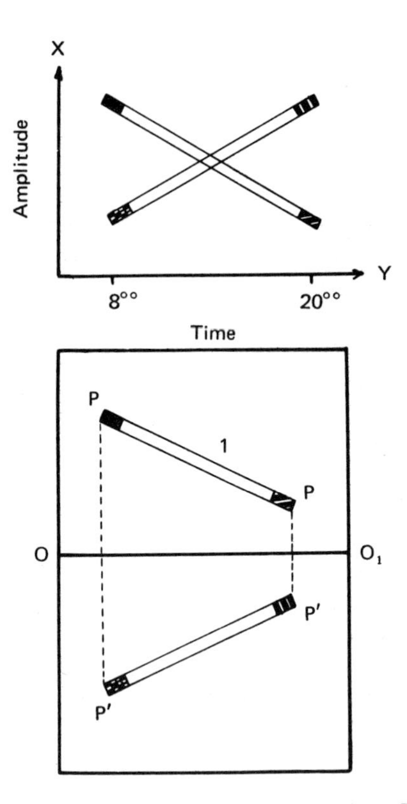

Fig. 45. Schematic diagram of mirror nature of "simple" rhythms. Explanation in text. From (135).

biosymmetry theory (591). Thus, in all types of rhythm curves (simple and complex rhythms) we detect assymmetry of the properties in time, and in some (the S form) there is symmetry of the properties in their translation in time.

It should be noted that this separation of rhythms into D, L, and S enantiomorphs is not simply a formal geometric operation, but provides a graphic representation of fundamentally different, specific parameters of the rhythmicity of individual biological objects. In practice this is manifested in the different responses of such functional rhythmic bioenantiomorphs to the action of the GMF and any other physical factors, e.g., drugs, and in the simultaneous manifestation of different physiological and biochemical properties. Studies of the biological action of the GMF must include due consideration of the division of biological objects into functional D and L enantiomorphs and their investigation by special programs, in view of the specific features of their responses. Examples illustrating the role of functional dissymmetry in the action of the GMF are discussed next.

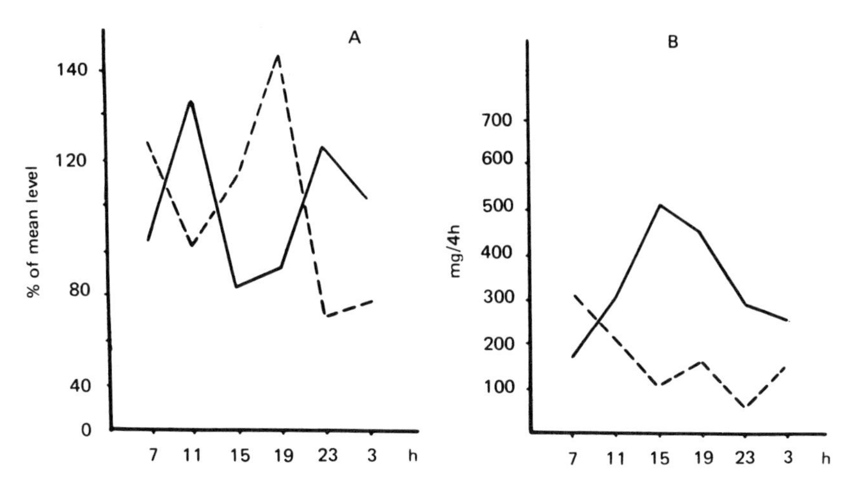

Fig. 46. Excretion of water (A) and 17-hydroxycorticosteroids (B) by patients with diencephalic pathology (187). Example of inversion of "complex" rhythm. From (135). Individual rhythm curves.

Experimental Investigations Confirming Conclusions on Functional Dissymmetry and Its Role in the Action of the GMF

The analysis presented above shows that the individual responsivity, differences, and physiological–biochemical variation of organisms express their dissymmetric structure, in which the GMF plays an important role both from the moment of formation of the organism and throughout its entire ontogenesis. This provides an explanation of many facts obtained in experimental biology, developmental biology, genetics, and ecology of man, animals, and plants. We shall discuss some of these facts to further clarify essential aspects of the question under discussion.

Numerous experimental investigations of biological rhythmicity show that the seasonal and diurnal variations of indices or characters in any kind of organism are represented by characteristic curves. These relations have been thoroughly investigated and these rhythmic characteristics have been called dynamic curves (285). The authors of this work present the results of their own investigations and experimental data of other investigators indicating the existence of *different types* of regulation of processes in organisms of the same species, and also the absence of uniformity in the manifestation of the diurnal rhythm in different individuals.

Their proposed typology of regulation is based on the different activity of the following systems: vegetative nervous, endocrinal, and cationic. The three types of diurnal rhythmicity distinguished by these authors represent different

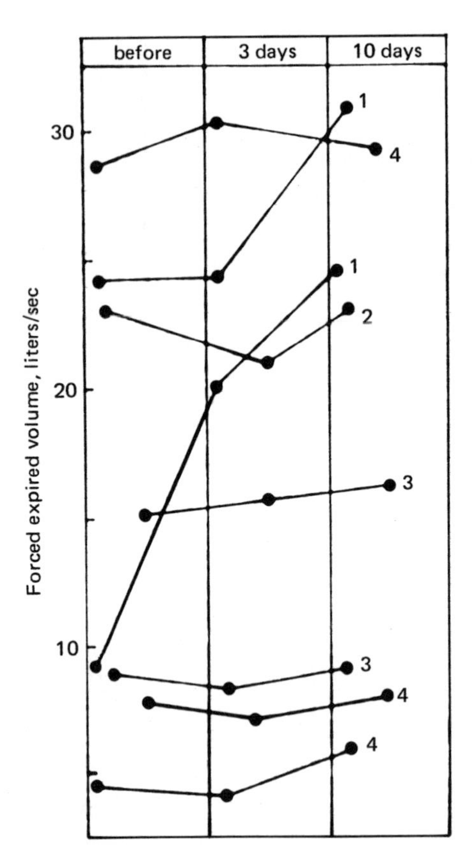

Fig. 47. Change in forced expired volume of air of individual patients with chronic bronchitis treated with preparation DTM 8 15. Data from (787) analyzed in (812). The graph shows left-handed (1), right-handed (2), symmetric (3), and intermediate (4) types of functional symmetry.

types of regulation: type 1, where the sympathicotonic system of nervous and humoral control predominates during the day; type 2, where parasympathetic nervous and humoral control predominates during the day; and type 3, where elements of the sympathicotonic and parasympathicotonic regulating systems are balanced to some extent during the day and night, or the sympathicotonic system dominates by day and night (285). These types of rhythmic changes in the organism are given above and, as we have shown, are based on the functional dissymmetry of the investigated objects (Figs. 42–44). Our exposition of the dissymmetric classification of biological rhythms gives a fundamentally new explanation of the indicated division into types. There is also confirmation of the conclusion of inversion of physiological, biochemical, and regulating systems of

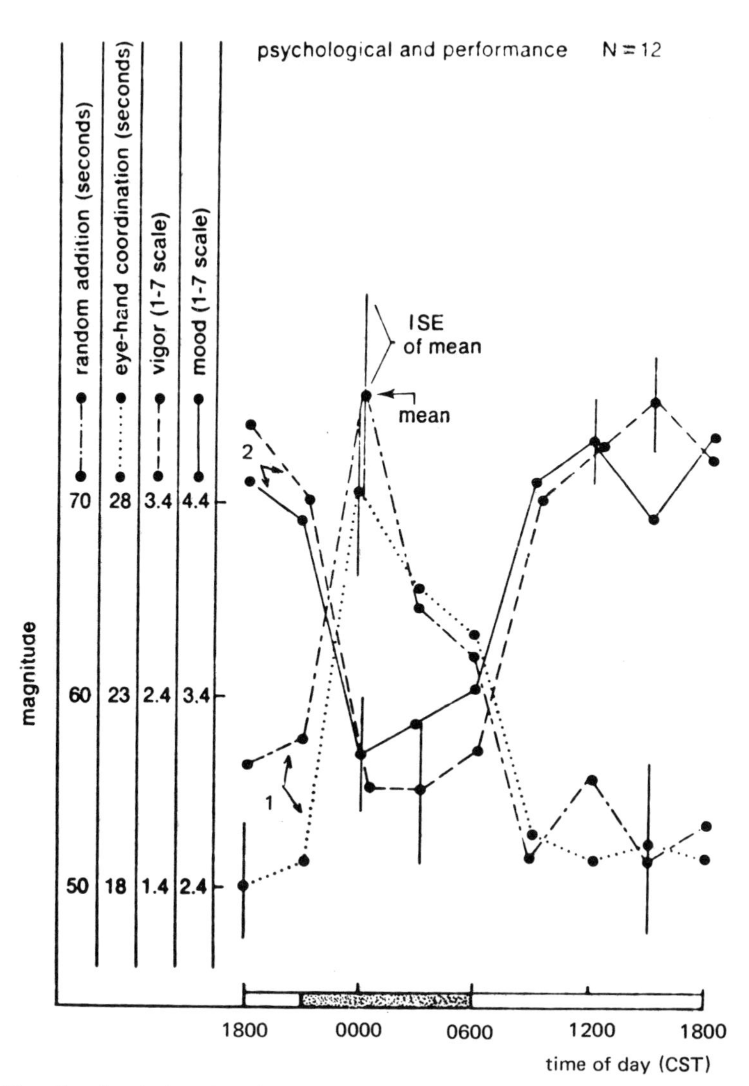

Fig. 48. Circadian rhythm of psychophysiological processes in man. The graph shows right-handed (1) and left-handed (2) types of rhythms. Data from (911) analyzed in (812).

dissymmetric objects, expressed in the corresponding biological rhythms, and the effect of the GMF on responses of D and L objects.

A graphic example of the role of functional dissymmetry in the action of the GMF is provided by work on the mechanisms of action of geomagnetic disturbances on the blood-clotting system of presumably *healthy* people (470, 472).

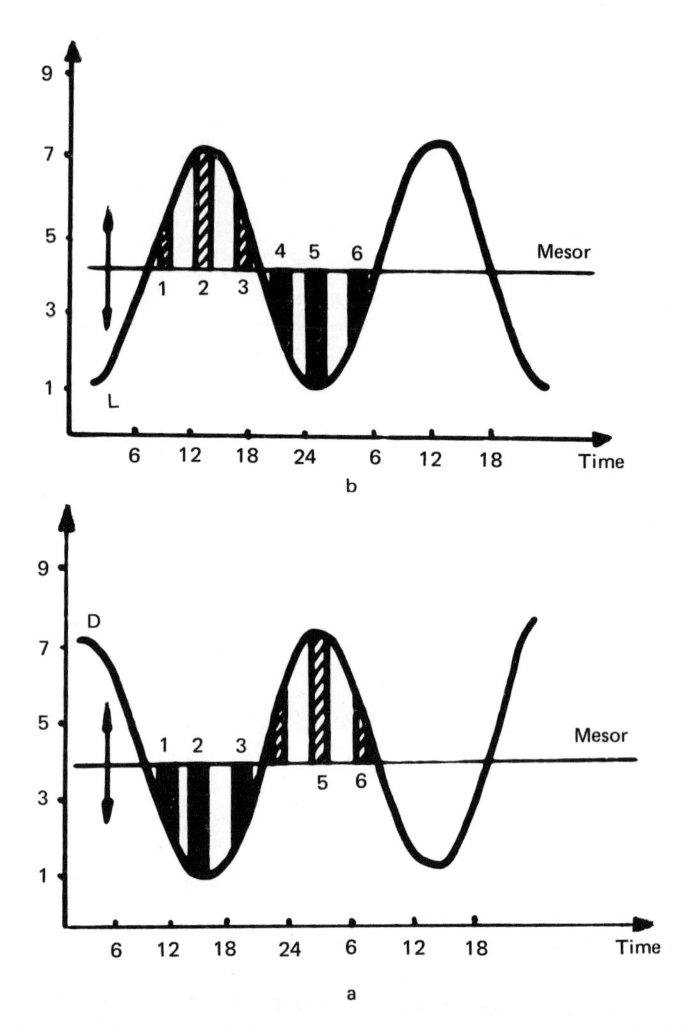

Fig. 49. Schematic diagram of individual components of D and L rhythm curves: (a) right-handed type of rhythm curve, (b) left-handed type of rhythm curve. Different curves indicate individual responses at particular points in time. Note that these points alter the amplitude of the D and L curves differently in relation to the median line (mesor).

These investigations showed *differences* in the responses of *healthy* people to the GMF. It was found that two-thirds of a number of healthy people investigated during a period of development of geomagnetic storms showed a moderate *increase in the sympathetic tonus* of the vegetative nervous system, while the other one-third showed an *increase in tonus of the parasympathetic* system.

Thus we see that two completely independent researchers arrived at the

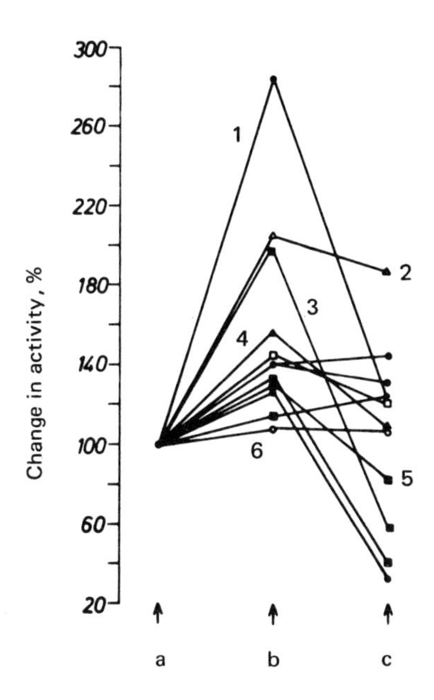

Fig. 50. Individual responses, based on measurement of motor activity, of animals to addition of amphetamine to food. Responses before (a), during (b), and after (c) administration of preparation to animals. Different curves indicate responses of different animals. 1–5, DL functional types; 6, symmetric. Data from (727) analyzed in (135).

same conclusions from an analysis of individual biological responses, and in one of them the effect of the GMF in these processes was clearly demonstrated.

For a clearer and more convincing demonstration of our views on functional dissymmetry we give several more examples that illustrate these functional differences well. Italian chronobiologists have recently reported the results of treatment of a number of pulmonary diseases with corticosteroids (787) (Fig. 47). It is clear from Fig. 47 that there is a distinct division of the investigated patients into left-handed (L), right-handed (D), symmetric (S), and intermediate (DL) forms according to their response to treatment with drugs. A similar situation has been observed in a study of physiological rhythms in *healthy* people (733).

Similar data can be cited from another experimental study (187). The authors studied endogenous rhythms of excretion of potassium, sodium, water, and hydroxycorticosteroids by patients with various brain diseases. They found in the course of these investigations that in addition to the exceptional diversity of the rhythms of investigated processes there were cases where the rhythms were completely inverted, i.e., directly opposite in different individuals with the same disease (Fig. 46, example of inversion of "complex" rhythm).

The authors of many similar works confirm that men (and animals) exhibit

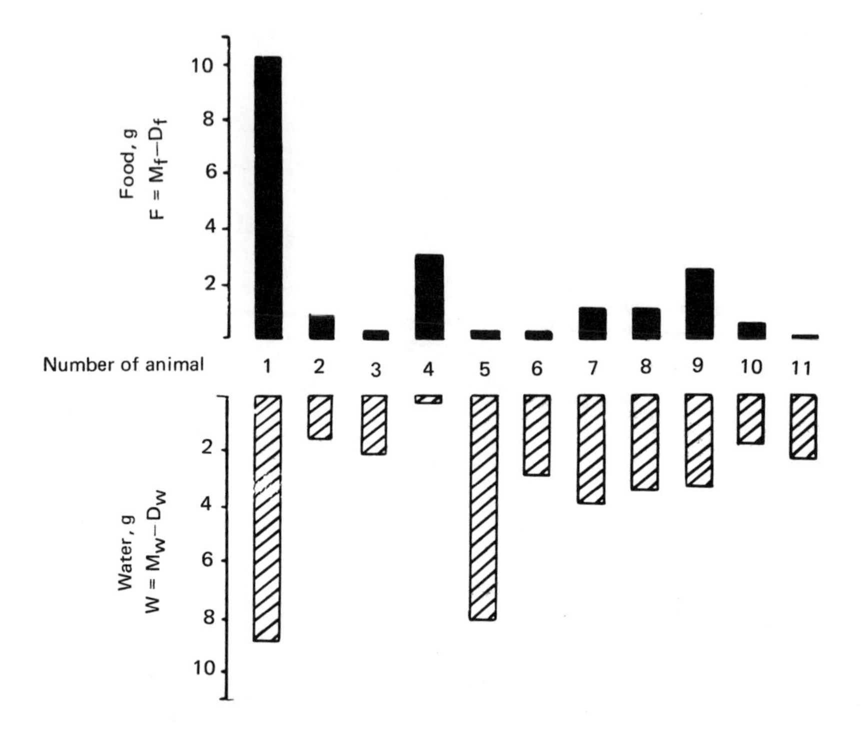

Fig. 51. Individual responses of different animals with respect to consumption of water and food in the presence and absence of a magnetic field. Explanation in text. (1–11) Different animals (1104). From (135).

completely different responses to the same physical factor, but they either fail to give any explanation and ignore the observed differences, regarding the matter as nothing unusual, or draw obviously erroneous conclusions.

For instance, in the above-cited study of rhythmicity in patients with diencephalic pathology (187) the reported features of the rhythm, e.g., its inverse nature and so on, were attributed to the main disease—damage to parts of the brain. This conclusion, however, may be erroneous for the following reasons: first, the data of the investigators themselves, who found *different* types of rhythms in patients with the *same* form of disease; second, inversion of rhythm is a feature of normal organisms and is a very common biological phenomenon (285).

From the viewpoint of functional dissymmetry inversion of rhythms is merely a reflection of the modifications (L or D) of biological objects, and their discovery depends only on the frequency of occurrence of the particular rhythm. A good illustration of inversion of biological rhythms in man and their interrelation with the GMF is provided by the results of a study of the flicker fusion rate of subjects in a compensated GMF (760) (Fig. 14). This figure clearly shows inversion of the biological rhythm and the role of the GMF in the formation of the response: Before the change in the GMF (days 0–5) no differences were found in the test response of the subjects, but when the GMF was compensated (days 5–15) one subject immediately showed an inverse type of rhythm! Thus, it is apparent that there are profound differences in the functional responses of biological objects, and that the GMF plays an important role in this. It should be borne in mind that the functional enantiomorphic type of biological object depends also on the dissymmetric factor itself, i.e., the index, property, or character that is being investigated: The same object can be left-handed for one character (index) and right-handed or symmetric for another, and so on. As an illustration of this we give an example from a study of the circadian variation of the psychological state of healthy people and their ability to perform tasks (911). Although the authors of this paper give averaged group rhythm curves, and not individual results of the study of each subject, the dissymmetric differences in the investigated indices are very clearly manifested (Fig. 48).

Thus, we find that a single population of any biological objects has an internal, definite, functional dissymmetric division based on the action of the GMF and dependent on it.

Individual Single Responses of Biological Objects as an Expression of Functional Dissymmetry

For a more thorough understanding of the problem of functional dissymmetry we consider the question of individual biological responses. If we divide the D or L rhythm curve into individual time elements, taking the median line between them (mesor) as a reference line, we can easily see that any D or L rhythm curves consist of separate individual responses at a particular instant (Fig. 49). These individual changes in the investigated indices, like the rhythms, have a common characteristic feature: depending on the rhythm curve to which they belong, they lie above or below the median line (mesor). This establishes a fundamental dissymmetric difference in the instantaneous (single) individual response of different biological objects—D, L, DL, and S forms. Hence, individual responses of organisms, like rhythms, can be divided into dissymmetric modifications according to the rhythm to which they belong. This is distinctly manifested, for instance, in the response of the organism to pharmacological preparations (727). It was reported that the addition of amphetamine to water and food led to differences in the locomotor activity of mice (Fig. 50). Figure 50 shows that the administration of amphetamine to one mouse (No.6) led to no changes in motor activity whatsoever either at the time of administration or even after administration of the preparation—this is an example of functional symmetry—the S form of response. In other individuals, however, the level of motor activity increased by as much as 300% on the day preceding the experiment (this is the L form of response or, more precisely, the DL intermediate type, when we consider reference points a, b, and c).

Thus, in the separate individual responses of an organism the corresponding dissymmetric modification is revealed by the presence or absence of a response. A similar picture of different degrees of responses was found in an investigation of the synchronizing effect of temperature on the rhythmic activity (Table 3) of various animals and birds (892).

The data given in the table clearly show differences in the functional dissymmetry of the investigated objects; these differences are manifested in the presence or absence of a synchronizing effect of temperature. These facts have been inexplicable hitherto. From the table we can estimate approximately (since it was a very small and unrepresentative sample) the frequency of occurrence of functional D and L enantiomorphs and forms for such a diss-factor as temperature synchronization of locomotor activity, since different individual animals and birds were used for investigation of the effect of temperature cycles (personal communication of author). Features of functional dissymmetry can also be seen in (1104), where the consumption of water and food by mice in the presence and absence of a magnetic field was investigated (Fig. 51). Figure 51 shows individual responses of different animals (Nos. 1–11) in different locations in the experiment. It is apparent also here that the animals are divided into functionally dissymmetric forms in relation to two diss-factors—water and food consumption. Mice Nos. 1 and 9 are functionally symmetric, whereas the others represent various dissymmetric forms: Number 4 is left-handed (L), Nos. 3, 5, 11 are right-handed (D), and Nos. 2, 7, and 8 are intermediate (DL). A similar division into functionally dissymmetric forms has been observed in an analysis of such different objects and characteristics as the psychophysiological features of the state and behavior of normal and corpulent people (135) and the growth charac-

Table 3. Comparison of Synchronizing Effect of Temperature in Lizard, Mouse, and Sparrow[a]

	Fluctuations of temperature cycle, °C								
	7.2	3.6	1.6	0.9	26	15	4	3	6
Presence of synchronization		Lizard				Mouse		Sparrow	
Distinct	17	6	6	5	—	—	3	2	—
Dubious	—	1	1	5	3	3	—	2	—
Absent	1	1	1	5	6	8	—	8	—

[a] A dash indicates no experiment for this temperature cycle.

teristics of plants (293). The author of the latter paper found that the populations of beans and cucumbers that he investigated consisted of three groups: the first developed more rapidly in the GMF conditions in the northern hemisphere, the second developed more rapidly in the GMF conditions in the southern hemisphere, while the third group showed a neutral response to the GMF. The different reactivity of people to the GMF, due to functional dissymmetry, is clearly illustrated by the example of investigations of skin potential (432, 433, 436, 437). These investigations, made on healthy and sick people (and on animals), showed that according to their response to helio- and geomagnetic factors human organisms can be divided into three main groups: magnetotropic, magnetostable, and intermediate. In magnetotropic people changes in skin potential are found three to four days before and on the day of the magnetic storm, whereas in magnetostable people they are found two days before and on the day following the storm. We should mention that individual differences in galvanic skin responses in humans have been thoroughly investigated and have been the subject of a special examination (899). Thus, in human bioelectric responses we can again detect the role of functional symmetry and its close relation with the GMF.

F. Brown in his studies of various biological objects (742–745) has shown that clockwise or counterclockwise rotation of the objects themselves, like the action of a rotating weak magnetic field on objects, leads to opposite rhythmic responses in comparison with a stationary control. These responses are either positive, or negative, which indicates the existence of left- or right-handed forms of objects in the experimental material and the decisive role of the GMF in the creation of these forms. We should mention that an important biological role of a rotating weak magnetic field was shown in very interesting investigations on mice (1037, 1040, 1042, 1044, 1045). The dissymmetric differences of individual single responses of biological objects, like rhythms, can be established only if there is some set of reference points.

Furthermore, an analysis of any single rhythm curve shows that throughout the whole period of investigation of an object there is a continuous alternation of left- and right-handed reference points, if the adopted basis is the median line (mesor, in the terminology of chronobiologists), which represents the symmetric type of point or reaction. This is the essence of the homeostasis or the dynamic equilibrium of the living object, and this variability and rhythmicity are related to the action of the GMF. This is well illustrated by the works of German researchers on the orientational ability of bees in the GMF and in the compensated GMF (951) (see Fig. 67). Thus, in all dissymmetric functional responses of objects we see the leading role of the GMF.

Possible Causes of Biological Symmetry and Dissymmetry

In previous chapters we presented evidence that the individual response, physiological–biochemical qualitative diversity of objects, and variability of rhythmic process in living organisms reflect their dissymmetric structure, and the GMF plays an important role in these processes. Questions of symmetry in nature and the nature of symmetry itself have been thoroughly and comprehensively examined (591). So far, however, there have been very few studies that throw light on the question of the main *causes* and/or factors responsible for the symmetry and dissymmetry of biological (and organic) objects (842, 1001, 1016).

From an analysis and summarization of studies of various kinds and our own experimental data for plants and crystals we have concluded that the GMF and gravitation (ordinary and rotational) play an important role in the formation of these fundamental properties of living and nonliving objects (125). A common basic feature here is the Curie principle, according to which symmetry and dissymmetry owe their origin to external conditions—to physical factors creating these properties. In succession we shall consider evidence of the leading role of the GMF in the dissymmetry of objects, and we shall discuss the important role of gravitation in another chapter.

First, it should be noted that the GMF is implicated in the formation of ''dissymmetrifying'' factors themselves, i.e., factors responsible for the left- and right-handedness of objects. For instance, in the case of dissymmetric organisms or organs, the predominant development of parts in one particular direction determines the symmetry of the object.

If we turn to paleomagnetobiological research, we find information that GMF reversals caused a change in the type of symmetry of biological objects: The direction of twist of the shells of the *Globorotalia menardi* complex was reversed (851), and the symmetry of the frontal commissure of rhynchonellid brachiopods was destroyed (224, 225). In a recent paper the authors reported that this change in the type of symmetry could not be attributed to any external environmental factor apart from paleomagnetic reversals. In their opinion the increase in the number of rhynchonellid species with the asymmetric S-form of frontal commissure and the increase in the number of individuals of this type, until they clearly predominated in the populations, were associated with reversals in which the polarity changed from normal to reverse (225) (Fig. 52).

We can also observe the effect of the GMF on the dissymmetry of biological objects at the present time. For instance, the symmetry of a root in a number of simple features depends on the development of the lateral roots, whose disposition is definitely correlated with the GMF (398, 1055–1063), and hence the symmetry of the root is also determined by the GMF. As was shown above, a similar situation is found in the case of biological rhythms, which have a left-handed, right-handed, or symmetric functional form, depending on the GMF. Thus, the first conclusion that can be made is that the GMF is probably one of the main causes of the formation of diss-factors for various living and nonliving objects and hence is a cause of the dissymmetry of objects.

The GMF is not only involved in the formation of diss-factors, however, but

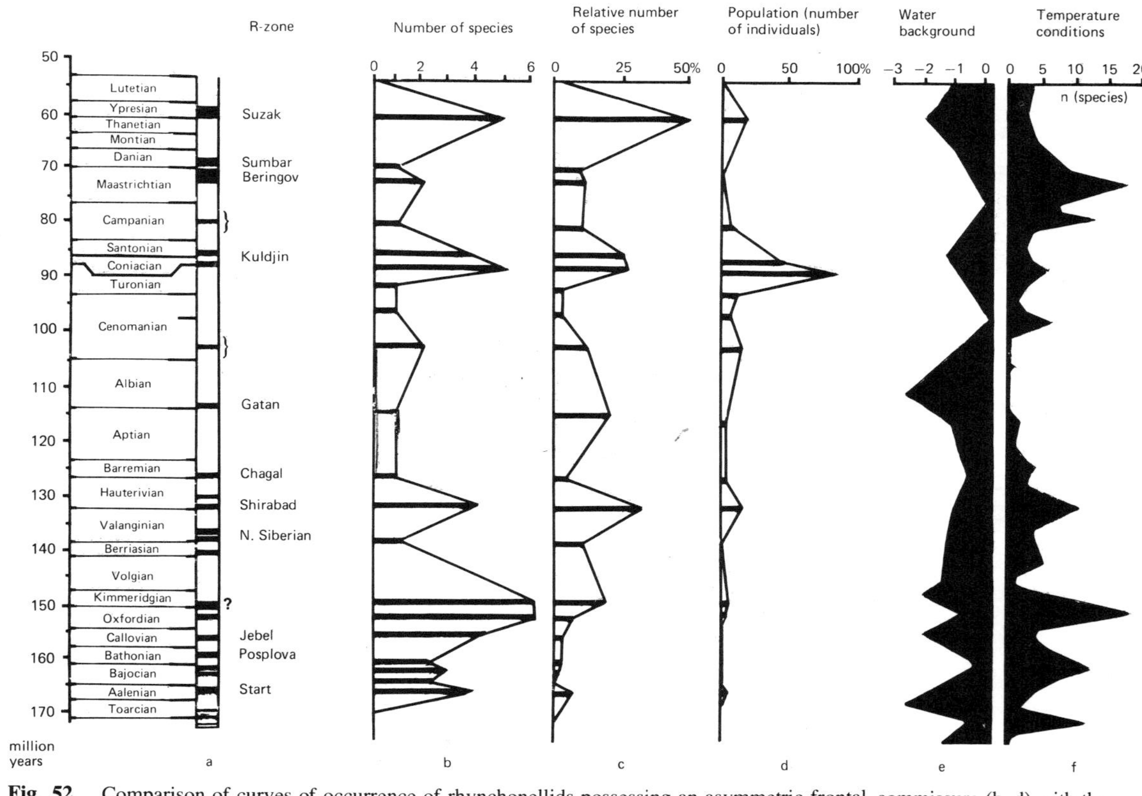

Fig. 52. Comparison of curves of occurrence of rhynchonellids possessing an asymmetric frontal commissure (b–d) with the paleomagnetic scale (a) and curves of variation of the water background (e) and temperature regime (f) in relative units according to biogeochemical tests. From (225).

continuously acts on the dissymmetry of objects. As an example of this we can cite data on the dissymmetry of florets in plants of the same species in relation to the GMF (794). The author found that the dissymmetry of the florets of the red silk-cotton tree *Bombax ceiba* L. at different geographical points on the earth varied regularly according to the change in dip of the GMF at the places where this plant was investigated. A leading role of the GMF in dissymmetry phenomena in plants has also been reported in other studies, e.g., on sugar beet (376, 377, 379, 398, 399), corn and wheat (552), and radish (394). It was reported that alteration of the horizontal or vertical component of the GMF by Helmholtz coils led to a change in the orientation of the root creases (394) and the response of the plants, *and to their separation into new functional types* (293), which is further confirmation of the role of the GMF in biodissymmetry.

We must single out especially the thorough and well-planned investigations of the dynamics of biosymmetry in plants (552). From many years of research on the dissymmetric properties of corn, wheat, rye, and other plants, Sulima found various kinds of cyclicity in the succession of bioenantioforms (Fig. 53), which he attributes to solar activity and the GMF (552). He investigated 279, 453 wheat grains in 1960–1968, and 93, 389 corn grains in 1962–1964, and determined the relative numbers of D, L, and S forms of these plants. This led to the establishment of important laws in the succession of D, L, and S forms. First, the symmetric fraction of the biomodifications is internally heterogeneous, plastic, and mobile, and second, the symmetric S fraction is more closely related functionally to the D form than to the L form. He also showed that there is a considerable excess of frequency of changes from the D form to the S form, and that this is accompanied by annual alternations of the value of these D to S changes. Sulima established empirically that left-handedness prevailed in even years—1960, 1962, 1964, 1968—when north-polar geomagnetic conditions predominated, while right-handedness dominated in the odd years—1961, 1963, 1965, 1967, 1969—during the manifestation of south-polar geomagnetic conditions (552). It is quite possible that the established cyclicity in the succession of D, L, and S forms and the specific effect of north and south geomagnetic conditions are

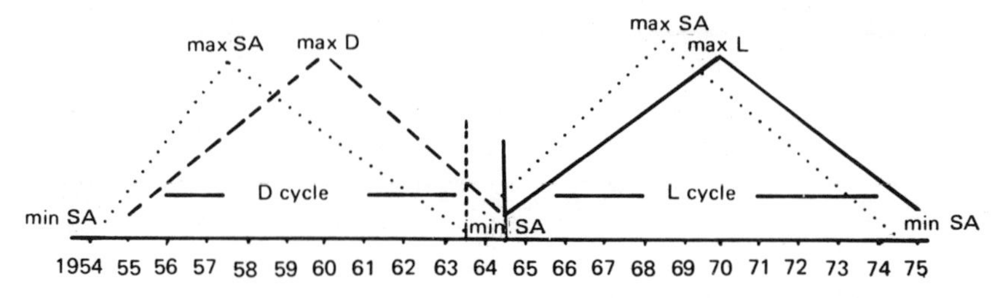

Fig. 53. Provisional scheme of relation between solar activity and dynamics of variation of L : D ratio in wheat and corn. Analysis made from grains and shoots. From (552).

related to solar activity, viz., to change in polarity of magnetic fields of bipolar groups of spots, and unipolar magnetic regions, on the sun and consequently to the sectoral structure of the interplanetary magnetic field, which is responsible for the biological effect.

The data presented above show what an important role the GMF plays in the formation and manifestation of the specific properties of the symmetry of biological objects and to what complex influences it is subjected on the part of other physical factors. We shall now discuss the possible underlying causes of this effect.

It is generally accepted that the symmetry of the magnetic field is axial and cannot engender dissymmetry in any objects (514). Here, however, we must recognize the difference between the physical features of an artificial magnetic field and the GMF, whose action on biological and other objects is homogeneous and total, because of the vast space that it occupies (536). Another point of fundamental importance is that the GMF has a special feature indicating its internal asymmetric structure, despite its apparent homogeneity.

The north–south asymmetry of the GMF is clearly revealed in several geophysical phenomena, including the direction of the vector of the GMF vertical component and the movement of oppositely charged particles in the solar wind in the GMF.

The modulus of the total GMF vector can be decomposed into several components, particularly a horizontal and vertical component. If we investigate the direction of the vector of the GMF vertical component, we find that it has a different direction in the northern and southern hemispheres. This leads to differences in the interaction of charged particles with gravity in the northern and southern hemispheres.

An investigation of the movement of oppositely charged particles arriving at the earth with the solar wind showed that they are specifically sorted according to sign and are directed into different sectors—morning and evening—of the magnetosphere. The GMF separates the particles according to charge: the protons go to the evening side, and the electrons to the morning side, i.e., the direction of particle drift depends on the GMF, and the picture of this separation is clearly displayed in the polar aurorae.

In speaking of the asymmetry of the GMF we must bear in mind a special feature of its manifestation. The variations of the GMF are due to current systems in the earth's ionosphere and magnetosphere. The ionospheric current systems are responsible for the S_q variations and baylike disturbances, and they have different directions at different local times of day. The current systems of the S_q variation consist of four current vortices, which have their centers at a geomagnetic latitude of 35° and have opposite directions in the northern and southern hemispheres, and also on the night and day sides, which leads to the formation of an asymmetric field in the S_q variations.

Baylike disturbances are due to east and west polar electrojets flowing in the ionosphere at heights of 100–120 km. The eastern electrojet (geomagnetic latitude about 65–68°) goes along the oval of the polar aurorae and is situated mainly in the evening sector (1500–2200 LT); the current in it flows east, and hence it leads to positive bays in central latitudes, whereas the western electrojet is situated in the night and early morning sector (2200–0600 LT), and the current in it flows west, which consequently leads to negative values of the GMF horizontal component (reduction of the H component) in central latitudes. Thus, there is a diurnal distribution of baylike disturbances with negative and positive values during disturbed days. The presence of differently directed changes of the variable GMF, of *ionospheric* origin, can lead to dissymmetry of biological objects.

We also find characteristic features in the manifestation of *magnetospheric* sources of the variable GMF. The ring current situated at a distance of 3–5 earth radii always has a western direction (in the plane of the equator) and hence causes a reduction of the horizontal component during a storm (D_{st} variation), which is usually negative (up to 500 γ); positive values (5 γ) are very rare. It should be borne in mind that although this ring current is of one sign, it is formed by protons with energies of 1–100 keV on the night and evening sides. Hence at the time of formation of the ring current protons drift westward and are concentrated mainly in the evening sector of the magnetosphere. This leads to asymmetry of the ring current and consequently to a large reduction of the D_{st} variations in the evening sector at 60–90° geomagnetic longitude.

The currents along the field lines, which descend into the polar ionosphere or pass out of it, have opposite directions on the morning and evening sides. It is believed that on the morning side the current flows along the field lines into the polar ionosphere, passes through the electrojet, and flows out on the evening side from the ionosphere into the magnetosphere. It should be noted that at the same time the currents along the field lines form a thin current layer in which the current has a reverse direction on the opposite side. Thus, this feature of the magnetospheric current leads to asymmetry of the GMF in relation to latitude.

It has become known in recent years that the north–south and azimuthal components of the interplanetary magnetic field (IMF) play a leading role in GMF variations. Since the GMF strength vector at high earth latitudes has an opposite direction (upward relative to the plane of the equator in the north, and downward in the south) the process of reconnection of the GMF and IMF field lines is associated with different directions of the IMF in the northern and southern hemispheres, which affects the seasonal variations of the GMF.

It follows from the above account that the GMF, in view of its asymmetric structure, can be a factor causing asymmetry of biological objects, i.e., the general concept of the Curie rule is valid.

It will certainly be necessary in the future to determine the specific mechanism of the effect of the GMF and natural electric fields on the formation of dissymmetry in biological molecules and organisms. At present we can only put forward hypotheses and suggestions, which will require solid and abundant evidence. The above-mentioned special feature of the physical properties of the GMF may lead to dissymmetry of water molecules and consequently to the functional and morphological asymmetry of biological objects. We can postulate that the structure of biological membranes includes water molecules that have different orbital and spin moments of atoms and electrons, or have hygroscopic effects with a different direction of rotation around the axis. This hypothesis corresponds with expressed views and, in particular, with the view of the existence of D and L types of water molecule (453).

The GMF may also have a direct effect on the formation of enantiomeric biological molecules. In particular, it has been postulated that weak interactions and neutral currents are to some extent responsible for the chemical asymmetry of biological molecules (842). The authors believe that orbital electrons in chiral molecules have polarized spins associated with the direction of their motion. The direction of this motion may depend on the twist (left- or right-handed) of the helix of chiral molecules. The static charge in such a molecule is a helical potential field associated with electron motion, where an additional magnetic field formed around the electron interacts with the magnetic moment of the electron. Depending on the sign of the helix, the spin of the electron will be in the direction of its motion or in the reverse direction. The helical potential field, as a consequence of all this, will affect the coupling of spins and moments and alter the parity of the orbital electrons.

We have discussed this hypothesis so thoroughly because it is one of the best models indicating possible "points of application" of the GMF at the quantum-mechanical level. In particular, if we accept this hypothesis, we can postulate that the GMF affects π-electron systems, by altering the magnetic fields induced in chiral molecules, magnetic moments, spins and, ultimately, current systems in them, which leads to a whole complex of biological sequelae.

We should mention that there are several papers expressing the view of a possible leading role of natural magnetic rotating fields in producing asymmetry in biological molecules (859, 1001, 1016).

Thus, in the light of the above discussion we can conclude that the GMF determines the biosymmetric status of a living object. This probably occurs in the period of fusion of the gametes and coincides with the moment of untwisting of the chromosomes, when there is a favorable opportunity for action on single-stranded DNA. The template enantiomorphic characteristic of the DNA molecule formed at this moment subsequently determines the direction of asymmetric synthesis and the stereoisomeric configuration of all the subsequent replicating molecules. The enantiomorphic characteristic of the object formed in this way becomes the decisive factor in the functional response of the living creature to the action of the GMF and other physical factors.

On the basis of our conclusion we can understand why a genetic analysis of the combination of D and L characters in progeny does not correspond with the Mendelian laws of inheritance, as was established in (552), and why at the same time the response of bioenantioforms to the GMF may be genetically determined, as shown in (378). The decisive factor in this case is not the Mendelian laws of inheritance with their well-known numerical ratios in the F_1 and F_2 generations, but the *effect of the GMF* on the symmetric properties of living and nonliving matter at deep elementary levels—on genetic molecular systems.

In the next chapter we shall discuss possible hypothetical aspects of this effect.

Combined Action of the Geomagnetic Field and Gravitation on Biological Objects

We have repeatedly stressed that the GMF acts on living organisms in conjunction with the complex effect of many factors—gravitation, atmospheric electricity, cosmic rays, light, temperature, and other external environmental factors. The biological action of two of these factors, viz., the GMF and gravitation, is of a special nature in that it is unselective and acts on *all cells* of the organism *directly,* on the organism as a whole, and thus on all living organisms on earth. The direct, all-pervading action of these factors does not mean the absence of *specificity* in the biological action of the GMF and gravitation, but it is very difficult to demonstrate this.

The aim of the present chapter is not to give a full account of the complex interrelation of natural gravitational and magnetic fields in their combined effect on biological objects. We merely wish to draw attention to the existence of this effect, since we believe that the combined effect of the GMF and gravitation may be a common denominator for all vital activity in the biosphere, and for nonliving matter too. We need only mention that gravitation is responsible not only for sea tides, but also for tidal effects in the earth's crust and changes in the mass of the earth as a whole, to make the reader aware of the strength and significance of this factor (983). There is no doubt that living organisms are affected more by gravitation than nonliving matter, owing to their exceptional sensitivity to external fields, since they are colloidal systems in a state of unstable dynamic equilibrium (1049).

The correct approach is probably to discuss the following aspects of this problem: the effects of inertial and rotational gravitation on biological objects in stationary and rotating states. The grounds for such an approach are the investigations of physicists indicating that there may be two types of gravitational field (761, 834).

We regard the effects of the stars and different planets on man, animals, and plants (666, 831, 832, 844, 888, 908, 963) as a manifestation of gravitational action, although they may have a specific effect and emit special radiations. We can mention that one paper (724) has shown that the cosmic effect on biological objects may be mediated by disturbances of the GMF.

In our view, which is shared by other scientists (1148), the most fundamental discovery in the entire history of research on biological rhythmicity was the discovery of Brown and Park (753) that the GMF in its action on biological objects interacts with other vector forces, particularly the gravitational effect of the moon. This was the first experimental evidence of the existence of a common denominator for the biological temporal organization and spatial orientation of living organisms. In that investigation the authors studied the lunar rhythm of spatial orientation of the planarian *Dugesia* continuously for seven months. The action of a magnetic field of 0.05 or 4.0 G in opposition to the vector of the GMF horizontal component led to a phase shift of 180° in the monthly rhythm of geographical orientation of the planarians, and the weaker the acting magnetic field, the greater was its effect (Fig. 15). The response of the planarians to a magnetic field of 0.05 G was a function of the elongation of the moon. Thus, it was discovered for the first time that the combined effect of the GMF and gravitation is responsible not only for the spatial, but also the temporal, organization of biological objects.

In our opinion, studies of the ESR of man and animals shielded from the GMF (105, 408, 409, 540, 541) and subjected to magnetic action (105) are a first indication of how important it is to take into account the interaction of magnetic and gravitational fields in the reaction investigated. It was found that direction of the magnetic field vector, in opposition to that of gravitation slowed down the ESR (408, 409, 540, 541) and reduced the agglutinating power of immune sera (105).

The combined action of the GMF and gravitation will have a more perceptible effect on the entire organism, of course, owing to the higher sensitivity to external factors. Investigations have shown that subjection of plant seedlings to a vertical magnetic field of 20–25 Oe with the north and south poles in different positions either stimulated (clover) or inhibited (wheat) seedling growth (505). In most cases the best results were obtained with the south pole in the bottom position. In this case the differences in seedling growth reached 60–70%. The authors believe that these differences in growth can be attributed to the difference in direction of the magnetic and gravitational fields, but we think that this is a case of interaction of the biomagnetic field and the field of the magnet. This conclusion would be more convincing if the discovered effect of an artificial magnetic field could be obtained with objects shielded from the GMF and rotat-

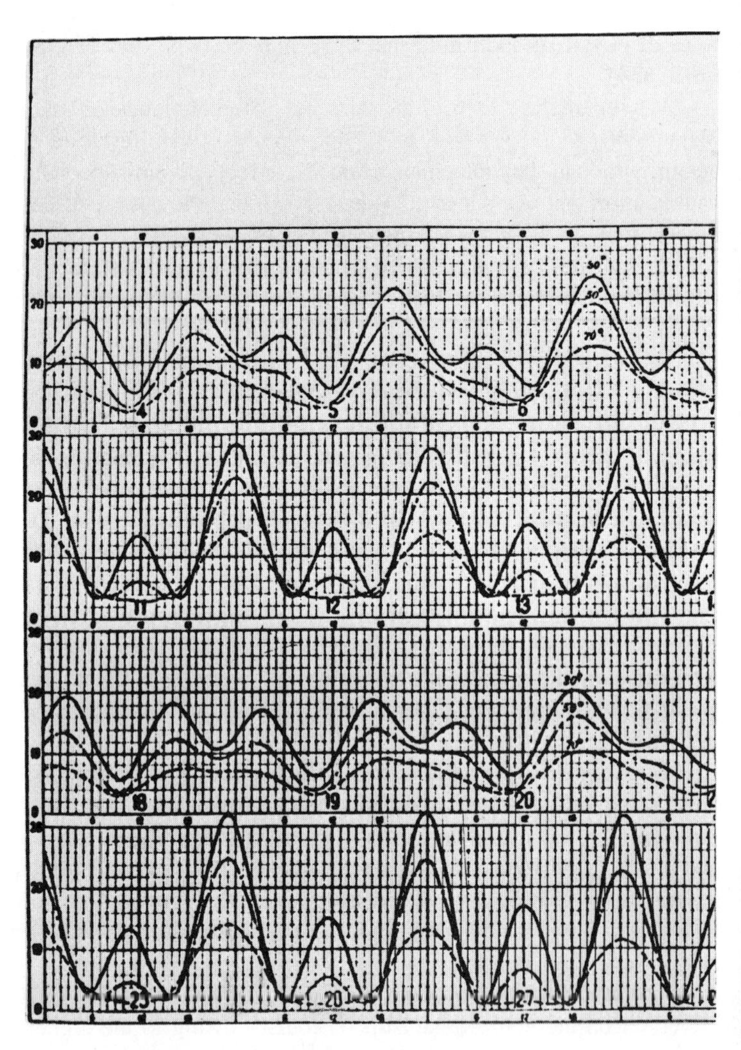

Fig. 54. Graphs of corrections to gravitational force for tidal changes due to motions of moon and sun. The corrections are calculated for each hour for latitudes 30, 50, and 70° and for local times of day for longitudes east of the Greenwich meridian. The x axis is the time axis, and the y axis gives the values in milligals.

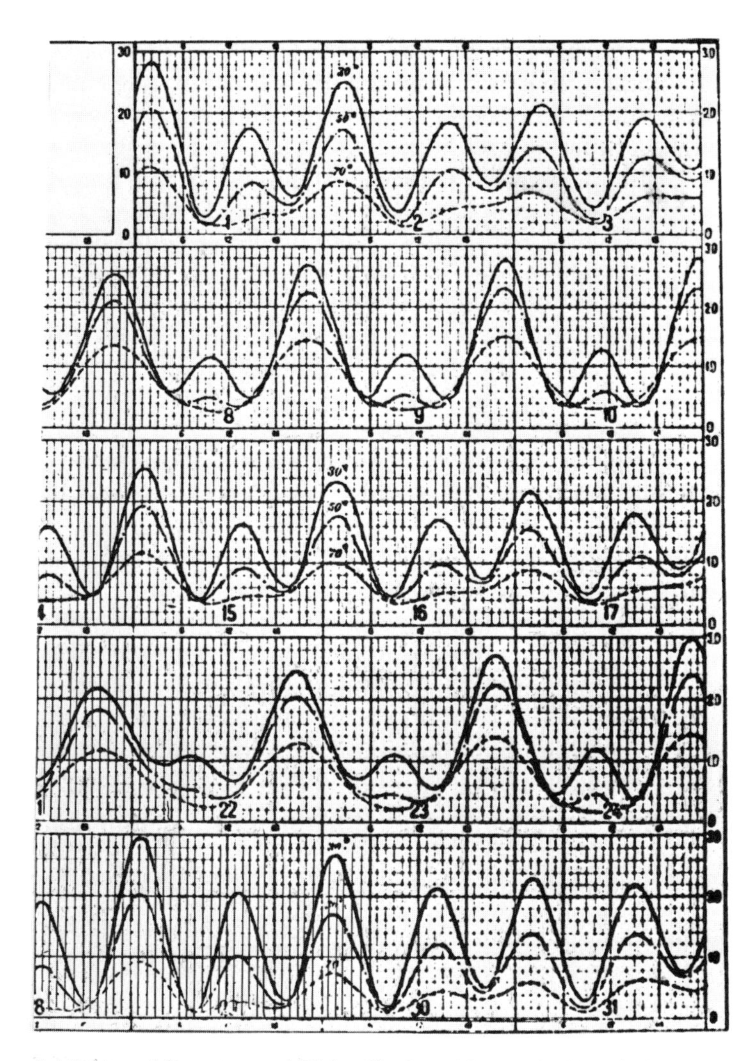

Publication of Department of Higher Geodesy, Moscow Institute of Geodesy, Aerial Photography, and Cartography, Moscow, 1975. The variation of gravitation for January, 1975, is shown.

ing, i.e., weightless. A similar approach must be made to studies in which the effect of the GMF vertical component is eliminated. It has also been reported that in experiments with cucumbers and beans in Kaliningrad, reversal of the vector of the GMF vertical component resulted in separation of a previously single plant population into two distinct groups with different rates of growth and development (289). In addition, the effect of a magnetic field of 18–25 Oe in the same direction as the GMF inhibited root growth in right-handed, left-handed, and symmetric "Vyatka" rye seeds ($p = 0.01–0.001$). When the field was opposed to the GMF the seedlings showed no response (396).

In investigations of the combined effect of gravitational forces on biological rhythmicity we found that the absolute level and the rhythm of root excretions by plants depended on the time of soaking of the seeds in relation to the lunar phase (134). Plants grown from seeds soaked before a change in any lunar phase showed differences in the rhythmicity and amount of the root excretions in comparison with plants that had been grown from seeds soaked exactly at the time of change of a given phase of the moon or some time after the phase change. Differences were found even in the case where the difference in times of soaking of the seeds in relation to the change in lunar phase was only one hour. Thus we find that the initial instant at which the living organism begins to function (moment of conception, moment of formation of embryo in animals and plants, moment of soaking of dry plant seeds, etc.) is of great importance, since the natural gravitational–magnetoelectric complex has its own characteristic features in each period of time. For a clear demonstration of the specificity of the changes in gravitational forces due to tidal effects (mainly the effect of the moon and sun) we give graphs of corrections to the gravitational field showing the diurnal course of tidal effects for several latitudes and longitudes on the earth (Fig. 54), and the relation between the diurnal course of root excretions and the variation of gravitation (Fig. 55).

The complicated nature of the interrelations of a living organism and the gravitational–magnetoelectric natural complex is shown by experiments involving the rotation of biological objects (742–745, 894–897) and the effects of rotating and stationary artificial magnetic fields (742, 745, 1028, 1029, 1037–1044, 1046, 1116). By using a special device capable of producing a brief period of simulated weightlessness on earth by rotation of the object in different planes, Hoshizaki *et al.* (896) showed that the particular time at which weightlessness was induced was of great significance for the entire subsequent course of the biological rhythm (897). If the simulated weightlessness was imposed early in the morning the biological rhythm was completely upset for almost three days (Fig. 56). Induction of weightlessness at noon led to a slow phase shift of 180° in the rhythm of the experimental plants in comparison with stationary control

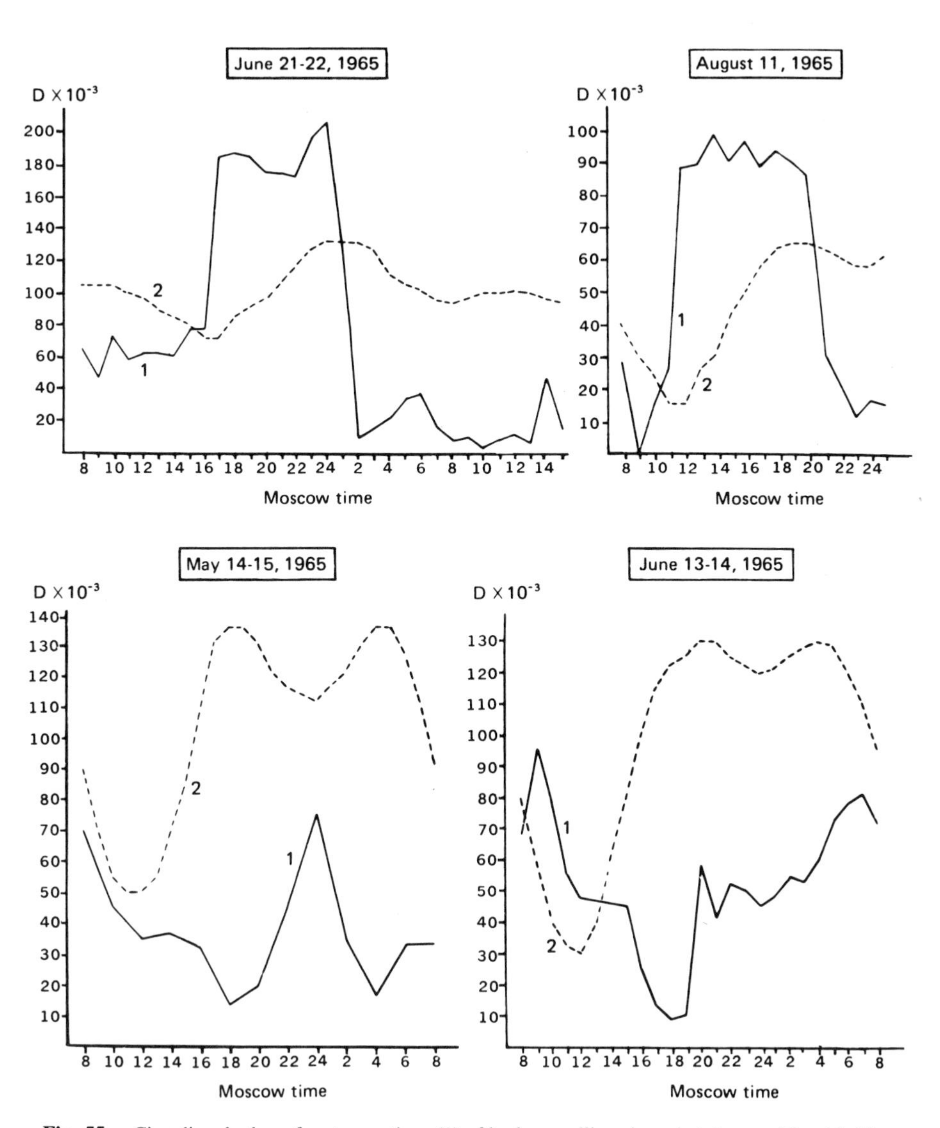

Fig. 55. Circadian rhythm of root excretions (1) of barley seedlings in a phytotron on May 14–15, June 13–14, June 21–22, and August 11, 1965 in Moscow and the diurnal variation of gravitation on these dates (2). The variation of gravitation on the indicated days was taken from the graphs of corrections to the gravitational force for tidal changes (see Fig. 54). From (134).

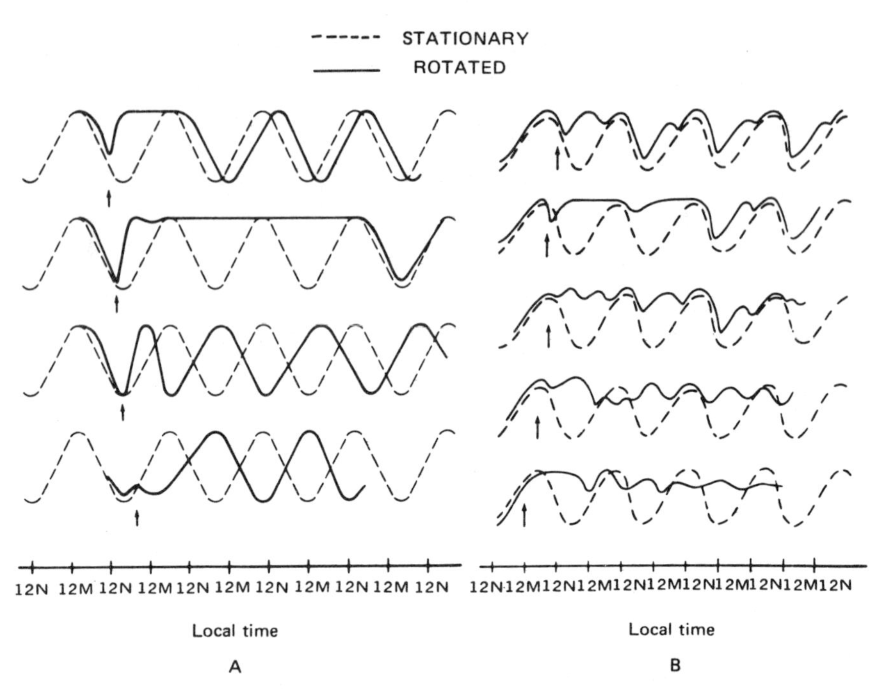

Fig. 56. Change in circadian rhythm of movements of *Phaseolus* (pinto bean) leaves due to induction of artificial weightlessness by means of rotation at different times: (A) day (B) night. Onset of rotation indicated by arrows. The curves for rotated and stationary control plants are mean curves for eight primary leaves of four plants. From (897).

plants. Induction of weightlessness during the night also led to phase shifts in the rhythm. Experiments in which we investigated the role of the combined action of the GMF and gravitation also revealed this effect. For instance, we observed a change in the rhythm of root excretions in plants in a special experiment in which the plants were suspended, like pendulums, on a fine steel wire 6 m long and 0.5 mm in diameter from a high ceiling (134). All the experiments showed that suspension of the plants increased the absolute amount of root excretions and altered the basic rhythm—the main maximum became more pronounced and an additional maximum appeared (134).

Thus, it is apparent that biological rhythmicity, controlled mainly by diurnal variations of the GMF, also includes the effect of the gravitational complex. In this complex the rotational gravitational field (761, 834), as well as the stationary gravitational field, plays an important role. Its biological role is revealed when biological objects are made to rotate in a clockwise and/or, in particular a counterclockwise direction (745, 826, 1055–1063). Such experiments showed that even a low rotation rate (1–6 rotations per minute, or even per day) led to a change in plant processes (394, 909), where the GMF, as has been shown above, is usually the decisive factor. For instance, when radish seedlings were rotated clockwise the root creases were not oriented in the GMF (389) and the root failed to swell (personal communication of the authors), the plant growth rate was reduced (909), and the uptake of water by dry *Phaseolus* seeds was altered (742). Similar changes were found in *Phaseolus* in different laboratories at great distances from one another (Marine Biological Laboratory in Woods Hole, Massachusetts, and Northwestern University, Evanston, Illinois).

Experiments involving rotation and the action of rotating magnetic fields have been carried out on living objects (743, 1028, 1029, 1037–1040, 1042, 1044–1047, 1116). These investigations confirm our conclusion that, although diurnal variations of the GMF play the leading role, there are other important factors governing the spatial orientation and temporal organization of biological objects. As investigations show, inertial gravitational and rotational fields are probably important, and there may be other unknown factors (1098). Incidentally, the Coriolis force, in our view, does not have a significant effect on biological processes, but the rotational gravitational field is important. This is apparent merely from the fact that rotations of biological objects in the northern and southern hemispheres lead to the same changes in plants (745, 910), whereas clockwise or counterclockwise rotation causes significant changes in processes occurring in biological objects (743–745). These changes, in our view, can be attributed to the different interaction of the physical vectors of the inertial and rotational gravitational fields, the GMF, and the vectors of the intrinsic biological magnetogravitational field of living organisms, which are different in right-handed, left-handed, and symmetric organisms.

The periodic, and also sharp and sudden, changes in the gravitational rotational field are probably due to particular positions of the planets (497) or to intense solar flares (410, 588, 1005). Hence, sharp changes in biological processes that show no correlation with the activity of the GMF or other important factors can possibly be attributed to sharp changes in the effect of gravitation on biological objects. The discovery of all the important acting heliogeophysical factors will definitely require the organization of special research in accordance with a comprehensive program. These investigations will have to be systematic, long-term, and standardized, so that the solar and cosmic nature of rhythms can be determined from their period. For instance, there is evidence that solar cycles have durations of 7, 8, 11.6, 12.6, 15, and 17 months, while cosmic cycles have durations of 26, 39, 53, 78, and 156 months (497). From the dominant period in the analyzed biological rhythm it will be possible to obtain information about the cosmoheliogeophysical factors controlling the particular biological process.

Fundamental Bases of Biological Effect of GMF (Hypotheses)

Sensitivity of Living Organisms to the GMF and Biological Superconductivity

The fundamental bases of the biological effect of the GMF present a problem in which only a few first steps have been made so far. There are more hypotheses and suggestions than conclusive answers. In presenting them, however, we refer to the words of Albert Einstein: "any scientifically based proposition that has internal perfection, i.e., a logical or mathematical structure, and external justification—correspondence with facts, can be, and should be, regarded as a plausible hypothesis."

In investigating the fundamental bases of the biological effect of the GMF we must bear in mind two essential features of this effect: first, the exceptional sensitivity of biological objects to the GMF, and second, the universality of the action of the GMF on biological objects. We shall consider these two features in turn.

The extraordinary sensitivity of living organisms to the GMF has led research workers to suggest that living objects have a special mechanism for detection of the GMF (706, 779, 971). In fact, it is apparent from the information that we have presented that biological objects and processes in them are directed and controlled by GMF variations that on quiet days lie in the range 10–20γ or a few tenths of angular minutes. Hence, such insignificantly small changes in the GMF determine the course of all biological processes—physiological and biochemical, genetic and biophysical—irrespective of the level at which they are considered.

Of all the currently known physical mechanisms only the Josephson junction effect, based on superconductivity, can provide an explanation of the sensitivity of biological objects to the GMF (910).

Superconductivity is the physical phenomenon of total reduction of the electrical resistance of a conductor at temperatures close to absolute zero (below 20°K). The states created at such extremely low temperatures represent the lowest energy level in the conductor, at which all, or a large proportion, of the conduction electrons are united by opposite spin and kinetic states (876, 910, 1096)—the so-called Cooper electron pairs, or paired electrons. A consequence of this unusual state of the electrons is the complete loss of resistance and complete exclusion of the external magnetic field from the interior of the superconductor. The currents produced in such a system can flow without any loss at all and require no electromotive force. At temperatures and magnetic fields greater than the critical T_c and H_c, superconductivity is destroyed and the conductor acquires normal properties (Fig. 57a).

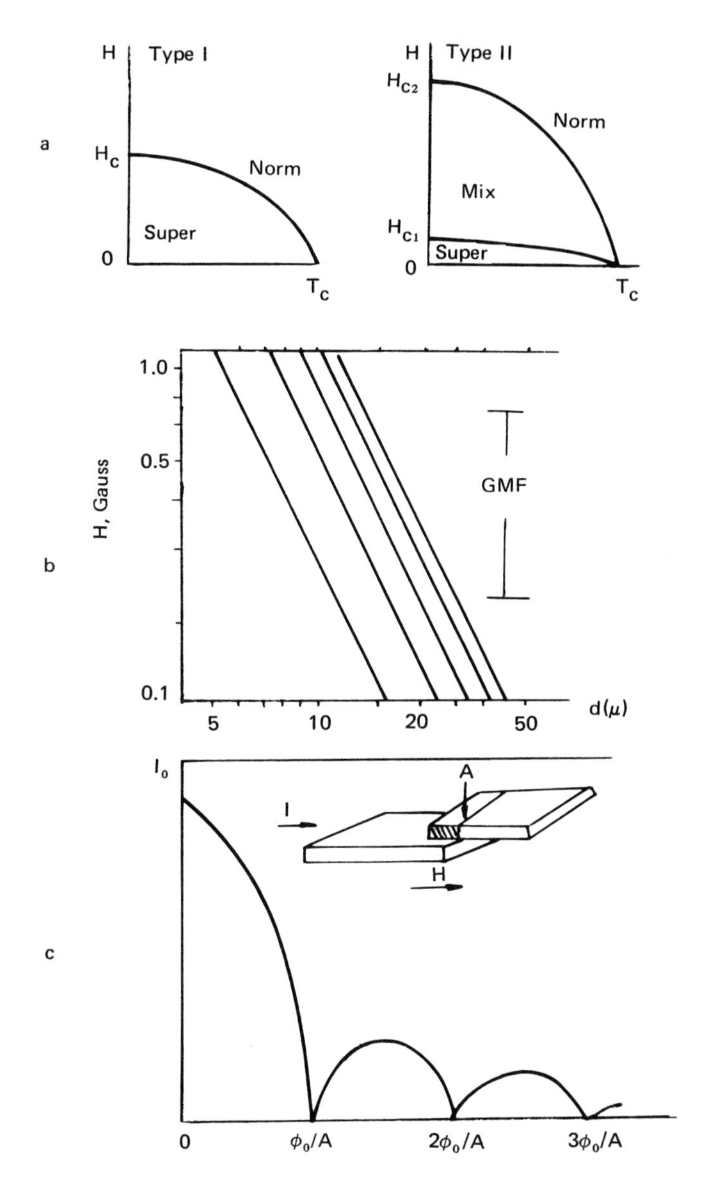

Fig. 57. (a) Relation between critical temperature and critical magnetic field for types I and II superconductors. (b) Lines of lowest free energy (greatest stability) of type II superconductor in mixed state in relation to external magnetic field for a specimen of diameter d through which a fixed flux ϕ_0 of quantized magnetic field passes. (c) Change in zero potential I_0 of Josephson current flowing through a bridge of area A, with an external field H directed along the junction as shown in the figure. ϕ_0 is the quantum flux. From (971).

There are two main types of superconductor: type I, where the critical magnetic field depends on the temperature, and type II (mixed), where the superconductive state can be accompanied by a normal state of the material and hence the material can transmit part of the external field. An important feature of type II superconductors is that in the mixed region they transmit a magnetic field only in multiples of unit quantum magnetic flux (multiplets). The magnitude of the quantum flux passing through the specimen depends on the external field and the size of the specimen. Unit quantum flux $\phi_0 \approx 2 \times 10^{-15}$ Wb (2×10^{-7} G).

The superconductivity current carried by paired electrons can flow not only in the semiconductor, but also between two semiconductors that are connected by an ordinary conductor or separated by a dielectric whose thickness does not exceed 10–20 Å. This is the so-called Josephson junction, named after the discoverer of this effect (910). The zero potential of the Josephson current is very sensitive to the strength of an external magnetic field and, in particular, can be completely destroyed by the GMF (Fig. 57c).

A detailed account of all aspects of superconductivity is given in several major monographs (876, 1096, 1134). For our discussion it is more fundamental and important that, according to theoretical calculations, superconductivity can occur in organic materials (852, 954) and biological objects (778–780, 971). These studies have been subjected to careful adverse criticism (937), but biological superconductivity is still regarded as a possibility by geomagnetobiologists.

What is the evidence in support of biological superconductivity and why is this hypothesis so attractive to geomagnetobiologists? First, there are theoretical grounds for believing that superconductivity can, in principle, occur at high temperatures (850, 938, 954, 955) and, in addition, there is experimental evidence of superconductivity in thin films of inorganic materials at room temperature (670, 955) and in organic molecules of the bile-acid type (876). Further evidence of the possible existence of biological superconductivity is the recently discovered superconductivity in a new class of thin films—biomolecular layers composed of a metal and a semiconductor separated by organic molecules (847). Geballe reports that their structure is almost a perfect analog of that of molecules in biomembranes (971). A molecular layered-membrane structure is a basic element in biological objects, and the lipids in it, together with protein and water molecules, play a decisive role. It has been suggested that the presence of cholesterol molecules in nerve-cell biomembranes may be the basis of biological superconductivity in processes occurring in nerves (780).

We can cite several more interesting facts indicating the semiconducting properties of biological objects and, hence, the possibility of tunneling effects in them (201, 1121). For instance, a study of the current–voltage characteristics of dry and swollen wheat seeds at a potential of 15 and 0.3 V, respectively, revealed *negative* resistance (250). This unusual property was observed only when the current passed in front of the germ region and indicates the existence of semiconducting (*p–n*) junctions in biological objects with all the resultant consequences.

In addition, it has been reported that the available temperature dependences of some processes in nerve and living cells (Fig. 58) indicate an electronic superconducting transition typical of electron pairs in a superconductor (Fig. 57b) and in these processes negative temperature coefficients are obtained (778–780). Furthermore, a recently published paper showed that superconductivity can be observed in the presence of high electric fields (670), and also that fields up to 100 kV/cm are present in nerve conductors due to the steep gradients of the electric fields in biological membranes of molecular dimensions.

We think that the hypothesis of biological superconductivity is also supported by the Kirlian effect (27, 1166). This is the emission of cold electrons when a biological object is placed in a high-frequency, high-voltage field. Such an emission of electrons from the surface of a biological conductor is only possible if superconducting regions are present. The existence of superconducting regions in living objects is also indicated, in our view, by the system of acupuncture points and meridians, which have high electrical conductivity (980, 1003, 1167). In the two cases mentioned above, the application of weak magnetic fields with the GMF completely shielded would give the same current dependence on the quantum magnetic flux as in a Josephson junction with a thin dielectric. We have grounds for such a conclusion, which we do not give here, and will merely indicate that our explanation of the Kirlian effect and interpretation of the system of acupuncture points differ from the explanations given by other investigators (980, 1003, 1167).

We must now discuss the objections advanced against the hypothesis of biological superconductivity (937). First, the effect of a magnetic field on the Josephson currents is the same, irrespective of direction (polarity), and then their use in orientation relative to the GMF is unclear. There is no contradiction here, however, since we observe a similar effect in the responses of living organisms, e.g., fish (1159) or birds (1216–1219), which make equal use of both geomagnetic poles for orientation in space.

Second, superconductivity is accompanied by a strong temperature effect (833) and can occur either at very low temperatures or when several other physical conditions are satisfied (670, 876).

Is the first of these conditions observed in the living organism? Yes, it is, but in very specific circumstances. It is a fact, as investigations on biological thermodynamics have shown (578, 1171–1174), that in the metabolizing cell performing internal work heat is not generated and hence entropy does not increase. This is due to the fact that the cell has a unique mechanism of heat removal involving continuous microphasic transitions of intracellular water and proteins from the liquid to the crystalline state. Their close interaction and the interdependent "liquid–crystal" or "disordered–superordered state" transitions constitute a unique mechanism for the creation of negentropy in living material.

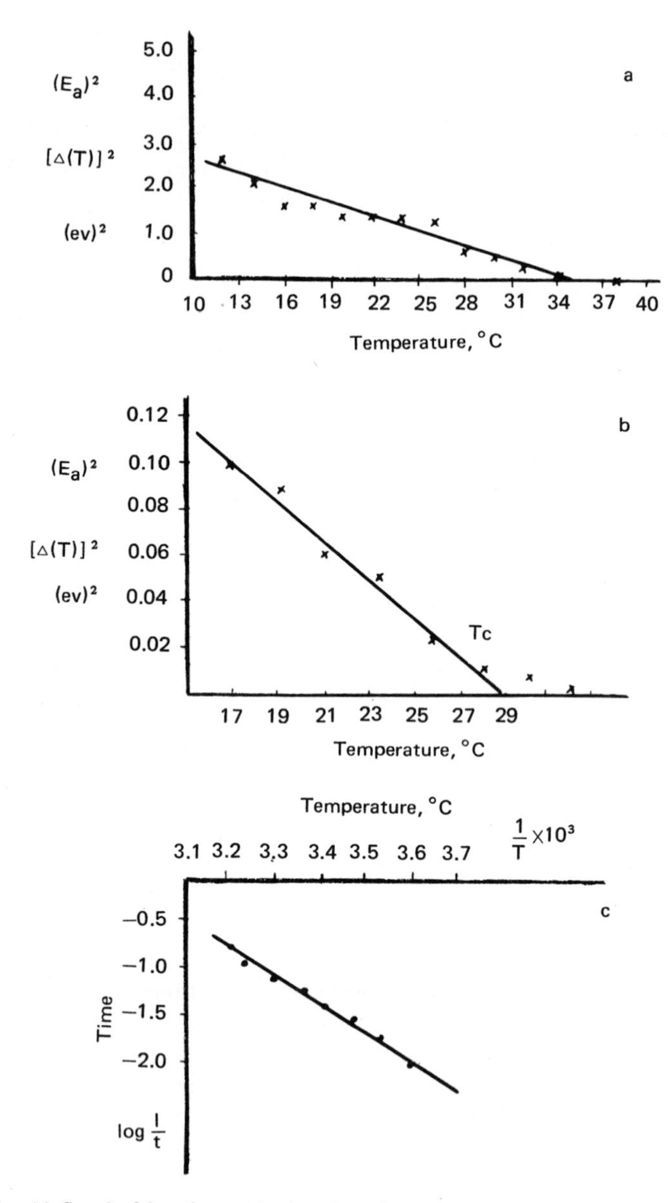

Fig. 58. (a) Speed of impulse conduction along frog nerve as a function of temperature. The linearity of this graph indicates similarity with the equation for the binding energy of Cooper electron pairs in a superconductor. The situation is similar in (b) and (c). From (778). (b) Growth rate of *E. coli* as function of temperature. T is half the binding energy of Cooper pairs. E_a is the activation energy. From (778). (c) Rate of erythrocyte hemolysis as a function of temperature. The duration of exposure to the hypertonic solution is plotted on the y axis. From (578).

Trincher states directly that "water inside the cell is in a state of *maximum order* [emphasis added]—in a state attainable in nonliving systems only at absolutely zero temperature" (578).

Thus, if we adopt as a basis the new thermodynamic concepts advanced by Trincher, we must assume either that biological processes take place, however paradoxical it sounds, at very large negative temperatures, or that in living systems, by the operation of a special mechanism of action of the biogravitational field, which creates specific conditions, there are effects dependent on molecular membrane biological superconductivity. This is not accomplished by the use of a complex cryogenic technique, of course, but by means of special processes in biological molecules, whose function is based on specific laws of living matter, which will be understood by science in the future.

Acceptance of the fact of biological superconductivity opens up pathways for the solution of many fundamental biological questions, including the hypersensitivity of biological objects to the GMF. In a Josephson junction the nonsuperconducting bridge is subjected to the effect of the external field and extremely small changes in the strength and direction of this magnetic field are sufficient to produce large changes in current. We need only mention that the sensitivity of special magnetometers based on the use of a Josephson junction can be as high as 10^{-11} G.

If we regard the living organism as a single pseudocrystalline structure (201, 1121), then particular links in this integral chain can be regarded as large Josephson superconducting junctions or loops. We know from the physics of superconductivity that the greater the area of a Josephson junction (or Cooper loop, which is the same thing), the greater its sensitivity to an external magnetic field. In view of published data indicating that the sensitivity of dowsers to a magnetic field is very high (1090, 1176), sometimes reaching exceptional values of 10^{-12}–10^{-14} G, as has been shown by the investigations of Dr. Harvalik, Director of the Research Center of the American Society of Dowsers (personal communication from Mr. Christopher Bird, Washington), we can propose two possible explanations of this hypersensitivity of man to magnetic fields. First, the human being may have a Cooper superconducting loop of very large area. This is indicated by simple calculations: For a minimum individual quantum flux of 2×10^{-7} G/cm^2 and a maximum sensitivity of man to a magnetic field of 10^{-14} G, the area of the Cooper loop is 10^7 cm^2. Second, the magnetic field may be detected in the living organism by a Josephson junction of smaller area, but there is a special system in the brain for amplification of the primary input signal by a factor of 10^6–10^8. In both cases extremely small anomalies in the ground will be detected with extraordinary accuracy by the human being, since they distort the structure of the external geomagnetic field in which the dowser is situated. The frame or pendulum in the dowser's hands is merely the indicating needle of this intricate living superconducting "sensing instrument."

Thus, the age-old problem of dowsing can be explained if we accept that *biological superconductivity* exists in living objects, although the actual phenomenon of dowsing includes many more physical effects that still do not lend themselves to a complete explanation (e.g., directed search, etc.). However, in discussing the hypersensitivity of man to magnetic fields, we stress at present the qualitative aspect of the difference between biological superconductivity and ordinary physical superconductivity, since this is of fundamental importance. Research workers who subscribe in their work to the view of a possible role of superconductivity in the detection of the GMF by man and other organisms (706, 778–780, 971) make an apparently slight, yet essentially very significant, omission. They fail to stress in their papers that the phenomenon under consideration can be called "superconductivity" only by analogy. In fact, in the case of biological superconductivity all the interesting quantum features of the physical effect have a completely different basis from the superconductivity known in physics.

The fact is that biosuperconductivity takes place in the conditions of a special conformational field that we call biogravitation (811). The presence of such a field is one of the conditions for the occurrence of biosuperconductivity in living cells and their membranes.

One must not infer, however, that biological superconductivity is an easy and simple accomplishment of the living organism. In fact, for the creation of biological superconductivity and its maintenance at the required level the organism pays a very high price—the life of the cells themselves. The short life span of the individual is the inestimable price that it pays for the possession of superstate effects—biogravitation, biosuperconductivity, the bioplasma, etc. The human organism maintains the *thermal homeostasis* of its body very accurately in the range 36.4–36.8°C. Apparently insignificant deviations of ±1–2°C from these values make us feel ill. The organism carefully and accurately maintains T_{crit}, the critical temperature for biological superconductivity. The organism responds equally sensitively to a change in the external magnetic field. The level of this field, as we know, is 0.5 Oe, and great changes in this strength (reduction or increase, reversals) lead to the death of living organisms if the mechanisms of *biomagnetic* homeostasis cannot compensate these changes. Thus, biological objects probably have a system of biomagnetic homeostasis whose purpose is also to maintain the level of biological superconductivity when sharp changes in the GMF occur. Hence, the study of the intrinsic magnetic fields of biological objects merits the closest attention of geomagnetobiologists, since it is the key to many enigmas of the living organism. The investigation of this question in the seventeenth and eighteenth centuries was inadequate, of course, owing to the low level of development of physical science as a whole, but the main concepts of the "science of animal magnetism," found empirically, were true (148, 149), including the "magnetic" polarities of parts of the human body.

Any hypothesis or theory of homeostatic regulation of functions in man must take into account, and provide an explanation of, such features as the system of acupuncture points present in living organisms. This system of points has two fundamental features and properties—their arrangement in the form of meridians, which constitute, as it were, an energy network for the organism, and their high electric conductivity in comparison with other points in the body. On this basis scientists have made the very valid suggestions that the acupuncture points are sites of connection of the living organism with the terrestrial air–ion, electrical, and magnetic fields (980, 981). We must consider the possibility that these points are sites of connection of the earth's magnetic field lines with the intrinsic biomagnetic field of the living organism.

Universality of Action of the GMF and Biological Symmetry

Since the GMF has an effect on absolutely all organisms and on all processes taking place in living organisms we must bear in mind the following features. First, the GMF exerts its effect through a weak-action mechanism, i.e., at extremely low energy thresholds; second, the GMF is one of the main causes of the rhythmicity of natural phenomena; and third, the GMF is one of the main causes of dissymmetry in nature. The globality and universality of the action of the GMF imply that the postulated mechanism of fundamental action must be all-embracing for all objects—both organic and inorganic. In a previous chapter, however, we noted that symmetry is such a universal property. The symmetry of objects is a fundamental phenomenon in nature, but it includes the opposites: left-handedness (L) and right-handedness (D), i.e., dissymmetry, or destruction of symmetry of the object.

We must draw attention to the fact that left- and right-handed objects, like dissymmetric objects, have *mirror* symmetry. This fundamental idea is clearly expressed in the universal formulation of dissymmetry (591). The most significant point for our discussion is that the formation of L and D objects involves mirror reflection (mirror ortho-transfer or ortho-rotation) by which one enantioform is converted to the other. Hence, L and D objects are essentially one, but differ radically in their properties. There is abundant experimental evidence of the differences in properties of L and D enantioforms (106, 107, 125, 135, 376, 377, 381, 591). Thus, we arrive at the important conclusion that mirror reflection (mirror ortho-rotation) leads to the formation of objects whose properties differ distinctly from those of the initial form. The importance of these operations for symmetry and the explanation of many fundamental effects in biology and physics has already been indicated (591, 682).

On the other hand, however, we have shown that the GMF plays a decisive role in the formation of L and D objects and in their subsequent responses during ontogenesis. The GMF is an extremely important dissymmetric factor in nature:

Its long-period components (of the order of hundreds of thousands and tens of millions of years) lead to reversals of poles, to an increase in the rate of mutational processes, and to the formation of new species of living organisms with *different* symmetry, while the short-period components (secular, annual, diurnal) are acting at present, and determine the complex dynamics of the homeostasis of living organisms, give rise to L and D bioforms, both morphological and functional, and thus lead to complex rhythmic oscillatory phenomena in the living organism.

Thus, it is apparent from the above that the GMF, in spite of the smallness of its diurnal amplitudes, has a decisive effect, by weak-energy action, on the state of biological objects in ontogenesis.

If what we have said above about dissymmetry and the role of the GMF in these processes is accepted, then another important conclusion inevitably follows: *The GMF is the factor that creates dissymmetric objects* by *mirror-transfer* (mirror ortho-rotation) *operations,* which it effects in some specific manner *at low energies*.

To understand how this specific mechanism of formation of L and D objects operates, we turn our attention to the phenomenon of transmutation of elements (919–922).

Transmutation of elements outside the radioactive series was discovered by the French scientist C. L. Kervran as a result of many years of original experimental research and a review of a vast amount of data from very diverse areas of natural science. He found that in living organisms (and nonliving matter as well) some elements are converted to others at *low energies,* e.g., magnesium is converted to calcium, iron to manganese, silicon to aluminum, and so on. In Kervran's opinion, transmutation of elements is effected by reactions of a special type—weak nuclear interactions. These reactions occur because there are some stable structures (^{1}H, ^{16}O, ^{12}C, ^{28}Si, ^{11}B) that can unite and separate at low energies. In these processes there is no change in the internal structure of the atoms, but only their union and separation. Some examples of transmutations are $^{12}C + ^{12}C \rightarrow ^{24}Mg$, $^{12}C + ^{16}O \rightarrow ^{28}Si$, $^{23}Na + ^{1}H \rightarrow ^{24}Mg$, $^{12}C + ^{12}C + ^{16}O + ^{40}Ca$. Although the phenomenon of transmutation is not in doubt, the mechanism of this effect has not been investigated, although various hypotheses and suggestions regarding transformations of elements at low energies have been put forward. One hypothesis is due to O. Costa de Beauregard, a world authority in the field of theoretical physics and Director of the Scientific Research Center in France. He thinks that an important factor in the transmutations of elements discovered by Kervran is the emission and absorption of "cold" and "hot" neutrinos, which are present in excess in the environment (781, 922).

In our opinion, however, the transmutation of elements (the "Kervran effect") can be attributed to a number of *symmetric transformations:* mirror reflection, translation + mirror reflection, rotation + parallel transfer (translation), etc. (for a fuller account of transformations see 591).

As evidence in support of our hypothesis we cite the following. First, the known transmutations, e.g., magnesium to calcium, and potassium to sodium, can be attributed to the fact that these elements have *mirror symmetry*. This follows from (109, 110), where it was shown that Mendeleev's entire periodic table can be represented as sets of dyads with mirror symmetry. In this form the set of elements is divided into two subsets: radial-even and radial-odd, differing in the sign of the configurational indices, magnetic quantum numbers, total-spin projections, spin magnetic moment projections, etc.

Second, the stable atomic structures (^{1}H, ^{11}B, ^{12}C, ^{16}O, etc.), to whose interaction the transmutation of elements is attributed, can be regarded from the standpoint of symmetry of atomic nuclei (454), and several transformations of them can give rise to the required transmutations.

Third, all the published work of Kervran (919–922) indicates that transmutations can be regarded from the viewpoint of symmetry group theory (271, 621). The characteristics of a group, accepted as axioms, are satisfied in the case of transmutation of elements: (1) there is an identity element (^{1}H, ^{11}B, ^{12}C, ^{16}O, etc.); (2) there are inverses (calcium–magnesium, sodium–potassium, etc.); (3) there is *closure:* $^{39}K + {}^{1}H \rightarrow {}^{40}Ca$; $^{24}Mg + {}^{16}O \rightarrow {}^{40}Ca$, and hence $^{39}K + {}^{1}H \rightarrow {}^{40}Ca \leftarrow {}^{24}Mg + {}^{16}O$ or $^{39}K + {}^{1}H \leftrightarrow {}^{24}Mg + {}^{16}O$ is possible; (4) there is *associativity:* $^{12}C + {}^{12}C + {}^{16}O \rightarrow {}^{40}Ca$.

We showed above, however, that dissymmetric objects are produced by the action of the GMF, and hence we can postulate that the cause of transmutation is the GMF, particularly since both factors are characterized by *weak* interactions. In addition, it follows that in L and D objects the main atomic structures are replaced by mirror-symmetric atoms, e.g., where the D form of the molecule has a D carbon, the L form has an L carbon. By combining as isomorphs they form different superatomic dissymmetric groups.

The preceding logical picture of the role of the GMF in the transmutation of elements has to be supplemented by factual data and experimental evidence. First, there are data indicating that magnetic fields (1000–5000 Oe) cause a change in the amounts of dry matter and ash in plants (249, 250, 319, 1064, 1065), and in the amounts of the same trace elements (copper, zinc, iron, manganese, etc.) in the animal organism (168, 373, etc.), and there are many such works. In addition, the chemical composition of underground stratal waters shows cyclic variation (V. I. Bratash, personal communication, 1960), which is similar in its dynamics to the periodic variations of the GMF. A decrease in the amount of one element is accompanied by a simultaneous increase in another: $NaCl \rightarrow MgSO_4 \rightarrow CaCO_3$, etc. It is possible that the pronounced depression and death of animals and plants due to thorough shielding from the GMF can be attributed to the absence in these conditions of transmutation of elements, a vitally important process for the metabolic and functional homeostasis of the living organism, which is usually controlled by the GMF. This may be one of the fundamental effects of the GMF on the metabolism of biological objects.

We have the following indirect evidence for an effect of the GMF on the dissymmetric properties of biological and inorganic molecules. The GMF affects physicochemical solutions and water (128, 420, 999, 1049), the orientation and possibly formation of L and D crystals of different substances (465, 466), and the transformation of DL tryptophan to the L form in solution (216, 353–355).

We can pick out one very interesting study that corroborates the proposed mechanism of a deep GMF effect. As we showed, above, the GMF plays a decisive role in genetic homeostasis: it gives rise to chromosome inversions, i.e., intrachromosomal structural changes involving the *rotation* of a chromosomal or chromatid segment through 180°, and to changes in the frequency of genes coding the synthesis of enzymic proteins. It has also been shown that the genetic code is based on pentameric symmetry, and the coding elements—the RNA triplets—have a specific arrangement in the form of an icosahedron (638, 639). The author reports that changes in the position of the codons, particularly their *rotations through* 180°, correspond to particular changes in the properties of the amino-acid residues. Hence, on the basis of the above account we can conclude that one of the possible ways in which the GMF affects genetic homeostasis and the code, in particular, is the transformation associated with *rotational symmetry* through 180° and with *mirror reflection*. Since such transformations in chiral molecules require alteration of the position of only one asymmetric atom, the GMF probably causes these effects, thus producing the *symmetric* properties of elementary particles.

It is possible that a change in spin and magnetic spin moments due to the GMF is the basis of these transformation operations. The high sensitivity of these parameters of elementary particles may be due to the fact that the GMF (and gravitation) is an essential characteristic of the "space–time" of elementary particles, in which they exist, and hence any changes in this "environment" are of decisive importance for elementary particles of any degree of complexity.

Thus we see that the apparently external, highly unusual, universal effect of the GMF on all *indices and properties* of living organisms on earth, actually conceals in itself the effect of the GMF on the fundamental parameters of the microworld, thereby establishing a very important connection—between the microworld and macroworld.

Is the GMF alone involved in the formation of this connection between the microworld and macroworld? This is certainly not the case, since gravitation also plays a very important role, and this connection can be attributed only to the interaction (possibly antagonistic) of these two important physical factors. From indirect arguments and observations we can infer that gravitation strives to maintain the equilibrium state, while the GMF strives to cause a change of state by rotation and dissymmetry. Sudden changes in mutations, the appearance of L and D forms of crystals, i.e., any asymmetric shifts, are probably caused by destruction of the equilibrium interaction of these two factors due to additional impulses from the magnetic and gravitational influences emanating from the cosmos.

Thus, we have considered two hypotheses that provide some answer to the important questions confronting geomagnetobiologists—the sensitivity of living organisms to the GMF and the universality of its action. Direct and precise experiments to confirm or refute the advanced hypotheses are essential, and until we have irrefutable evidence obtained in strictly controlled and reproducible laboratory conditions, these hypotheses will not survive the adverse criticism of physicists. In this connection, however, we can quote the very apt words of the world-famous American scientist—the cyberneticist Norbert Wiener:

> If a theorem merely looks grotesque or unusual and if your maximum effort cannot discover any contradiction, do not cast it aside. If the only thing that seems to be wrong about a proof is its unconventionality, then dare to accept it, unconventionality and all. Have the courage of your beliefs—because if you don't you will find that the best things you might have thought about will be picked up from under your own nose by more venturesome spirits; but, above all, because this is the only manly thing to do (Norbert Wiener: *I Am a Mathematician,* 1956, p. 359).

There is no doubt that progress in our knowledge of the biological action of the GMF and other important geophysical factors will depend on the progress made in many branches of physics, such as quantum mechanics and radioelectronics, cosmology, and geophysics, and also on the resolve of progressive physicists to take up this complex and major problem, as O. Costa de Beauregard, a world authority in the field of elementary-particle physics, has done in the case of transmutation of elements in biological objects (781), a cardinal problem posed in biology by C. L. Kervran.

CHAPTER 4

Specific Aspects of Geomagnetobiology

Man

Effects of the GMF on the Healthy Human Organism

Living organisms are closely linked with their environment. Hence the functional–dynamic properties of any organism depend on its adaptation to the conditions of existence, which is the basis of the homeostasis of living things. By this term we mean the dynamic unity and equilibrium formed between the living organism and the environment with its numerous factors. Until recently temperature, air humidity, and light intensity—in short, the set of factors included in the concept of meteorological environmental conditions—were regarded as the main external factors implicated in the homeostasis of living organisms. The effects of this set of factors on the biology of the living organism, particularly man, have been the subject of many scientific publications in the form not only of papers, but also of comprehensive monographs (100, 529, 672, 1110, 1178).

The question of the diverse biological effects of natural magnetic and electric fields, however, has still not been adequately investigated.

A probable reason for the lack of interest in other external factors just as important as meteorological factors is the fact that living organisms have no visible receptors, like the other organs of sense, for the detection of changes in the parameters of natural electromagnetic fields (1214). This has obviously led to a negative attitude toward the problem of the effect of natural magnetic and electric fields on man, animals, and plants.

General State of Organism. Facts accumulated over a long period compel us to consider the question of a direct effect of the GMF on the human organism. This was suggested, in particular, as far back as the 1930s. On the basis of an *a priori* assertion of the orientation of dipole molecules in cells Baumgol'ts (52) concluded that the GMF has an effect on the state of tissues and the induction of electromotive forces in them. He subsequently put forward the hypothesis that it was the simultaneous change in elasticity and electrical properties of tissues and organs that were the main result of the action of the GMF on living organisms, but he did not present any experimental data.

Another author (669), who measured the arterial pressure and determined the leukocyte count in 43 patients over the course of a whole year, reported that diurnal variations in the diastolic pressure and leukocyte content of the blood coincided with the diurnal variations of the GMF. A similar conclusion was drawn from a comparison of the heart rate and the GMF. The author (210–212) established a correlation between the heart rate and GMF indices from the results of more than 24,000 pulse measurements in presumably healthy subjects 20–40 years old. He reported that the effect of the GMF on heart rate was due largely to the change in direction of the field vector in a vertical plane—the dip—and not to the change in total field intensity. Data have also been obtained that indicate that the pulse is accelerated and the arterial pressure is increased in elderly people during magnetic storms (269). The cited observations indicate that the GMF plays an important role in the normal functioning of the cardiovascular system. There is also evidence of the effect of the GMF on other systems of the organism.

Higher Nervous Activity and State of Vegetative Nervous System. The GMF definitely has an effect on higher nervous activity (77, 203–206, 237, 239, 240, 243, 268). We have already discussed the response of the central nervous system to the removal of the GMF (by compensation or shielding). Compensation of the GMF rapidly alters the critical flicker frequency—an important characteristic of the operation of the visual analyzer and the higher divisions of the brain (703, 760). Chigirinskii (89) reports that the change in the level of dark adaptation of the retina, which is an important functional test, also depends on the daily rhythm of GMF indices. A comparative analysis of the curves of variation of the level of dark adaptation of the eye by Lazarev *et al.* showed a good agreement with the extreme values and maxima of diurnal activity of the geomagnetic field.

Special mention should be made of the monograph (724). Using a specially devised technique, the author distinguished categories of people with different types of higher nervous activity. These types were designated as cogitative and rational (A), artistic and sensitive (B), and passive and inhibited (C). The dates of birth (and, the corresponding presumed times of conception) of people belonging to these categories, and the geomagnetic conditions on these dates, were then analyzed.

A significant correlation was found between a particular level of geomagnetic disturbance and the time of birth (and, accordingly, conception) of each category of person. The author subjected the inheritance of paternal and maternal characters to a similar analysis and concluded that the GMF affected the acquisition of the attributes of a particular parent by the children.

Thus the level of magnetic disturbance or, more precisely, the state of the geomagnetic field at the moment of conception has its own effect on the nature of the nervous constitution, activity, and entire subsequent development of the human being. In a character analysis of children of large families from this standpoint Birzele noted the following. First, everyday experience shows that within a family children often have dates of birth close to those of their parents, grandparents, or great grandparents. Second, children whose dates of birth are close to those of their parents are very similar to them in character.

From an investigation of the level of magnetic disturbance within one or two days of the dates of birth, Birzele found that the similarity between the character of the child and one of its parents was correlated with the similarity of the level of magnetic activity during the birth of the child and the particular parent.

As an illustration we cite one of the cases described in Birzele's monograph. In the K. family, consisting of parents and seven children of different character, the children were questioned about their similarity in character to one of their parents. The results of two independent questionnaires are given in Table 4. As the table shows, all the children in the family could be divided, on the basis of the completed questionnaires, into three distinct groups according to their similarity in character to their parents: (1) children who definitely resembled their father in character (c,i); (2) children who definitely resembled their mother (e,h); (3) children whose resemblance in character to one parent was dubious (d,f,g).

Figure 59 shows a diagram of the level of magnetic disturbance on the 273d day before the date of birth, i.e., at the assumed time of conception of each member of the family. The diagram shows a remarkable similarity between the geomagnetic disturbance curves calculated for parents and children of similar character, whereas the curves for the dubious cases have an indeterminate, even course. Birzele's monograph (724), based on experimental study and retrospec-

Table 4 Difference in Character of Children in K. Family

Members of family	Designations	Results of questioning	
		First questionnaire	Second questionnaire
Father	A		
Mother	B		
Daughter	c	Like father	Like father
Son	d	Like father	Like mother
Daughter	e	Like mother	Like mother
Son	f	Like mother	Like father
Daughter	g	Like father	Like mother
Daughter	h	Like mother	Like mother
Son	i	Like father	Like father

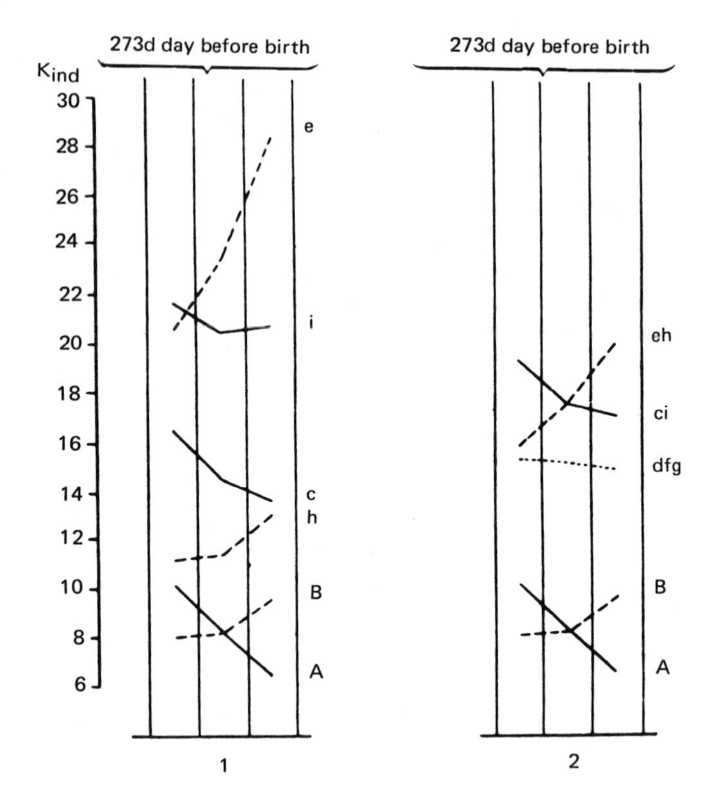

Fig. 59. Example of effect of geomagnetic disturbance on character of members of K. family. (1) State of magnetic disturbance on the 273d day before the birth of each member of the family. (2) Generalized diagram. The quantity plotted on the *y* axis is twice the mean value of the *K* index; the day before and after the date of the 273d day are plotted on the *x* axis. From (724). Explanation in text.

tive statistical analysis, contributes a great deal to our understanding of the role of the GMF in human life and makes an approach to the principles of biological action of the GMF.

Unfortunately, these very interesting data were obtained on a limited amount of material and cannot provide a basis for any generalizations or conclusions, particularly in view of the difficulty of interpreting them in the light of contemporary genetics.

From biomedical statistical research and special experiments (341, 625, 628, 642, 1041) we can conclude that the GMF has a direct effect on the central nervous system of man and animals. This confirms a previous suggestion that natural electromagnetic fields can affect higher nervous activity (90, 767). This would account for the results presented in papers giving accounts of the number of industrial injuries (946, 970), road disasters and accidents (331, 723, 1142), and the response of persons to signals (839, 1079, 1080), since the effects considered are directly related to the state of higher human nervous activity.

It should be noted that the vegetative nervous system of healthy people is very sensitive to variations of the GMF (51, 470, 472, 865). For instance, during geomagnetic disturbances (small and moderate storms) there is an increase in the tonus of mainly the sympathetic division of the vegetative nervous system and, in only 30% cases (usually in males), of the parasympathetic division of the vegetative nervous system. The authors of these papers think that the changes in the state of the vegetative nervous system are due to the effect of the GMF on the blood.

A study of the adaptation of healthy persons in the North European part of the USSR showed that out of all the investigated physical factors of the external environment in this region—atmospheric pressure, temperature and relative humidity of the ambient air, wind speed, and level of disturbance of the GMF—the last is of greatest importance for alteration of the physiological responses of the healthy human organism (41–43, 912). The results of a correlation analysis (Table 5) indicated the existence of moderate, close, and very close nonlinear relationships between changes in indices of vegetative–humoral regulation and the K index of magnetic disturbance in the above region [the results in all the investigated cases were highly significant ($p = 0.001$)]. The periods of high magnetic activity preceded the increase in excretion of neutral 17-corticosteroids and adrenalin by one or two days. The threshold aggregate diurnal value was $K = 15$, and when this level was exceeded the responses of the organism were enhanced (42).

Thus, changes in the GMF are associated with the same changes in man as those produced in animals by weak artificial magnetic fields: enhancement of inhibition in the central nervous system, retardation of conditioned and unconditioned reflexes, failure of memory, altered regulation of normal and pathological processes, etc. (240).

It has been concluded from numerous experimental studies that the nervous system is the first system to respond to electromagnetic fields, since it is the first to encounter any external stimuli, and it responds to the magnetic field in a direct and reflex manner (243). Kholodov stresses that the effect of electromagnetic fields on the central nervous system is nonspecific and is largely of a subsensory nature, and that their reception resembles the activity of interoreceptors. Of interest in this respect is Persinger's paper (1041) on a basic problem of parapsychology—information transfer from person to person without the use of known sensory models. It is suggested that the energy required for such extrasensory transmission does not come from the organism itself, but from its environment. Persinger concentrated his attention on three physical factors—extremely low frequency sound (infrasound) fields, high-voltage static electric fields, and

Table 5. Effect of Geomagnetic Activity (*K* Index) on Vital Activity of Healthy Persons in the European North [Modified from (42)]

Physiological index	Values of correlations	
	Confidence limits	Degree of correlation
1	2	3
Excretion of neutral 17-corticosteroids	0.912–0.922	Very high
Vitamin B_1 in urine	0.636–0.750	High
Weighted-mean skin temperature	0.600–0.663	High
Blood flow rate	0.598–0.670	High
Adrenalin excretion	0.256–0.401	Medium
Minute blood volume	0.353–0.456	Medium
Pulse pressure	0.258–0.356	Medium
Hemoglobin concentration	0.307–0.517	Medium
Oxygen capacity of blood	0.318–0.522	Medium
Erythrocyte sedimentation rate	0.327–0.463	Medium
Cholinesterase activity	0.227–0.411	Low
Pulse	0.242–0.346	Low
Blood pressure		
maximum	0.219–0.330	Low
minimum	0.168–0.466	Low
Thrombocytes	0.113–0.452	Low
Blood clotting	0.040–0.274	Very low

GMF disturbances. An analysis of daily data showed that there was a correlation ($r = -0.58$) between the monthly disturbance of the GMF (mean value for a period of 58 years) and monthly data for telepathic and clairvoyant effects.

Skin Electric Potentials. Processes that are closely associated with the electric characteristics of the organism probably react most distinctly to variation of the GMF. This is confirmed, in particular, by data relating to the electrical activity of the skin. For instance, during geomagnetic storms the electric potentials of human skin are altered and their distribution on the skin becomes asymmetric or more asymmetric (270, 432).

These responses of the human being in time are of a different nature during a magnetic storm (436, 437). Three groups of people can be distinguished: electromagnetically stable, electromagnetically mobile, and intermediate. It is obvious that these groups correspond precisely to the types of people that we distinguished on the basis of functional dissymmetry, which confirms the fundamental nature of our classification (136).

In electromagnetically stable people an increase in the static and bioelectric potentials and asymmetry in their distribution are observed one or two days before and one day after a geomagnetic disturbance, whereas in electromagnetically mobile persons the corresponding times are four days before and the day of the magnetic storm or disturbance. Thus, a change in the static potentials in different persons precedes the perception of the magnetic effect and also comes after it. There is no doubt that changes in static potentials and bioelectric activity are closely linked with the electrodynamic field discovered and investigated earlier (759, 1076).

Changes in skin electric potentials can have important consequences since, by affecting particular regions of the skin, which, as is known, are receptor zones acting as "external representatives" of many internal organs, the GMF can affect the functioning of these organs (167, 699). No objection can be raised to this from the physical viewpoint, since such interaction occurs between a direct electric current and a magnetic field—the so-called galvanomagnetic effects, which are particularly pronounced in semiconductors (Hall effect) (298, 592). Hence, it is quite possible that the GMF acts on the direct electric currents flowing in neurons and thus affects the central and peripheral systems of regulation of the living organism (596, 697, 698, 759, 1076).

Blood. We shall now discuss the changes observed in such an important internal medium of the organism as the blood, which is of immense importance for the functioning of all systems of the organism.

Heliobiologists have shown that solar activity is responsible for a number of changes in the blood of healthy people (330, 333, 334, 422, 515, 516, 1153). Researchers have discovered that the GMF probably plays a significant role in this case (24–26, 41–43, 72, 267, 275, 347, 370–372, 475, 476, 480, 580, 628).

We have already mentioned that direct comparison showed a synchronous variation of the human blood leukocyte count and the GMF (25, 26, 669). It was later found that the functional state of the fibrinolytic system of the blood of healthy persons is altered during geomagnetic disturbances (72, 470, 476). Fibrinolytic activity is reduced, which increases the probability of thrombosis (333).

The conducted investigations showed that the blood-clotting system of presumably healthy people responds very precisely to a change in geomagnetic activity. The close correlation found between the indices of the blood-clotting system and fibrinolysis indicates that stimulation and inhibition of blood clotting and its phasic nature depend on the GMF (472). Rozhdestvenskaya arrived at the following conclusions from an analysis of the functional state of the blood-clotting system and fibrinolysis in 260 presumably healthy humans. During a period of weak magnetic storms with a *gradual* commencement the blood-clotting time shows a tendency to decrease, whereas in a period of weak magnetic storms with a *sudden* commencement the blood-clotting time on the first day is increased. During the first day of moderate and severe storms with a *gradual* commencement there is a statistically significant reduction in the blood-clotting time and stimulation of fibrinolysis. During the first two days of moderate and severe magnetic storms with a *sudden* commencement, there are no significant differences in the mean clotting times. Rozhdestvenskaya attributes the last result to the actuation of adaptation mechanisms that protect the organism from excessive stimulation. The effect of geomagnetic disturbance on the blood-clotting indices of healthy humans was confirmed by the result of a correlation analysis. The correlation between the changes in blood-clotting indices and geomagnetic disturbance in the period of weak magnetic storms with a sudden commencement was $\eta = 0.795, p < 0.01$, and of moderate and severe storms with a sudden commencement was $\eta = 0.785, p < 0.01$. The thrombocyte count was also closely correlated with the geomagnetic disturbance in these periods: $\eta = 0.723, 0.680, p < 0.01$). The fibrinolysis indices were closely correlated with indices of geomagnetic disturbance during magnetic storms with a gradual and sudden commencement: $\eta = 0.808, 0.666, 0.832, p < 0.01$, respectively, for weak storms with a gradual and sudden commencement, and moderate and severe storms with a gradual commencement.

A prolonged investigation of 92 healthy people aged 17–30 years showed irregular variation of the ESR of the same person over a period of 24 h and on different days (287). There was a distinct correlation between the direction of the change in the ESR in a steady magnetic field (0.8 Oe) and the change in the vertical component of the GMF during the same period of investigation. From daily investigations of the erythrocyte count and the hemoglobin content in healthy people (over a period of four months), Koval'chuk (275) found that when the geomagnetic activity was low or moderate the blood indices varied in time with the global variation of geomagnetic disturbance. He also reported that a very sharp change in geomagnetic activity (an increase in the A_p index by more than 100 in one or two days) was accompanied by a drop in the erythrocyte count and hemoglobin content.

It has been reported that variation of GMF activity is of definite significance in the seasonal dynamics of leukocytes, blood-clotting time, and thrombotest results (25, 26). It was found that the leukocyte count and the nature of the changes in the coagulogram indices, such as tolerance of the plasma to heparin, thrombotest result, and fibrinolytic activity showed a tendency toward 27-day recurrence during the year. The indices of the blood-clotting system—clotting time, plasma recalcification time, prothrombin time, fibrinogen concentration, factor KhSh, and also ESR and hemoglobin content—showed no tendency toward 27-day recurrence.

In healthy young people (19–22 years old) during magnetic storms the leukocyte and thrombocyte counts were lower, clotting was slower, the change in ESR was accelerated, and the fibrinolytic activity was higher (24).

In Koval'chuk's opinion the course of GMF variation is reflected best in the count of peripheral blood erythrocytes, but the degree of correlation is different at different times (276). By using a combination of mathematical methods of revealing latent periodicity, Koval'chuk showed that in the everyday course of the most important functional parameters of man, such as the maximum and minimum blood pressure, the number of blood elements formed, heat rate, muscle strength, changes in the local weather situation, and fluctuations in the level of disturbance of the general planetary magnetic field, there is a set of very close frequencies—about 7, 9, 12–13, and 30 days—which are characteristic of the dynamics of solar activity (278). Koval'chuk believes that the endogenous biorhythms of the human organism can anticipate impending changes in the external environment. His study of the causes of the fluctuations in the daily values of the blood hemoglobin content of people living at different latitudes in the USSR, including more than 800,000 primary clinical analyses in 1959–1972 (278, 279), showed a highly significant inverse correspondence ($p = 0.001$) between the daily hemoglobin content of most subjects and the degree of disturbance of the GMF (A_p index) on the preceding day or two. He reported that the value of the reliable positive cross-correlation coefficients indicating the degree of similarity to variations of the sunspot number was significantly higher ($p = 0.001$) than for the magnetic index A_p.

Koval'chuk also found that the mean monthly and, in particular, mean annual values of 18 functional indices of the human organism—blood hemoglobin, various blood-clotting indices, blood phospholipid and cholesterol levels, heart rate, general excitability of somatic nervous system, etc.,—investigated in a very large number of subjects at different places on the earth and reported by different authors, showed a significant correlation with the heliogeophysical situation (279). In a series of special experimental investigations by several methods he also found that a constant magnetic field (20–2000 Oe) had a stabilizing effect on the spatial structure of DNA in solution. Using a special technique he succeeded in showing that the spatial structure of the DNA of living cells and in solutions, maintained in sterile conditions, responds exceptionally precisely, like the integrating self-regulatory systems of man, to impending changes in the GMF and corresponds to them (281).

Koval'chuk infers that the primary general biological coupling mechanism responsible for the unity of living systems and the environment could be the universal direct sensitivity of the cell genetic apparatus to such effects and arrives at the general conclusion, like Brown (748–750), that cosmic factors are mainly responsible for synchronization of the long-period biorhythms of living systems. Koval'chuk particularly stresses the fact that the cross-correlation coefficients in some cases indicate reliable preperception of changes in the GMF three or four days beforehand. This preperception is a common feature and has been described for many biological processes (92, 121, 436, 437).

Growth and Sexual Development. The GMF, which has such a pronounced effect on the state of healthy adult people, cannot fail to affect the development of children. Although there has been no direct evidence of this effect so far, some facts discovered in recent years necessitate a reexamination of this problem (286). One of these facts is "acceleration"—the increase in growth rate and speeding up of the sexual development of the human organism. The average growth of children of all age groups, from newborn to adolescents, is increased, and girls mature earlier (223, 880).

Acceleration is of global nature and its features, indicated above, are observed in normal healthy children in different climatic regions of the planet, in urban and rural conditions, in all national and social groups of the population, and in children growing up and brought up in diverse conditions. The universality of acceleration suggests that the reasons for it are planetary. It has been suggested that the reason could be the increase in the mean level of electromagnetic fields in the biosphere throughout the considered acceleration period (447, 1072).

There are grounds, however, for believing that acceleration is directly due to a change in the GMF. As the above account indicates, acceleration shows the same features of global synchronism and cophasing as various other processes in the biosphere. The acceleration process itself, despite a general trend toward increase in growth rate and other changes, is of an irregular nature (1033), but, as Fig. 60 shows, the rises and falls are synchronous in all age groups.

The curves of acceleration and variation of the GMF during the same period are similar (131, 133, 136), which is indirect evidence of a possible connection between these phenomena. The dip and declination of the GMF play a particular role in acceleration.

The results of our investigations show that acceleration and retardation processes depend on the different effect of individual GMF elements (*H, Z, D, I, X, Y*) on the development of the regulating systems of the organism (Fig. 61). Since these elements are subject to very diverse changes at different geographical points, the global picture of acceleration is exceedingly diverse, although it shows common features, e.g., those associated with a 5% reduction of the earth's magnetic moment (428). It is difficult at present to give a full explanation of our discovered biological role of individual GMF elements, since more thorough investigations of acceleration processes on a worldwide scale in accordance with a common program (139) will be required for the discovery of general trends and detection of specific relations between GMF elements and particular developmental systems of the human and animal organism. We think, however, that the many facets of the effect of individual GMF elements on the living organism can be satisfactorily attributed to the above-described effect of the GMF on genetic homeostasis. As mentioned earlier, individual blocks of genes respond to changes in different GMF elements, which indicates the complex architectonics of the internal structure of the gene. Since individual systems of the organism are

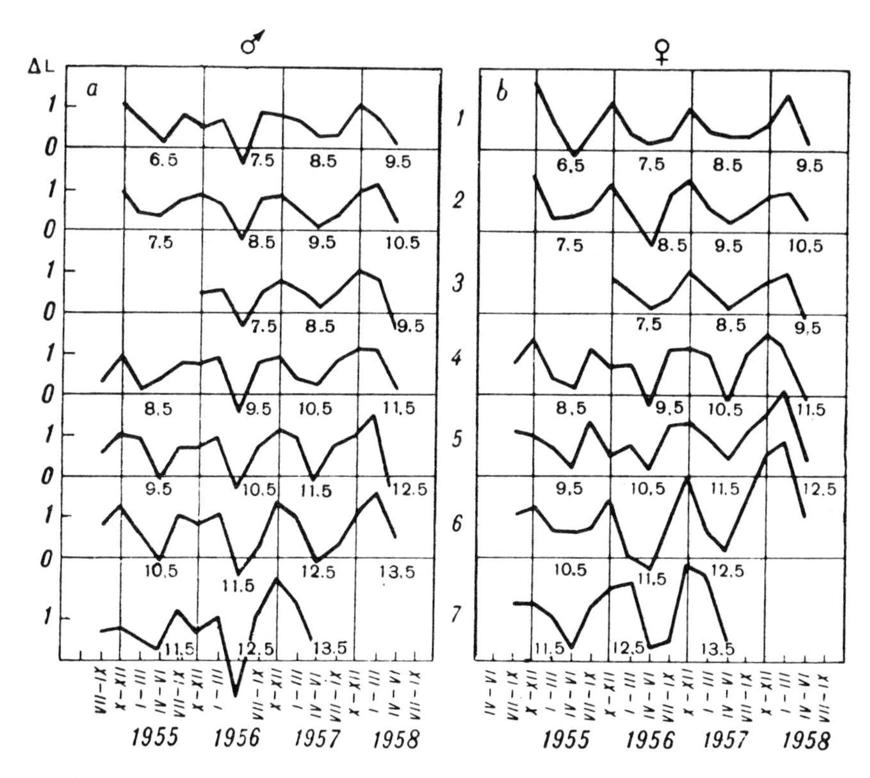

Fig. 60. Change in body height increment in different age groups of boys (a) and girls (b) in rural conditions in Poland (1033). (1–7) Age groups from 6.5 to 11.5 years.

controlled by different blocks of genes, it is quite understandable that in acceleration conditions this will be manifested in a different relationship between these systems and different elements, viz., those which control the activity of a particular system of genes.

A different opinion is held by Vasilik, the author of many papers on the effect of the GMF on acceleration (600–611). While sharing our view on the role of the GMF in acceleration processes, he believes, nevertheless, that the main factor is the variation of intensity of the GMF. He bases this conclusion on a comparison of the available data on changes in types of individual development (times of sexual maturation, changes in body size) in different countries with the variation of GMF intensity. In his opinion, this relation may have a wavelike nature. At low GMF intensity maturation is speeded up and there is a slight reduction of body size, whereas at high GMF intensity maturation is retarded and the growth period extended, which leads to an increase in body length of the adult organism, and so on. Thus according to this idea an increase or reduction of

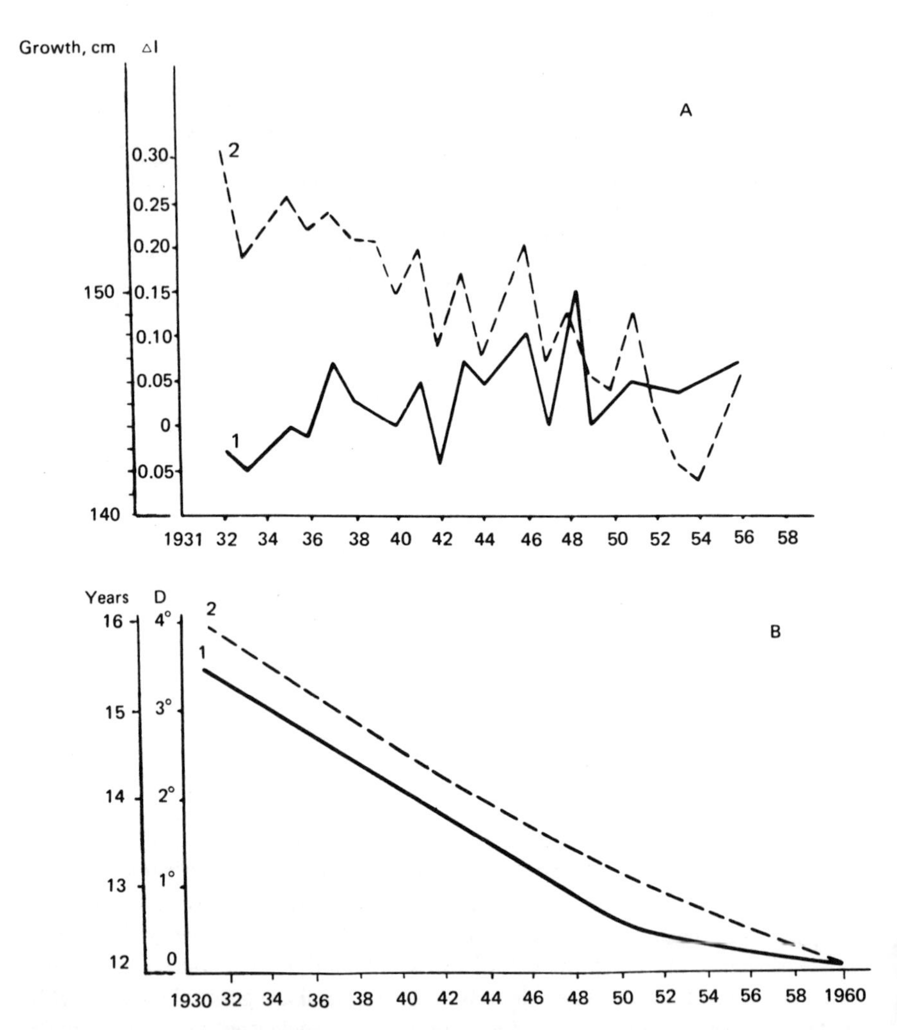

Fig. 61. Acceleration processes in Sweden and Norway in relation to different GMF elements. (A) Acceleration of growth of children in Sweden and role of GMF. Change in growth of 12- to 13-year old boys (1) and change in GMF gradient (ΔI) during same period (2). Tromso Observatory, Sweden. (B) Acceleration of development of children in Norway and the role of GMF. Change in times of sexual maturation of girls (1) and change in declination (D) of GMF in same period (2). Tromso Observatory. Curve (1) shows age of subjects at onset of first menstruation. From (880), taken from (131).

the GMF intensity can lead to synchronization or desynchronization of the development of individual systems of the organism.

Vasilik corroborates his view with the results of a specific analysis of acceleration processes (603).

The lowest value of GMF intensity is found in South America. The investigations established that in Chile, for instance, there was some reduction of the body length of males between 1920 and 1960. American children living in Rio de Janeiro differed from their peers in the U.S.A. in their smaller body length. In Africa the GMF intensity is a little higher than in South America. This affects the growth and development of the local population—children from Dakar at the breast-feeding stage differed in weight from European children, while the body length of adult males—the Tutsi from Ruanda—exceeded the body length of europoid males, and so on.

Since the physical development of man on earth has become a major global problem, and acceleration is attracting the intense interest of scientists of most diverse specialties, we must discuss in more detail the question of the role of the GMF in acceleration processes.

As our above analysis shows, variation of the total intensity of the GMF is an important factor in acceleration and retardation processes. This GMF parameter, however, operates over long periods of time and has a decisive effect on the long-period components of acceleration processes. This is clearly revealed by the results of archeomagnetobiological research.

Acceleration at present, however, depends on the biological action of the individual GMF elements controlling the complex links of hormonal development of man and animals. Since the seasonal, annual, and secular variation of these elements at different geographical points is different, there is great diversity in the acceleration and retardation processes in different countries, although there is a general trend toward acceleration of growth and development. This is why even adjacent countries show different trends in the variation of times of sexual maturation, body size, weight, etc., since these parameters depend on the local specific variation of the GMF elements. The examples we cited [acceleration processes in two adjacent countries (Sweden and Norway)] clearly illustrate the relation between individual GMF elements and different acceleration indices (Fig. 61).

The hypothesis of a wavelike effect of the GMF on biological growth and development (605, 606) is very simple and attractive, but there are serious objections to it and it is inconsistent with facts cited by Vasilik himself. For instance, the hypothesis fails to indicate what criterion should be adopted as a guide in selection of the GMF intensity responsible for acceleration or retardation, if the aim is to predict these effects. It is not clear, according to the hypothesis, why equal GMF intensities, in Europe and Australia, for instance, should be associated with opposite variations in human physical development, as the author himself reports, and so on.

It should be noted that, irrespective of the views of scientists on the basic mechanisms of acceleration, the role of the GMF is very distinct, and even social aspects of the effect of the GMF become apparent. Being a global factor, the GMF leads to extensive changes in the population structure by accelerating or retarding the sexual development of the young and thus affecting the number of marriages contracted in different countries, the state of psychic activity, and the state of health of the population, since it is closely related to the hormonal state of man. This diverse effect of the GMF on people is mediated, of course, by a complex chain of socioeconomic and climatic conditions, which in several cases play a decisive role in human social life. It is certainly now time to investigate, within the framework of UNESCO, the role of geophysical factors, primarily the GMF, in the global changes occurring in the physical development of man in different countries, particularly since there are long series of observations, extending over decades, of various functional indices of man (1177), physicochemical indices (420, 999, 1049), the state of plants and animals (751, 752), and various geophysical factors (1191).

It has also been suggested (599) that the western drift of the GMF is reflected in a change in body size fluctuations and migration of the population.

Although the hypothesis of a decisive role of the GMF in acceleration is a very attractive one, there are as yet no direct experimental investigations confirming this hypothesis.

All the data discussed above and the analytical comparisons indicate that the GMF is a very important factor in the evolutionary development of living organisms on earth and that the modes of action of the GMF on the human organism are extremely diverse. These are, first of all, alteration of the most diverse functional features and processes in the human organism. The GMF would appear to have a special effect, however, on the most important centers of nervous and humoral regulation in the living organism—the hypothalamus and the cerebral cortex. This could account for the diverse relations observed between the GMF and the various manifestations of functional activity in man. We shall first discuss the interrelationship between some functional states of the female organism and the GMF.

Effects of the GMF on the Female Organism. As a result of long-term medical observations and comparisons of data indicating the periodicity and synchronism of effects and processes occurring in the female organism, and also of investigations of large groups of people, midwives and gynecologists have directed their attention to a possible role of the GMF in these phenomena. Such stressful and important periods in the physiology of the female organism as menstruation, labor (196, 966, 1152, etc.), and pregnancy disorders (675, 1170) have been subjected to careful analysis. Statistical treatment of a large amount of data has established an effect of the GMF on the menstrual cycle. The number of commencements of menstruation in women on particular days depends on the geomagnetic activity. A calculation from data on the course of 7420 menstruations, obtained by questioning 810 girls 14–18 years old (pupils in a Prague nursing school), showed that during reduced geomagnetic activity the frequency of onset of menstruation was higher and, conversely, during increased activity onsets of menstruation were less frequent (966). The variation of the duration of the menstrual cycle is correlated with the variation of the strength of the GMF (724).

From a retrospective statistical analysis, Birzele (Graz University, Austria) established a correspondence between the duration of the menstrual period and geomagnetic disturbance. On the basis of Knaus and Ogino's maximum and minimum types of menstrual periods, he showed a close correlation between geomagnetic disturbance and this extremely important physiological index of women (Fig. 62). We tested these data on independent material and obtained similar results.

Birzele's work is confirmed by the work of E. Jonas, the well-known Czech specialist in astrobiology (908). From several thousand statistical investigations, he established that cosmic factors have an exceptionally pronounced effect on the physiology of the female organism. According to Jonas's data, this effect includes such important aspects as fertilizability, sexual differentiation (formation of male or female fetus), and viability of future child. The author of these highly interesting investigations, which are very important from the social viewpoint, believes, like several other scientists (831, 832, 844), that the key factor in all the above effects is the position of the planets and stars. One must remember, however, that in addition to their specific effect, their action is mediated by the GMF, as we showed in previous chapters. This is also distinctly shown in Birzele's book (724), where he traced the effect of geomagnetic disturbance on the specificity of formation of types of persons according to astrobiological calculations. This applies equally to periods of enhanced and reduced fertilizability in woman. The synchronism of this process in different provinces in Japan and different countries of Europe was shown by Moriyama (999) and, as was mentioned above, Birzele established its correspondence with geomagnetic disturbance (Fig. 63). Thus, we come to the important conclusion that long-standing astrobiological observations can be largely accounted for by the effect of the GMF on the human organism. We think it important that the effect of cosmic factors on specific features of manifestation of the GMF should be investigated. Approaches to this problem have already been made by scientists of various specialties (117, 118, 467, 497, 509, 618, 619, 1143, 1191). A thorough investigation of the role of cosmic factors in the life of living organisms on earth will help to reveal features inherent to living organisms as specific biomagnetic objects (756, 836, 1015, 1193) and to explain the most important role of weak

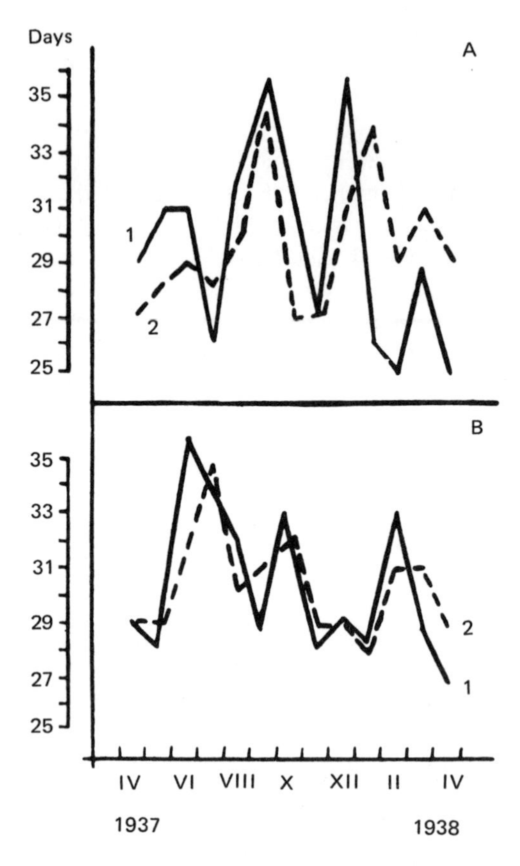

Fig. 62. Relation between duration of menstrual period and cyclic variation of geomagnetic activity. The graph shows the correspondence between the geomagnetic *K* index (1) and the change in duration of the menstrual cycle (2) for maximum (A) and minimum (B) type (according to Knaus and Ogino). From (724).

electromagnetic fields in their vital activity (445, 450, 835, 967, 1121). Retrospective statistical analyses show an important role of GMF activity in the perinatal period of human life (1043). The relation between the IPAT anxiety scores of students and the state of GMF activity (*K* index) in the seven days before and seven days after their dates of birth was investigated. In 82 female students a correlation was found between the above indices with $r = +0.29$–$0.31, p < 0.001$, for the day of birth ± 1 day, but no such relation was found in male students (83 subjects). When the males (38 subjects) and females (45 subjects) who showed great anxiety, i.e., whose anxiety scores exceeded the mean level, were considered separately, the degree of correlation increased to r

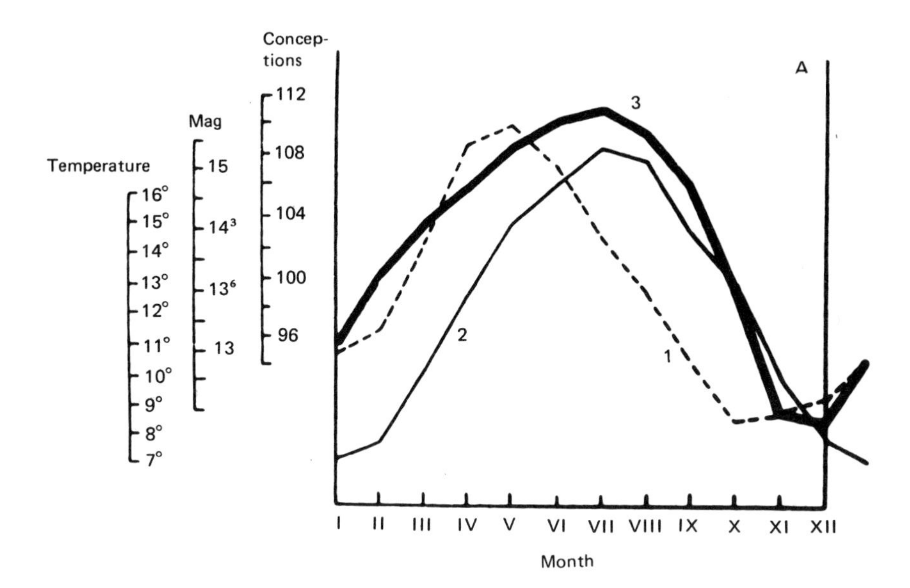

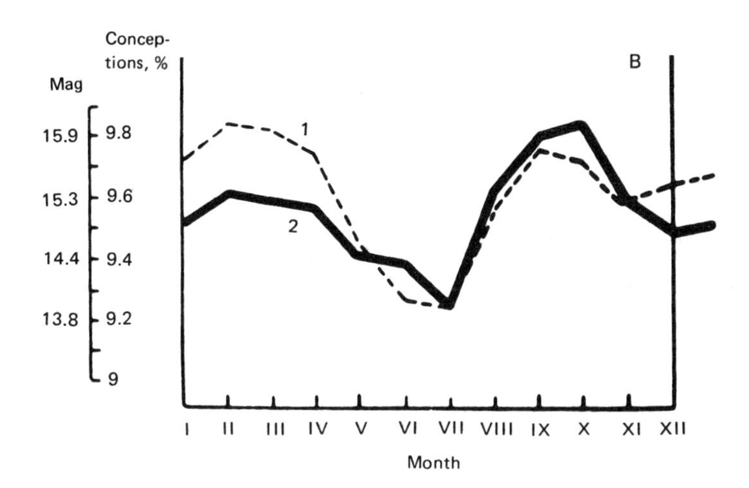

Fig. 63. Relation between number of conceptions and cyclic variation of geomagnetic activity. (A) Annual variation of extramarital conceptions in Germany (1872–1883) (1), annual variation of temperature (2), and variation of K index of geomagnetic disturbance (mean for 40 years) (3). (B) Annual variation of extramarital conceptions (1) in 1948 and variation of geomagnetic activity (2) in same year. From (724).

$= +0.40–0.51$, $p < 0.001$, for the time from one to three days before birth. These investigations again show an important role of the GMF in the vital activity of man at different stages of life.

Some authors (196) believe also that not only the onset of menstruation in women, but also the course of labor, depends on geomagnetic activity. This conclusion was based on an analysis of 5424 menstruations in 1046 healthy women and an investigation of the course of 8103 births in Lvov Region. It was found that the daily rhythm at the start and end of the births depended on the daily rhythm of the geomagnetic field. At times of the day when the GMF was stronger labor was more intense. The curve of daily rhythm of births reproduced the curve of daily variation of the GMF with a lag of six hours; the coefficient of correlation $r = +0.827$. This led the author to suggest that the latent period of the effect of the GMF on labor is about six hours. There was also a direct relationship between the frequency of births and the strength of the GMF, but it was most appreciable on the second, ninth, and thirteenth days after the disturbance of the GMF. The relationship between the frequency of menstruations and the strength of the GMF was most pronounced six days after the onset of large variations of the GMF ($r = +0.652$).

Special note should be taken of the observations of the author of (196) on the effect of magnetic storms on labor. Severe storms cause definite disturbances in the rhythm of the birth rate. On the first day of a storm cases of onset of births are more frequent, whereas on the second day the number of births is reduced, but increases again on the third to fourth day, and by the end of the storm decreases to the initial level. Premature births begin more often during a magnetic storm, and toward the end of a storm the number of rapid births is appreciably increased. The dynamics of the onset and the changes in amplitude and frequency properties of the short-period variations of the GMF associated with magnetic storms correspond in many ways with the dynamics of the onset of processes in the female organism.

Of great interest are the investigations of the effect of the orientation of the GMF on the sex determination of the embryo in embryogenesis (1–4, 6–9, 12), which is inexplicable, of course, from the standpoint of contemporary genetics.

Geneticist V. Abros'kin concluded that a particular orientation of the embryo in plants and animals in the GMF in particular periods of embryonic development is important for determination of the sex of the developing organism (1–12).

Orientation of the radicle of plant embryos during seed germination toward the GMF north promotes subsequent female sexuality—the predominance of female florets in monecious species and the predominance of female individuals over males in diecious plants. Orientation of the embryo radicle toward the GMF south promotes predominance of the male sex. The author reached this conclusion on the basis of experiments on the germination of hemp and cucumber seeds

Table 6. Effect of Orientation of Embryos during Embryogenesis in GMF on Sex Ratio (%) after Birth [from (1–7)]

Objects of investigation and number in experiment (not control)	Orientation of head end of embryo and sex of offspring				Control individuals oriented differently[a]	
	Toward north		Toward south			
	Male	Female	Male	Female	Male	Female
Drosophila eggs 2600	55.3	44.7	38.8	61.2	47.3	52.7
Hen eggs 1173	73.5	26.5	28.3	71.3	50.6	49.4
Cows 19,000	52.5	47.5	47.8	52.2	50.0	50.0
Horses 79	71.5	28.5	36.3	63.7	50.0	50.0
Women 824	81.5	18.5	13.7	86.3	50.0	50.0

[a]In the control individuals the head of the embryo was oriented at random in the case of *Drosophila*, hen, and horses, and toward the west and east in the case of cow and women.

(1, 3). These results were subsequently confirmed directly or indirectly by the experiments of other scientists (106, 379, 381, 521, 565) with the same and other plant material (corn, beet, etc.).

In animals, birds, and insects, orientation of the embryo with its head toward the GMF south promotes the appearance of female individuals among the newborn, and orientation toward the GMF north promotes the formation of males (4–12).

The author investigated the effect of predominant orientation of the fore end of the main body axis of the animal embryo on its sex determination in experiments on the incubation of oriented *Drosophila* and hen eggs, and also in studies of oriented embryogenesis in cows, horses, and man. We summarize the results of these investigations in Table 6.

The position of the head end of the *Drosophila* embryo was determined from the presence of processes (filaments) on the fore end of the egg (4). The position of the hen embryo is related to the shape of the egg: If the egg is placed with its blunt end to the left, the embryo is situated on the yolk with its head end almost always forward of the observer (5). These features enabled Abros'kin to orient the eggs in the GMF during incubation.

The head end of the embryo's body in the cow and horse is directed in most cases away from the head end of the body of the pregnant animal, and the position of the human embryo in the mother's body is, in principle, the same (2).

In comparisons of the orientation of mammalian embryos in the GMF with the sex ratio of the newborn, the position of the embryos in the GMF was determined from the predominant orientation of the mothers in the GMF: Cows and horses spend a considerable part of their time in stalls whose position after erection is constant from year to year; hence orientation of the animal's body in relation to the geomagnetic poles is also constant throughout the years.

The orientation of a pregnant woman relative to the earth's magnetic poles also effects the sex of the future child. As Table 6 shows, the relation between the sex ratio of newborn humans and their orientation in the GMF during embryogenesis is much more pronounced than in animals. The author attributes this to the more definite and constant position of the human embryo in the mother's body in comparison with animals. He writes: "The statistical significance of the obtained results confirms that if women in the first two months of pregnancy sleep with their head toward the north [in which case the embryos have their head toward the south] there is a predominance of girls among the children born, and if they sleep with their head toward the south, boys predominate" (6).

Abros'kin also established that orientation of adult organisms in the GMF could affect the expression of some secondary sexual characteristics, particularly the development and function of the mammary glands in mammals (7).

The author attributes the effect of a particular orientation in the GMF on the organism to the different rates of flow of biological fluids (which he showed experimentally) when the organism is oriented in different ways in the GMF (14). This leads to differences in metabolic rate, which affect the viability of the embryos and can lead to selective embryonic mortality of individuals of a particular sex. This, in Abros'kin's opinion, alters the sex ratio of the newborn without contradicting modern genetics and embryology.

We can conclude from the data cited that at early stages of embryonic development the hormonal systems controlling the final differentiation of sexual dimorphism of the fetus are probably under the controlling influences of the GMF.

All the data presented above indicate that the state and functional activity of the normal healthy organism is closely related to the state of the GMF. Hence, we can expect that in the sick organism, where the compensatory functions are seriously affected, there will be an even closer relationship with the GMF and its disturbances.

Effects of the GMF on the Sick Human Organism

As yet there have been no detailed investigations of changes in the interrelationship between the GMF and the human organism when disease is present. The available data show that this association is equally important at the initial and later stages of disease. In the second case, however, sharp changes in the state of the GMF often cause pronounced effects. This is exemplified by such states as preeclampsia and eclampsia—pregnancy toxemia (1170). An even more illustrative example of such a relation is the group of cardiovascular diseases. This kind of disease has attracted the interest of doctors and heliobiologists in connection with the wide occurrence and frequent discovery of synchronous onset of complications in this group of noninfectious diseases (176, 389, 481). Many recently accumulated facts indicate that geomagnetic disturbances affect the course and aggravation of these diseases, especially in their later stages. An association between the GMF and the course of diseases of this group has been discovered by various investigators. A correlation between the increase in number of complications of cardiovascular diseases and solar activity has been found (533–535, 547, 766, 767). There has been a recent strengthening of the view that this correlation is due mainly to the modulating effect of solar activity on the GMF, whose sudden changes are one of the direct causes of cardiovascular accidents (158, 229, 458, 533–535, 1069). This is clearly indicated by the high value of the coefficients of correlation between the frequency of cardiovascular accidents during the year and the mean monthly K index of geomagnetic activity (498).

Table 7 shows how closely the group of cardiovascular diseases is associated with geomagnetic activity—the correlation coefficient in almost all cases exceeds 0.50. Incidently, the data for angina pectoris confirm the close relation between this kind of disease and the sunspot situation, particularly the transit of the main group of spots across the central meridian of the sun. Scientists have shown interest in this for a long time, particularly since the work of French investigators, who could indicate precisely the time of transit of spots across the central meridian of the sun from the state of their patients (1107).

One should note that in the study of heliogeophysical correlations the authors, usually cardiologists, thoroughly investigated the severity of the diseases and their incidence in relation not only to the GMF, but also to other environmen-

Table 7. Correlation between Cardiovascular Diseases and Heliogeophysical Factors (Irkutsk, 1956–1964)

	Correlation coefficients ($r \pm m_r$)		
Disease	Relative sunspot number W	Area of sunspots S	Geomagnetic activity ΣK
Angina pectoris	0.75 ± 0.12	0.69 ± 0.14	0.59 ± 0.11
Myocardial infarcts	0.48 ± 0.22	0.28 ± 0.26	0.68 ± 0.14
Strokes	0.36 ± 0.25	0.20 ± 0.27	0.70 ± 0.15
Hypertensive crises	0.26 ± 0.27	0.30 ± 0.26	0.50 ± 0.22

Table 8. Correlation between Cardiovascular Accidents and Geophysical Indices (Kiev, 1966, 1585 Cases)

Index	Correlation coefficient $r \pm m_r$	Level of significance p
Disturbance of GMF	-0.89 ± 0.09	<0.02
Drops of atmospheric pressure	$+0.54 \pm 0.32$	>0.1
Temperature drops	$+0.61 \pm 0.28$	>0.1

tal factors—pressure, drops in air temperature, precipitation, wind speed, cloud, ionization, radiation conditions, etc. A significant and high correlation, however, was found only when exacerbations of cardiovascular and other diseases were considered in relation to chromospheric flares and geomagnetic activity (273, 481, 830, 1018).* As an example we reproduce Table 8, from an investigation in which the effect of three factors on cardiovascular diseases was compared (265).

The data of Table 8 provide an excellent illustration of the significance of the correlation between the level of geomagnetic disturbance and the dynamics of cardiovascular diseases.

The number of cardiovascular diseases differs distinctly on magnetically quiet and disturbed days. For instance, in Sverdlovsk in 1964 the mean daily index of the incidence of strokes was 3.5 on magnetically active days and 2.8 on quiet days (389).

Similarly, in Leningrad in 1960–1963 the number of requests for first aid for patients with myocardial infarct in a day with high magnetic activity was 6.6, and on a magnetically quiet day was 3.4. There was an increase in the incidence of various complications of cardiovascular diseases, including an increase in the number of cases of sudden death by a factor of 2–2.7 ($p < 0.001$), with increase in the intensity of geomagnetic disturbances (177). This was observed simultaneously in different towns. It should be noted that the data of all the investigators for the correlation of diseases with the GMF agree, but the information on the course and outcome of diseases in relation to the onset of magnetic disturbances differs.

There are data indicating that the incidence of cardiovascular accidents and deaths is highest on days of commencement of magnetic storms and the day or two after (38, 177, 229, 265, 288, 309, 389, 406, 460, 533–535, 546, 657). The highest incidence of disease on the first or second day after a storm could be due

*Correlations with several meteorological factors, mainly particular types of weather, were also significant [B.R.].

to the responsiveness of the organism itself, the latent period in the development of the particular complication, and the fine structure of the electromagnetic field accompanying the geomagnetic disturbance (presence of short-period fluctuations, their amplitudes, the frequency range of their changes in the solar cycle) (33, 360, 361, 627, 1106).

It should be noted that the days of highest incidence of disease coincide with the days when geophysicists engaged in observations of the state of the GMF report the appearance and increase in amplitude of short-period fluctuations of the geomagnetic field (427, 509), which is further evidence of the biological significance of this frequency range.

There is evidence, however, that geomagnetic storms with a sudden and gradual commencement have a biological effect on other days, particularly the third and fourth days after the storm (183).

Recent data on the biological role of short-period fluctuations of the GMF account for the different effectiveness of storms with a sudden and gradual commencement, and the dynamics of diseases in different phases of the storms. In particular, it was reported in (427) that on the third and fourth days after the occurrence of a storm with a sudden commencement there are maxima in the frequency of appearance of short-period fluctuations of the pc1 type (pearls), and on the second through fourth days of maximum variations of the pc3 type (509), the biological role of which is now quite definite. The increase in the number of infarcts on these days after a magnetic storm with a sudden commencement was previously inexplicable, but we now have an explanation, particularly if we take into account the high level of E_z of short-period fluctuations of the GMF (86).

Different Kinds of Disease

In turning now to a consideration of the cyclic nature of epidemic diseases we must point out first of all the complexity of this process and the effect on it of social and natural factors, and also of preventative measures, which determine its development as a whole (650).

A number of facts, however, indicate that the analysis must include a consideration of the effect of heliogeophysical factors and, in particular, the effect of the GMF. A basis for this is provided by the data obtained by epidemiologists (118, 647, 648, 650). Phthisiologists have shown a connection between the GMF and pulmonary hemorrhages (366, 392, 417, 1073), and deaths from tuberculosis (817, 818). Similar conclusions were drawn from an investigation of the effect of the GMF on the course of a number of noninfectious diseases of man (cardiovascular diseases, eye diseases, cancer), and on the state of the blood system (220, 221, 302, 333, 457, 472, 476, 480, 546) and nervous system (382, 721, 837, 838, 1018–1020).

Different diseases have common features that indicate their connection with

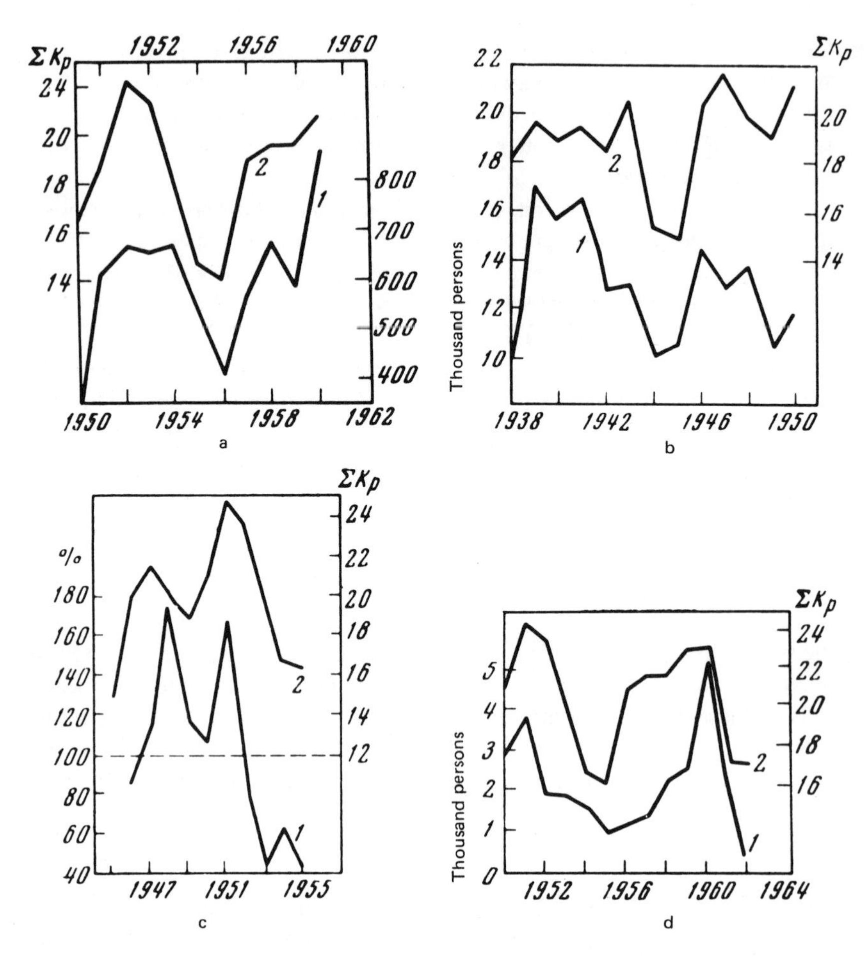

Fig. 64. Dynamics of epidemiological process and variation of planetary index of magnetic disturbance (ΣK_p) (650). (a) (1) Incidence of dysentery in the world, thousands of persons; (2) ΣK_p. (b) (1) Mortality from smallpox in Australia, 1:100,000 persons, (2) ΣK_p. (c) (1) dynamics of severe cases of scarlatina in Leningrad in relation to the number of mild cases, taken as 100%, (2) ΣK_p. (d) (1) incidence of poliomyelitis in Japan, thousands of persons, (2) ΣK_p.

the GMF. For instance, the level of infectious diseases (scarlatina, diphtheria, dysentery, etc.) fluctuates considerably, whereas the efficacy of counterepidemic measures is much higher, sanitation and hygiene are improved, and so on (650). A direct effect of geomagnetic factors on disease agents and the susceptibility of man to infection could provide an explanation of this fact. Medical statistics (on a regional and global scale) confirm this conclusion (Fig. 64). Increased disturbance of the GMF is accompanied by an increase in the incidence of and mortality from smallpox, dysentery, whooping-cough, poliomyelitis, and tetanus (647, 650). When the relation between diseases and individual elements of the GMF was analyzed, however, it was found that acute attacks of glaucoma, for instance, occur mainly on days with weak ($\Delta H \leqslant 60\gamma$) variations of the horizontal component of the GMF (662).

The scientific medical literature contains extensive and thorough studies of Soviet scientists on the effect of the GMF on various groups of diseases (51, 164–166, 207, 208, 219, 351, 390, 391, 473, etc.). A detailed examination of these major investigations would require considerable enlargement of the book not only for an account of the investigations themselves, but also for a thorough discussion of the underlying causes of the diseases, and we would then need to describe the functioning of various systems in the normal healthy human. Hence we will confine ourselves to a brief account of the main results obtained by researchers in the course of their major studies.

Cardiovascular Pathology. We have already indicated the important role of the GMF in the clinical aspects of this disease. In addition to the important role of atmospheric pressure and temperature, authors report that the state of magnetic activity on the day of occurrence of mycocardial infarct affects the outcome and incidence of various complications of this disease. On days characterized by unfavorable heliogeophysical conditions patients show an increase in such infarct complications as ventricular fibrillation, arrhythmia, atrioventricular block, and low cardiac output (165, 166). From a study of 2143 cases of acute macrofocal myocardial infarct, the authors concluded that the incidence of complications of this disease is greater when the GMF intensity deviates appreciably to either side of the mean. This feature of the GMF effect has been reported by other investigators in the case of eye and kidney diseases (207, 208, 351), gastric hemorrhages (272), epilepsy (112), etc. When there is a considerable reduction of GMF intensity ($K_p = 0$) there is a highly significant increase in the incidence of ventricular fibrillation and atrioventricular block in man, paroxysmal tachycardia in men and women, myocardial ruptures, and contractile inadequacy (165). An increase in GMF intensity also leads to an increase, though less pronounced, in the incidence of complications in acute infarct. It will probably be necessary to carry out future research with the aim of ensuring that medical institutes make use of the forecasting service to take into account the changes in the GMF intensity (158, 288, 459, 547, 788, 865, 1027).

The effect of GMF intensity on the state of patients with rheumatics and rheumatic heart disease has been thoroughly analyzed (390, 391). Geomagnetic disturbances, in the authors' opinion, are a significant factor aggravating the state of stress in the mechanisms of human adaptation. This conclusion applies both to low latitudes, where the work was conducted (390, 391) in the resorts in the south of the USSR, and to high latitudes, where the state of adaptation of healthy people was investigated in regions of the Far North (41, 42, 370–372, 912). In rheumatic patients on magnetically active days, the normal diurnal course of the heart rate was disturbed, ventricular asynchronism was increased during the day, the phase of isometric contraction was prolonged, and the stroke volume was altered—it increased in the second period of treatment, although there was a distinct reduction initially in the period of adaptation and treatment. Thus, the GMF has a significant effect on the adaptation of rheumatic patients—on the state of their cardiovascular system.

Since the course and outcome of many diseases, and also human adaptation processes, depend on the state and function of the pituitary–adrenal system, the investigation (51) that analyzed the effect of the GMF on this important link in neurohumoral regulation of patients with ischemic heart disease is of particular interest. An investigation of the dynamics of excretion of catecholamines in the urine of 135 patients with myocardial infarct and 94 patients with atherosclerotic cardiosclerosis and angina pectoris revealed a reduction in the activity of the neurogenic link of the sympatho-adrenal system in the acute and postinfarct state. In patients with angina there was also an increase in activity of the hormonal link of this system with increase in magnetic disturbance.

The use of the method of superposed epochs enabled the authors to establish that on the day of development of small, moderate, and large storms with a gradual commencement, and also small storms with a sudden commencement, the level of noradrenalin excretion was reduced in the acute and postinfarct period; on the day of development of moderate and large storms with a sudden commencement it was the same as on the preceding quiet days. The level of adrenalin and noradrenalin excretion decreased with increase in GMF intensity. The functional relations of the physiological indices and geomagnetic activity were very close: the correlation ratio was 0.601–0.824 for $p < 0.01$–0.05. Healthy persons also showed a close correlation between the K index of magnetic disturbance and the excretion of 11-hydroxycorticosteroids. The maximum excretion of corticosteroids was observed 32 h after the geomagnetic disturbance (108).

Psychic Disorders. The effect of the GMF on the central and vegetative nervous systems indicates that there is a relation between geomagnetic activity and several psychic diseases (112, 203–206, 219) and neurotic reactions. The cyclicity in this group of diseases has been known for a long time and hypotheses on their possible cause, implicating environmental factors, have been put forward. Some features of the clinical signs of psychic diseases indicate that solar activity and geomagnetic disturbances, i.e., increases or reductions of the GMF intensity, are direct causes of this kind of disturbance of human psychic activity. The number of schizophrenic patients shows cyclic fluctuations with a period of about 10 years, and there is a significant high correlation between incidence of the disease and solar indices ($r = +0.7$–0.9). The sunspot numbers in the prenatal period of the patients were significantly higher (203–206). The most sensitive months of the prenatal period, making the greatest contribution to the possible subsequent disease, are the first three months after conception (203–206, 724).

A comparison of geomagnetic activity in different years and clinical manifestations of epilepsy showed a definite correlation between the indices investigated (219). For instance, in years of high geomagnetic activity (1960, 1961) there was a large number of twilight days in epileptic patients but, on the other hand, there was a reduction in the number of epileptic fits. The ratio of the number of fits to twilight days was 2.13 and 1.82 in years of high geomagnetic activity (1960, 1961), and 10.58 and 11.07 in years with minimum geomagnetic activity (1964, 1965). Other indices also revealed differences in the nature of manifestations of epilepsy in years of maximum and minimum geomagnetic activity. For instance, in 1960 the mean period spent by an epileptic patient in the hospital was 56.4 bed days, and 1965 it was 45.3 with similar treatment. Calculations showed that in 1960 there was one epileptic fit every 11.16 days, and one twilight day every 23.78 days, whereas in 1965 the ratios were 3.64 and 40.26 days, respectively. The authors of the cited study concluded that clinical manifestations of epilepsy are different in years of maximum and minimum geomagnetic activity: In a period of high geomagnetic activity there is an increase in the number of epileptic psychoses and their equivalents, and in a period of reduced geomagnetic activity the number of twilight states is smaller, but the number of epileptic fits is greater. This conclusion has been confirmed on independent material and in another investigation (112), where the number of fits of epileptic patients on quiet magnetic days was higher than on magnetically disturbed days ($r = -0.670$–0.750).

It has also been reported that an increase in geomagnetic activity leads to an increase in the number of neurotic difficulties, reactions, and complaints in neurotic patients even in cases where resort treatment is beneficial (865). What is very characteristic, however, is that the author, like the other investigators cited above, also noted a significant deterioration in the state of neurotic patients on days with minimum geomagnetic activity. Thus it is apparent that the views expressed in Chapter 2 are confirmed by medical data, viz., that the organism reacts distinctly to any change in the geomagnetic environment, whether it is a sharp reduction or increase, since the deep connecting mechanisms are based on the biomagnetic homeostasis of the living organism and involve current systems in the membranes, and modification of their quasi-crystalline structure (64) and their permeability (121–123, 809). This is particularly well exemplified by a serious eye disease such as glaucoma, where destruction of the visual functions of patients is related not only to ophthalmotonus, but also to the state of the vascular system of the eye (207, 208).

Eye Diseases. The primary mechanisms of the glaucomatous process were subjects to thorough investigation in the papers cited above (207, 208). The author of this investigation correctly paid great attention to a study of the general vascular permeability and metabolic processes in patients with primary glaucoma and to the relation between these indices and variation of the GMF. He investigated the dynamics of general vascular permeability in 140 patients in the period 1969–1972 by the Lendis method (percentage loss of protein). The investigations showed that geomagnetic disturbances definitely affect the state of vascular permeability. In a period of maximum geomagnetic activity there were the greatest deviations in the general vascular permeability of patients and an increase in frequency of decompensation of the glaucomatous process ($p < 0.01$–0.05). The nervous response to geomagnetic disturbance was followed by wavelike changes in vascular permeability and an increase in the number of acute glaucoma attacks on the second, fourth, sixth, eighth, and tenth days, and a sharp increase in permeability on the sixth day. In contrast to (662) the author of the cited paper found an increase in the number of acute glaucoma attacks on days of increased geomagnetic activity: The mean daily indices of the number of acute glaucoma attacks were 0.05, 0.06, 0.104, and 0.154, respectively, for completely magnetically quiet days, magnetically quiet days, magnetically active days with a C index of 0.5, and those with a C index of 1.0 or more. Geomagnetic storms with different characteristics have a different effect on the frequency of occurrence of acute glaucoma attacks. Potentially dangerous times are the day of onset of the storm, and also on subsequent days—the sixth day for small storms with a gradual commencement, the fourth, seventh, and ninth days for moderate and strong storms with a gradual commencement, the first, third, sixth, and ninth days for moderate and severe storms with a sudden commencement, and also two days before the onset of small storms with a sudden commencement. The author of the cited work came to the important conclusion that the increase in GMF activity during storms can probably be regarded as the trigger that actuates one of the most important pathogenic mechanisms, such as vascular permeability, whose modification leads to the development of decompensation. Thus, like the several studies of cardiovascular diseases (164), the study of the role of the sympatho-adrenal system in ischemic disease (51), and work on ophthalmology (207, 208), completely confirms our earlier conclusion that the main mechanism of the biological effect of the GMF is its effect on the permeability of biological membranes (119, 121, 122, 124, 133, 134, 809).

Urolithiasis. The formation of stones in the urinary pathways (urolithiasis) is a very common, worldwide disease. The dynamics of this disease shows several characteristic features indicative of a role of the GMF (351). First, as for many other groups of diseases, there is a characteristic periodicity of incidence during the year and the correlation noted between the GMF and other diseases is again found. For instance, an increase (or reduction) of magnetic activity is accompanied by a reduction (or increase) in the incidence of cases of renal colic. The coefficient of correlation between these indices is $r = -0.78$ and increases to 0.96 and 0.87, respectively, for periods from 0800 to 2000 h and from 2000 to 0800 h if the *u* index is taken as an index of magnetic activity.

The author of the work cited made a very interesting comparison of the effect of the GMF on the distribution of urolithiasis in the world. He found that there are long-lasting foci of urolithiasis in countries of Central Asia, India, the Balkans, Egypt, Iran, whereas in countries of Southern Africa, North America, and the Far East this disease is rare. A comparison of the map of the distribution of urolithiasis in the world with the map of continental currents compiled by Vestin (651) showed that the foci of urolithiasis are situated in places where the zero current line passes through regions of minimum GMF intensity, whereas in regions of high GMF intensity—the Kursk magnetic anomaly, Northwest Siberia, and the other locations mentioned above—there are no foci of urolithiasis. This is good confirmation of the important ecological role of the constant component of the GMF (131, 136, 141, 263, 400, 603, 607).

The very wide range of groups of diseases investigated by the various authors is remarkable and is probably due to the fact that the GMF and its disturbances have a direct effect on living organisms, and on their main substances and internal processes. Natural electromagnetic fields play an important role in the functioning of the human organism (243, 263, 400, 536, 622, 624, 627, 756, 759, 835, 836, 932, 967, 1015, 1027, 1120, 1121, 1203–1207).

The Geomagnetic Field and Occupational Activity

There are many aspects of this important and extensive problem, but we shall consider only those that have not been well discussed in the scientific literature. We showed previously that the GMF has a pronounced effect on all human functional characteristics. Hence changes in the GMF will affect any occupational activity and will be revealed by the investigation of any processes. The examples we give, relating to medical work (865) and road accidents (1142), show the great range of effects involved.

It is reported that sudden changes in GMF amplitude, particularly against a background of low ambient air temperature, lead to a significant increase in the number of industrial accidents (946). Of particular interest in this connection are the data indicating an effect of the GMF on the performance of civil aviation pilots (214, 215). This question is very important, since flight safety depends on many factors, including the effectiveness of preflight medical checks of pilots. The author of the papers cited investigated the effect of variation of the GMF, viz., the relative variation of the GMF horizontal component, expressed as a percentage of the mean, on the working efficiency of flight crew personnel in tests to decide if they should be allowed to fly. The performance of the crew members was investigated over a period of 45 days with a photographic model of the instrument panel of an IL-18 aeroplane on the usual working days of the pilots. The indices investigated were the mean time taken to read the panel with unlimited time, and the number of errors made when the reading time was limited (15 sec). The investigation showed that on days when the variation of the GMF horizontal component was greater, the number of errors made in complicated tasks by pilots was 30.56% higher, and by copilots 23.9% higher, and the time to carry out some special procedures was 6.2% and 20.0% longer than in a quiet period. There was a direct correlation between the course of the geomagnetic disturbance, the time to read the panel dials by pilots, and the number of errors made ($r = +0.883 \pm 0.098$ and 0.929 ± 0.061 for $0.05 > p > 0.02$), but there was no correlation for the performance of tasks in a limited time ($r = -0.612 \pm 0.28, p < 0.1$). There was no correlation between the results of reading of the panel dials and the state of GMF disturbance in the case of copilots. The author attributed this to the specific nature of the job.

Of interest in connection with these investigations are the studies of Maxey, who attributes importance to the effect of low-frequency electromagnetic fields of natural origin and ions of opposite charge on the work of pilots (981). This is a matter that requires serious and thorough attention, particularly when one considers the high values of the vertical electrical component in a field of low-frequency geomagnetic pulsations (87, 765), and that its changes at any heights can attain large values, even as high as several kilovolts (1024).

We must draw attention to other aspects of the effect of the GMF on occupational activity. In particular, the suggestion that the state of GMF disturbance should be taken into account in conducting clinical laboratory analyses (50, 51, 473) merits attention. The authors report that an investigation of patients showed great variation in homogeneous groups of patients and in the same subjects investigated on different days. In addition, a doctor investigating a patient often encounters sharp fluctuations in the results of laboratory analyses, which do not correspond with the clinical course of the particular disease. For instance, a sharp reduction in the level of urinary excretion of catecholamines in patients with acute myocardial infarct, an atypical feature of this period of the disease, is sometimes observed. The results of electroencephalographic and ballistocardiographic analyses of patients (374, 390, 391) and hematological analyses of healthy persons (25, 26, 347) may also depend on the state of the GMF. Hence, the possible effect of the GMF should be taken into account in the laboratory practice of medical institutions and in industrial work. For instance, it has been reported that there is a seasonal and diurnal variation in the adhesiveness of chromium films evaporated in vacuum onto quartz and silicate sodium glass and that this variation is related to solar activity and the GMF (353). Since GMF disturbance causes an increase in the radioactive background (507, 508), it is obvious that the level of geomagnetic activity must be taken into account in radiobiological studies and radiation-dose measurements, which are extensively carried out. Thus even this brief account of facts reveals the importance of consideration of the state of the GMF in any occupational situation.

In the following sections we shall discuss experimental facts indicating that animals and plants are not only affected by geomagnetic activity, but their homeostasis depends on the GMF.

Biological Effect of Anomalous Geomagnetic Field

Investigations of the effect of regional and local anomalies of the GMF on living organisms are of particular interest. In anomalous regions there is a pronounced change in the gradients of the GMF components which can reach several hundred and, in some cases, even several thousand, gammas per kilometer. A unique case of this is the Kursk magnetic anomaly, where the vertical component of the GMF reaches 1.5 Oe, which is three times greater than the vertical component of the normal field.

The effect of the anomalous GMF in Kursk and Belgorod Regions on the incidence of disease among the population has been investigated (101, 264, 576, 577). The investigation of incidence of disease was based on the following nosological units: hypertensive disease, rheumatism, neuropsychic diseases, malignant neoplasms, eczema, nephritis, vascular diseases of the central nervous systems, and other diseases. The incidence of disease in the populations of Perm Region and Primorskii Krai, situated at the latitude of the Kursk anomaly (101), and in anomalous and normal parts of Kursk and Belgorod Regions during the same years (576), were compared.

For almost all the nosological units the correlation between the degree of anomaly of the GMF and the incidence of disease in the population was significant at a high level ($p = 0.05$–0.0001). In anomalous regions the incidence of hypertension, rheumatism, and neuropsychic diseases was 120–160% higher than in normal regions. A careful subsequent analysis (570) showed that only the correlation between vascular diseases of the central nervous system, hypertensive disease, eczema, climacteric disorders, and active rheumatism in normal and abnormal regions was statistically significant. In the case of vascular disorders the correlation was positive, and in the case of the other groups of diseases it was negative.

The dynamics of growth and yield of crops in the region of Kursk magnetic anomaly was also investigated (263, 566, 570, 574, 575). According to the mean data (1964–1969), the yields of winter wheat, rye, corn, sunflower, and annual grasses in regions with an anomalous field were lower than in regions with a normal field. The potato harvest, however, was greater in the anomalous regions than in the normal regions. In anomalous regions the growth of sugar beet roots lagged behind the growth of roots in normal regions (the differences were significant at the $p < 0.05$ level), but no differences in sugar content were found.

These investigations are of great interest, since the biological effect of the constant GMF and, in particular, the role of regional anomalies are analyzed. An important factor in the vital activity of living organisms is the variable GMF in its whole range of variation, since investigators believe that this factor is responsible for the change in many functional properties of organisms, which we will discuss in the following sections.

Animal World

The actual existence of an effect of the GMF and its role in the life of animals have been investigated by various methods. There are three main areas of research on the biological effect of the GMF on living organisms: (1) vital activity, (2) position and orientation in space, (3) long-range migration. The choice of object has frequently been dictated by considerations of methodology: ease of collection, suitability for laboratory cultivation, availability of pure genetic lines, and also the classic and long-standing nature of the research problem itself—orientation of birds in flight, homing and migration factors, position of body in space, etc.

Microorganisms and Viruses

The close correlation between the dynamics of infectious processes and the variation of the GMF implies that the GMF causes direct and diverse changes in the vital activity of microorganisms (rate of multiplication, pathogenicity, hereditary characters, etc.). For instance, shielding of microorganisms from the GMF greatly reduces their growth (15, 21, 412–415, 685) and gives rise to mutant strains (85, 95, 530). An analysis of changes in the spontaneous phage titer in a lysogenic *E. coli* K_{12} (λ) culture showed that it was correlated with changes in the horizontal component of the GMF (53, 312, 313).

It has been suggested that changes in the cultural and morphological characters of microorganisms are connected with magnetic activity (312, 729), and that the strength and polarity of the GMF affect the marine microbial flora and its distribution in marine deposits (1129, 1130).

If we accept that the GMF has a direct effect on bacteria and viruses and take into account the very high rates of growth and multiplication we can understand the rapid evolution of these organisms, the succession of parasitocenoses and viral strains, and as a consequence, the fluctuations in the incidence of various diseases. At present, however, the effect of the GMF on insects has been more thoroughly investigated.

Insects

Experimental and statistical investigations have revealed distinct regular features in insect behavior, which have provided a basis for a thorough investigation of the effect of the GMF on them (31, 84, 573, 652, 686, 691, 692, 764).

Sex Ratio and Mutations. Laboratory experiments have shown that the orientation of *Drosophila* larvae with the head of the embryo to the north (by the compass) leads to a larger number of male adults, while a southward orientation leads to a larger number of females (4). This implies that the orientation of the eggs in relation to the magnetic poles at oviposition is important for the development of individuals and that the natural sex ratio can be altered by altering the orientation in the GMF. This indicates that the GMF acts directly on the genetic apparatus of insects. This conclusion is confirmed by the results of an experiment with *Drosophila* in hypomagnetic conditions (1155), and analytical investigations of the dynamics of the natural mutation processes (122, 126, 136).

Subjection of *Drosophila* to an artificial magnetic field, or to the natural magnetic field reduced to ~0.05 Oe, gave rise to mutations leading to a deviation of the sex ratio from the expected normal 1:1 ratio (1155). Insects, especially of the F_2 and F_3 generations, kept in hypomagnetic conditions (0.05 Oe, Helmholtz coil) for a long time after mating showed a very statistically significant change in the sex ratio toward a predominance of males. A brief subjection to such conditions (0.05 Oe, 48 h) did not produce such an appreciable effect, although F_2 individuals in some tubes of the experimental series showed a considerable deviation from the expected sex ratio.

Analytical investigations have also shown that the GMF plays an important role in the mutational changes occurring in insects in natural conditions (122, 123, 809). Direct comparison showed that the seasonal course of chromosome inversions corresponds exactly with the variation of the GMF during the same period (Fig. 65).

It is of interest that the frequency of appearance of ST and TL gene inversions depends on the variation of different elements of the GMF—the dip and declination. This may indicate a specific relationship between the GMF and the molecular structure of the genetic apparatus of living organisms. These hypotheses, however, require thorough testing and very reliable experimental corroboration.

Rhythmic Activity. Observations in the field confirm that the attraction of insects to a quartz lamp at night depends on the state of geomagnetic activity (78, 79, 81, 82). An analysis of a large amount of material (75,336 individuals) revealed that the coefficient of correlation between magnetic disturbance and increase in the number of attracted insects was $r = +0.926$ ($p < 0.001$). On days with a magnetic storm the diurnal rhythm of activity of the beetle *Trogoderma glabrum* was greatly altered (80). This insect in laboratory conditions showed a distinct rhythm of activity by day and rest by night, but motor activity was often observed day and night. Such disturbances of rhythm were observed on 69 days out of 284. An analysis showed that the rhythm was altered much more frequently on disturbed days than quiet days. The proportion of days with altered

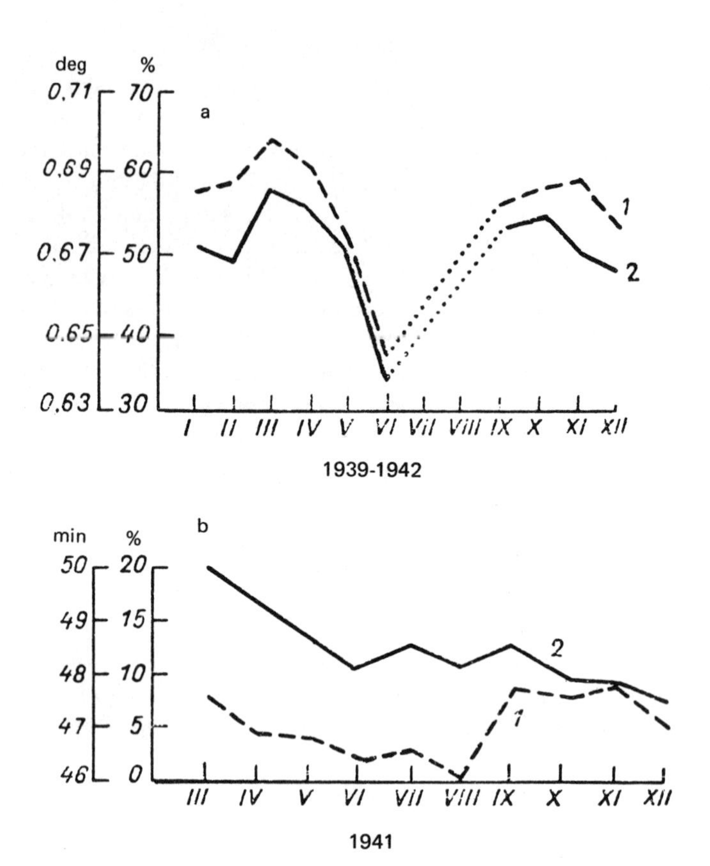

Fig. 65. Dynamics of mutation process in *Drosophila* in natural conditions (California, U.S.A.) and variation of GMF (130). (a) Change in concentration of order of gene ST: (1) dip of geomagnetic field (deg); (2) inversions of gene ST (%). (b) Change in concentration of order of gene TL: (1) declination of GMF (min); (2) inversions of gene TL (%).

rhythm for magnetically disturbed days was 0.59, whereas for magnetically quiet days it was only 0.16. The difference in these figures was statistically very significant ($p < 0.001$). The disturbances of rhythm had a 27-day period and were much more frequent in spring and autumn. These times are characterized, as we know, by manifestations of magnetic disturbance associated with the position of the sun on the ecliptic (see Fig. 4). It has been reported that different insects react differently to geomagnetic disturbances (83).

Insect Orientation. A large series of careful and precise experiments has shown that other physical environmental factors besides the usual ones (light, food, and other ecological factors) have a decisive effect on the orientation of the insect body in space (969).

Beetles, Termites. "Ultraoptical" orientation was recorded for the first time in laboratory experiments with the cockchafer (1112–1115). Adult beetles (previously cooled to render them immobile) were returned to normal conditions. When the cold torpor wore off, the beetles began to move and take up particular positions. The direction finally chosen by the beetles was not random, but was the result, according to the author's terminology, of ultraoptical orientation. The insects selected a particular orientation relative to the vectors of the acting physical fields, which had a particular compass orientation. It was subsequently discovered that ultraoptical orientation depended on the action of vectors of unidentified physical factors that can pass through walls, glass, and the wire grid of a Faraday cage, and can induce geographic orientation.

The results of the investigations (1112–1116) of insect orientation were statistically very significant. Although the author did not draw any direct or final conclusion, all the data indicate that the GMF plays a major role in the "ultraoptical" orientation of insects. This is clearly indicated by the choice of particular geographic directions by the insects, the effect of a magnetic field of 10 Oe on this choice, the penetration of the unknown factors responsible for ultraoptical orientation through any materials, and so on.

Observations in field conditions have also confirmed the important role of the GMF in the life of insects (691, 692, 969). For instance, it has been established that termites orient their underground galleries and entrances to termitaries mainly in the direction of the magnetic meridian (685, 686). There are data indicating that the female (queen) takes up a position along the magnetic meridian in the termitary (169, 797). Such observations have led research workers to investigate in detail the sensitivity of other insects to the GMF and their response to changes in the GMF.

Flies. The orientative ability of flies has been investigated in very diverse experiments: in square boxes with glass walls, and in tall round Petri dishes shielded from light by paper. The dishes were rotated through 60° after every test. In addition, the insects (*Calliphora erythrocephala* Meig. and *Musca domestica* L.) were photographed in special cubical containers under artificial illumination in a windowless room, where the positions of landing and rest of the insects on an illuminated square or circle were investigated. When required, the angle of inclination of the container walls and the light intensity could be altered within prescribed limits (694). The direction of each insect's body relative to the geomagnetic poles was measured to within 1°. The experiments showed that, irrespective of the experimental conditions, the insects preferred sectors bounded by angles of 20° on either side of the north–south and east–west axes of the horizontal component of the GMF. We should mention that the frequency distribution of the selected angular directions relative to the poles of the insects on landing and at rest was greatly altered when the GMF was compensated to approximately 5% of its natural value. The effect of compensation was revealed by photographs of the insects in normal conditions and with the GMF compensated. It was found that compensation of the GMF led to less preference for the north–south and east–west directions than in normal conditions (Fig. 66).

We have mentioned that the preferred directions for insects in the GMF were north–south and east–west or directions close to these lines. The objectivity of photography, the large samples that the investigators used, and the thorough mathematical analysis of the data guarantee the correctness of the results and rule out the psychological errors that can be made in experiments of such kind (694). For instance, in the case of the resting position of *Drosophila melanogaster* the differences between the main north–south and east–west orientation sectors and the northeast–southwest and southeast–northwest directions were statistically very significant ($p < 0.0005$).

It was mentioned above that compensation of the GMF by Helmholtz coils altered the natural orientation of the insects. The same effect occurred when only the horizontal component of the GMF was compensated, or when an artificial field produced by a rotating magnet was applied. It is true that the GMF had a very different effect on the orientation of individual flies. This is quite consistent with the proposed principle of functional dissymmetry, which postulates a diversity of responses of biological systems to the action of the GMF (125, 126, 135).

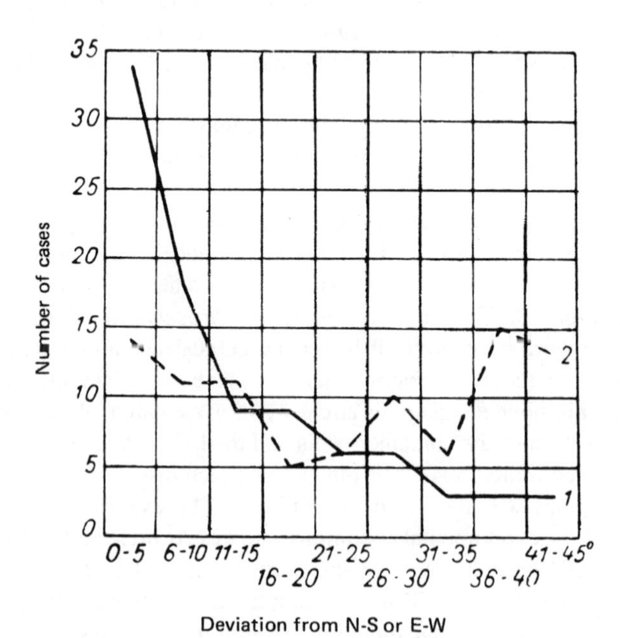

Fig. 66. Effect of compensation of GMF on direction of body of flies on landing (694). (1) Geomagnetic field; (2) GMF compensated.

It should be noted that in the study of the effect of the GMF on insect orientation particular care was taken to ensure a correct statistical assessment of the results. The reliability of the data indicating a preferred orientation of insects in different experimental conditions was assessed by means of the Kolmogorov–Smirnov tests, the Fisher test, Student's *t*, the Kuiper–Stephens test, etc.

Bees. The problem of the orientation of bees in space in search of the hive or food is one of the oldest and most traditional scientific problems. The role of the GMF in the orientation of bees has attracted the interest of research workers. Early work with bees, however, failed to reveal any effect of the GMF (840, 1086).

Research conducted in recent years has led to a reassessment of the role of the GMF in the life of bees. As is known, the dance of bees in a vertical plane is a special way of imparting information about the location of plants from which pollen and nectar can be collected (840). By the motion of its body, particularly the tail end, the dancing bee communicates the azimuthal direction of the food supply. These indications of direction, however, always contain some error (*Missweisung*), i.e., contain certain deviations from the correct route. It has been established that the reason for these "incorrect indications" of the true position of the food is variation of the GMF. Research workers (951) found that when the GMF was compensated (the hive was placed in a modified Helmholtz coil system) the fewest incorrect indications of the direction of the food were obtained. For instance, in the control experiments the deviations from the correct direction were $+4.8$ to $-10°$ with a spread of ± 1.4 to $3.8°$, whereas on compensation of the GMF to a level of 0–5% they were $-3.8°$ with a spread in the range ± 0.8 to $2.5°$ (Fig. 67). The statistical treatment of these data and the criteria used have already been mentioned.

Thus, by correct experimental procedure a definite indication of the role of the GMF in the life of bees was obtained for the first time. It should be noted that the effect of the GMF on bees and, in particular, the compensation or shielding of the GMF do not take effect immediately, but after an interval of time of about 0.5–1 h. In addition, adaptive responses are possible, i.e., the insects can become accustomed to an altered electromagnetic environment. This is in good agreement with the data of other magnetobiologists who have investigated the adaptation of various groups of animals and birds to magnetic fields (236, 1215).

Drosophila. Although the first results were promising, the question of the role of the GMF in insect life is still an open one. It has now been definitely established that the GMF has an effect on the position of the insect's body in space, and there is a definite interrelationship between the perception of gravity by insects and the GMF. This conclusion, derived from experiments with bees (951), has been confirmed in experiments with *Drosophila* (1196). The negative geotactic orientation of *Drosophila* on a plane with a slope of 30° was definitely

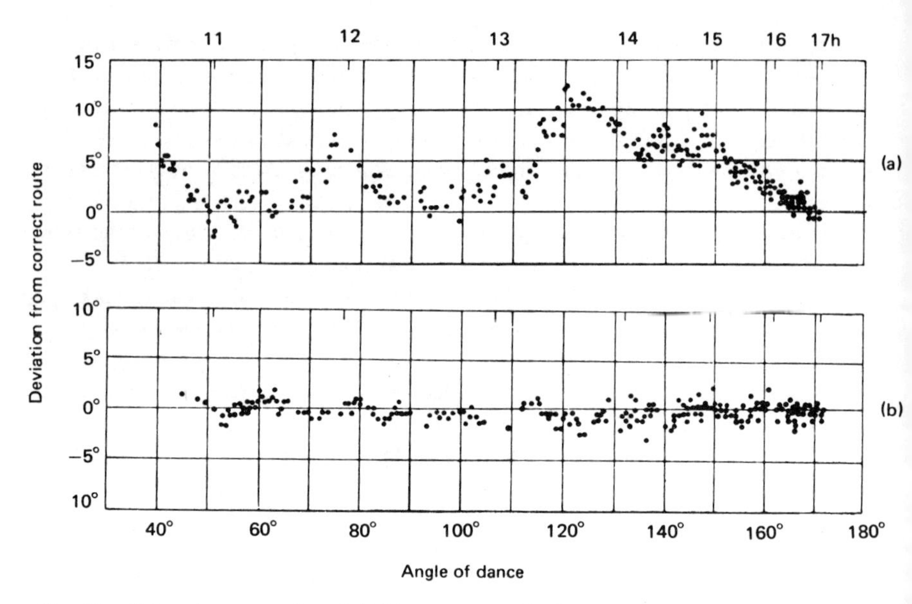

Fig. 67. Effect of compensation of GMF on accuracy of indications of direction by bees (951). (a) Geomagnetic field, March 31, 1967; (b) GMF compensated, April 2, 1967.

reduced by 5.5° if the lines of force of the GMF ran in the direction of motion of the insect. It is still not clear if the GMF affects the mechanism of gravireception or the insect nervous system as a whole.

Investigations on insects, however, have not yet provided specific evidence that they use the GMF in any way, in their flights for food, in their behavioral orientation, or in migrations. The GMF affects the deviation of the indications from the true direction in bee dances, but the direction of flight to the feeding area, the landing, and the orientation of the body on landing at the feeding site are independent of the GMF (840, 1086).

It should be noted that the effect of the GMF on gravity perception is of very great importance and provides a basis for further investigations of the role of the GMF in the orientative ability of different groups of animals and birds.

Birds

A possible role of the GMF in the life of birds has attracted the interest of investigators for a long time. In particular, hypotheses implicating the GMF in the ability of birds to orient themselves in space and to navigate have been advanced since the middle of the last century (762, 763, 791, 792, 979, 990, 1146, 1160–1162, 1186). At the present the GMF is regarded as one of the main factors underlying these special features of bird behavior (45, 97, 114, 446, 447, 677, 829, 914–917, 1136, 1138–1140, 1188, 1189, 1215–1219, 1221, 1222, 1227). It has been known for a long time that birds can undertake short and long migrations and find their way home to their nesting sites ("homing").

We can distinguish two types of travel by birds—homing and seasonal migrations. The first type of travel usually entails flight over short distances (100–200 km) and return to the nesting site in a day or two. The second type—seasonal travel—occurs twice in the year: the spring arrival and the autumn departure. Seasonal migrations involve prolonged flights over distances of hundreds or thousands of kilometers. In both cases the birds unerringly find the route that takes them to their nesting sites. For instance, carrier pigeons are capable of homing even when there are no visible terrestrial or stellar reference points. The birds can find their way home in solid cloud, in thick fog, or in localities where terrestrial landmarks are unfamiliar or absent, as in the case when the birds are released over the open sea. Observations of such kind are confirmed by recent investigations, where the migrational flights of birds in dense cloud have been investigated by radar (708–710, 807, 913, 1013). Birds care capable of homing even when they are transported to an unfamiliar location in a drugged state or in covered rotating cages (861, 927).

Observations of seasonal migrations have shown that young birds, which have had no previous experience, are capable of seasonal flights by the same route as that followed by experienced, fully grown birds. Statistically reliable data indicate that the only environmental factor that can affect the orientative ability of birds is GMF disturbances (1137, 1138). The selection of the main direction of autumn migration (ESE) by juvenile gulls is less accurate, the greater the disturbance of the GMF on the day of the experiment. On magnetically quiet days the gulls select the right direction. These facts indicate that birds can "remember" the position of their nesting sites and find their way to them, and that disturbance of the GMF can probably affect their navigating mechanism.

Although the sun and terrestrial landmarks (landscape features, nature of coast line, atmospheric conditions, e.g., prevailing winds) are of great importance for bird orientation (864, 978), these factors cannot account for the extraordinary homing ability of the choice of routes in seasonal migrations (258, 259, 653, 962, 1036). One of the real physical guides could be the GMF. The theory that the GMF is the main orientative signal in bird migrations, however, still encounters serious difficulties. First, researchers report that so far no sensory organs receptive to the GMF have been found in birds (864), although some birds have peculiar "magnetic" stones in their digestive organs (799), and some investigators have failed to develop conditioned reflexes in birds to a magnetic field (1026) or to a rapid change in its direction (829). In addition, apart from rare exceptions (914, 915, 1139, 1140, 1188, 1189), investigations show that artificial magnetic fields created around the head, wings, or bodies of birds do not prevent homing (487, 488, 726, 763, 854, 940, 973, 1088, 1091, 1223).

In particular, the artificial magnets, which (in the opinion of the authors who used them) ought to interfere with the orientative ability of birds, weighed from 0.5 to 1 g and had strengths of 1–100 G. The experimental techniques, evidence, and testing, and the theoretical premises for the construction of a bird navigation theory based on the GMF have roused criticism and objections (517, 829, 940, 1022, 1213).

Nevertheless, there has been a gradual accumulation of data indicating a possible role of the GMF in the vital activity of birds. First, it has been experimentally confirmed that birds can respond to weak artificial magnetic fields (236, 1135, 1140, 1188, 1189), show changes in activity (156, 988, 1010), and respond to a change in the GMF (517, 917, 985, 986, 997, 1138). It should be especially noted that birds have been found to be highly sensitive to a range of magnetic field intensities close to that of the GMF (321, 1078, 1215).

An investigation of bird orientation in laboratory conditions by the round-cage method (936, 984) showed that different kinds of birds in the migration season (even when kept in cages) make directional movements—"migrational restlessness." For automatic recording of the migrational movements the whole perimeter of the round cage is divided into 8–12 equal parts by perches. Pressure on the perch by the bird closes the contacts of an electric relay controlling the recorder, and hence each movement of the bird in the corresponding direction is accurately recorded. Since each perch corresponds to a geographical direction (every 30–45°), a count of the total number of jumps made by the bird over a long period can reveal the direction of activity of the birds in particular periods of time.

The round-cage method showed that when the birds were shielded in a steel chamber the previously distinctly oriented migrational movements of the birds were upset (841, 986, 1215), and the birds themselves, in the state of migrational restlessness, were affected by the GMF. This conclusion was tested by putting the round cage with the bird inside large Helmholtz coils (2 m in diameter). One pair of coils, set in the north–south direction, weakened the GMF, while the other pair, perpendicular to the first, created a new field (0.47, 0.88G) with a different "north" (Fig. 68). When the new field was applied and the north direction artificially altered, the main migrational direction of the birds was also altered (985, 986).

Experiments with birds were also conducted in a planetarium. The round cage was placed in a rectangular solenoid, which created an artificial magnetic field (from 0.2 to 3.46 Oe). This method allowed the use of artificial astronomic reference points in conjunction with the magnetic field. On the basis of these experiments the authors (321) concluded that birds determine the direction of migration by astronomic reference points and the direction of the horizontal component of the GMF. If the guiding stars are not visible or their position is

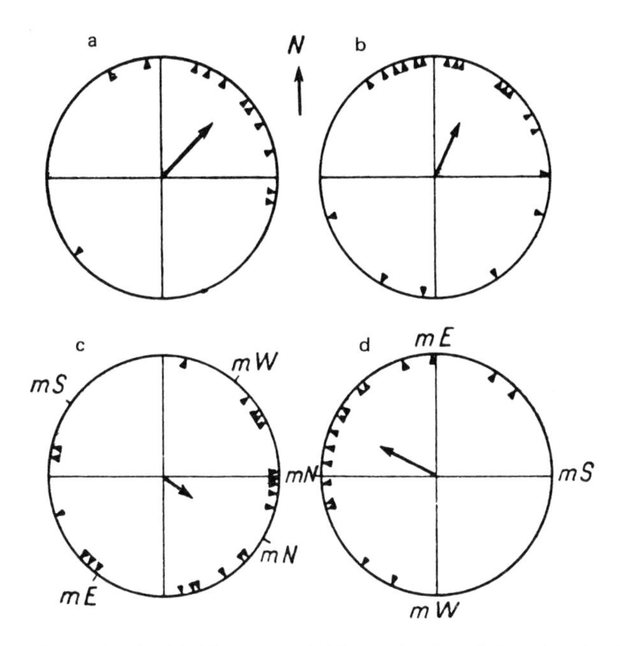

Fig. 68. Effect of natural and artificial magnetic fields on selection of migration direction by birds (987). Position of cage: (a) in a faraday cage; (b) in a timber house (control); (c) in Helmholtz coils with north in the direction ESE: (d) the same, with north in the direction W.

unusual, the birds orient themselves by the direction of the GMF horizontal component.

We note in conclusion the need for caution before rejection of the role of the GMF in bird life. The main reason for such caution is that laboratory investigations, even by the most refined techniques, but in a limited space (cage or planetarium), can give only a partial answer to the question of the effect of the GMF on orientation and navigation. A complete answer to the question will require conditions in which the birds fly freely, or telemetry experiments, where they have unlimited freedom of movement. The sensitivity of birds to the GMF is becoming more and more definite. There are data indicating that the GMF affects the life of birds from the moment that the eggs are laid in the nest (252). For instance, orientation of the eggs with a particular end toward the north alters their rate of growth and the sex ratio of the progeny (5).

Observations can help to reveal the reasons why birds select different migration directions—north in spring, south in autumn—according to their physiological state (828). A major factor here may be their special sensitivity to the GMF, which corresponds almost exactly in time with the above-noted increase in geomagnetic activity in these periods of the year. The discovered cases of contradictory results can also be satisfactorily accounted for by functional dissymmetry, when the same factor, in this case the application of an artificial magnetic field, has a different effect on different enantiomorphs (135). In addition, the successful conducting of experiments on the orientative ability of birds may depend largely on the geomagnetic activity during the period of observation (1138, 1139), the selected magnetic field strength, which is particularly biologically effective in the range of GMF values (1215), and other factors.

Possible Mechanisms of Orientation. The ability of birds to find their way to their nests, the orderly seasonal migrations of birds, and other orientation phenomena have for long interested scientists of different specialties—ornithologists, ecologists, physicists, and bionicists. The last group is interested in the navigational ability of birds.

Birds can migrate over several thousand kilometers and arrive within a few kilometers of their destination. For instance, the bristle-thighed curlew *Numenius tahitensis*, which nests in Alaska, migrates in winter to the islands of Tahiti and Hawaii, a distance of about 10,000 km, of which 3000 km are over the sea (329).

Although the mechanism of bionavigation in birds has not been discovered, research has tended to follow two lines, which we will call "functional" and "physical." In the first case the oriented flight of birds is attributed to an internal physiological mechanism involving instinct and remembrance of nesting sites, seasonal migration routes, and so on. The other trend, the "physical" approach, is based on the idea that birds have special means of perceiving physical factors of the external environment (the Coriolis force due to the earth's rotation, the position of the sun above the horizon, guiding stars, the GMF, etc.). Some researchers think that such reference points enable birds to determine the coordinates of their nesting and wintering sites and their migration routes.

The spatial orientation of birds is certainly a highly complex problem, which has been the subject of hundreds of theoretical and experimental scientific studies (44, 653, 805, 916, 978). We will not discuss the problem as a whole, but will consider only studies in which a possible role of the GMF in bird orientation is discussed and the validity of such a hypothesis assessed.

A possible role of the GMF in bird orientation was first suggested in the middle of the nineteenth century (990) and was subsequently subjected to thorough discussion and even to tests, although with artificial magnets. The results of the investigations were so contradictory that at certain times scientists completely rejected any implication of the GMF in the navigational ability of birds. In the 1960s research on magnetobiology and bird orientation again evoked interest in the problem and led to the advancement of interesting new hypotheses. For instance, it was suggested that an important element in the mechanism of the GMF effect on bird orientation could be a special structure in the bird's eye—the "comb" (97, 1154). In the opinion of Danilov *et al.* (99), the combined effect of the GMF and light on the comb creates conditions for manifestation of the Kikoin–Noskov photomagnetic effect, and hence the comb can be regarded as a kind of biological magnetometer. Continuous illumination of the comb may give rise to magnetic concentration in the GMF and the consequent change in field intensity affects the degree of separation of electric-charge carriers in the parts of the comb. These effects may give rise to diffusion currents capable of acting as stimuli for the optic nerve fibers.

Other investigators attribute special significance to the GMF horizontal component (446, 447, 791, 792, 1071). An analysis of numerous maps revealed that the migrational routes of birds are perpendicular to the isodynamic lines, i.e., they run in the direction of the gradient of the intensity of the GMF horizontal component. Hence, it was postulated that birds might remember their nesting sites by the associated intensity of the GMF horizontal component (447). The author of this hypothesis also believes that birds, by assembling in flocks and "post-nesting forays" (group flights in the direction of the future migration flight) form "a group conditioned reflex" to the GMF as a conditioned stimulus. The geomagnetic navigation hypothesis is supported by the behavior of flocks during migrations: After resting and feeding at a number of points on the migration route the birds perform circular flights, which become ellipses with their long axis in the direction of the migration route. This may be confirmation of a group-conditioned reflex to the GMF and selection of the azimuth for continuation of the flight.

Another interesting hypothesis on bird orientation is based on perception of the GMF dip and gravitation (45, 1221, 1222). The two hemispheres of the earth have only one closed line on which the magnetic dip has the same value (isocline), and on the corresponding isoclines in different hemispheres the dip differs only in sign. At the same time, the earth's gravitational force varies with latitude owing to the irregular spherical shape of the earth. In interesting papers the German researchers Wiltschko and Wiltschko not only advanced the hypothesis of a role of GMF dip but also tested it experimentally and fairly well confirmed it (1215–1219, 1221). The experiments were carried out by the same round-cage method and with the same kinds of birds that were used in the Helmholtz-coil experiments (Fig. 69) (1215, 1220).

Different factors affecting the natural geomagnetic complex and its individual elements (horizontal and vertical components, dip) led to statistically significant changes in the choice of direction by birds. These experiments showed that the choice of migration direction by birds depended on their ability to determine the position of the GMF poles (particularly the "north" pole) and they did this by also using the gravitation vector. The authors of these papers concluded that birds can determine the north direction from the smallest angle between the direction of the GMF lines of force and the gravitation vector. An analysis of the experimental data showed that in all cases the birds selected a direction in which the angle between the GMF and gravitation vectors was least far "north." In the case where the artificially created magnetic field had no

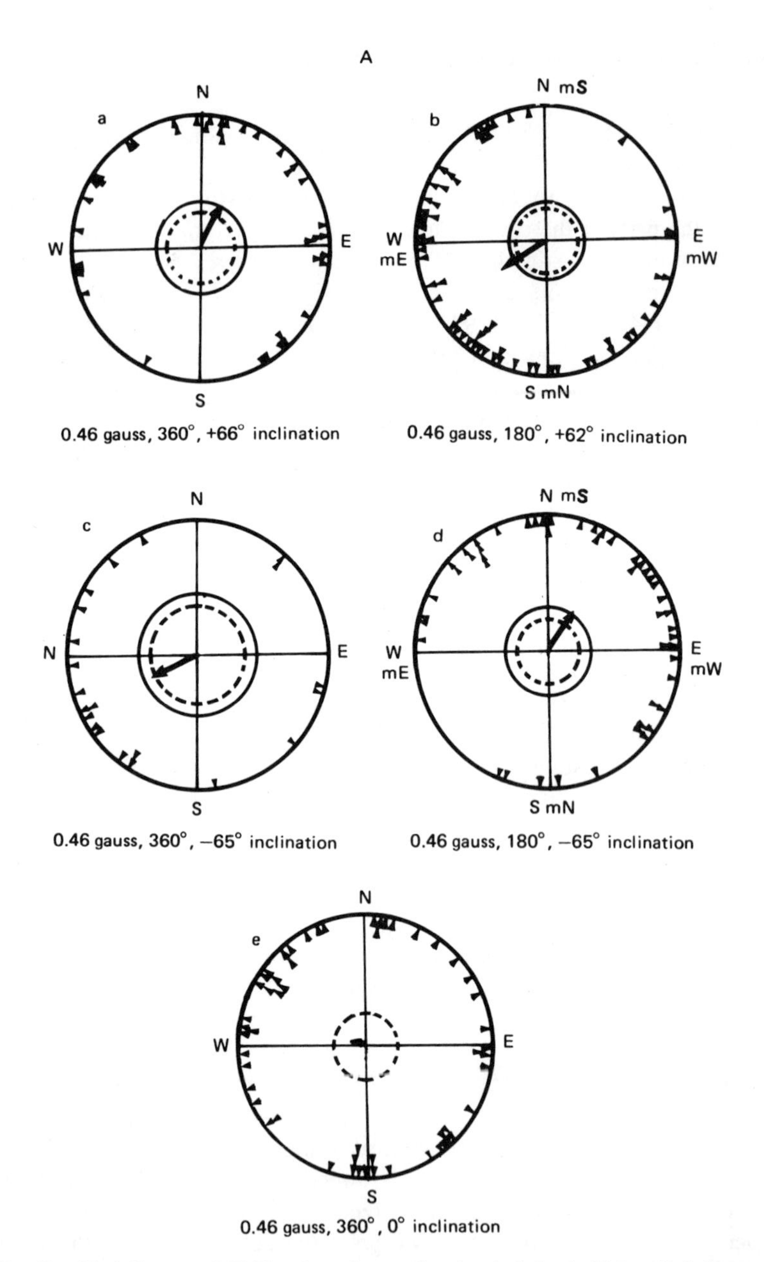

Fig. 69. Possible influence of GMF and gravity on directional choice in birds. (A) Influence of artificial magnetic field on directional choice in birds. Experiments in spring, 1971. Mean of single test nights (triangles); mean vector comprising single test nights (arrow); 5% (dashed) and 1% (solid) significance border of the Rayleigh test (inner circles). (a) Tests in earth's local magnetic field; (b) tests in arrangement with horizontal component reversed; (c) tests in arrangement with vertical component reversed; (d) tests in arrangement with polarity (both components) reversed; (e) tests in arrangement with only horizontal magnetic field present. (B) Section through magnetic field in plane parallel to force of gravity and north–south axis. g, force of gravity; H_e, local GMF vector; H_h,

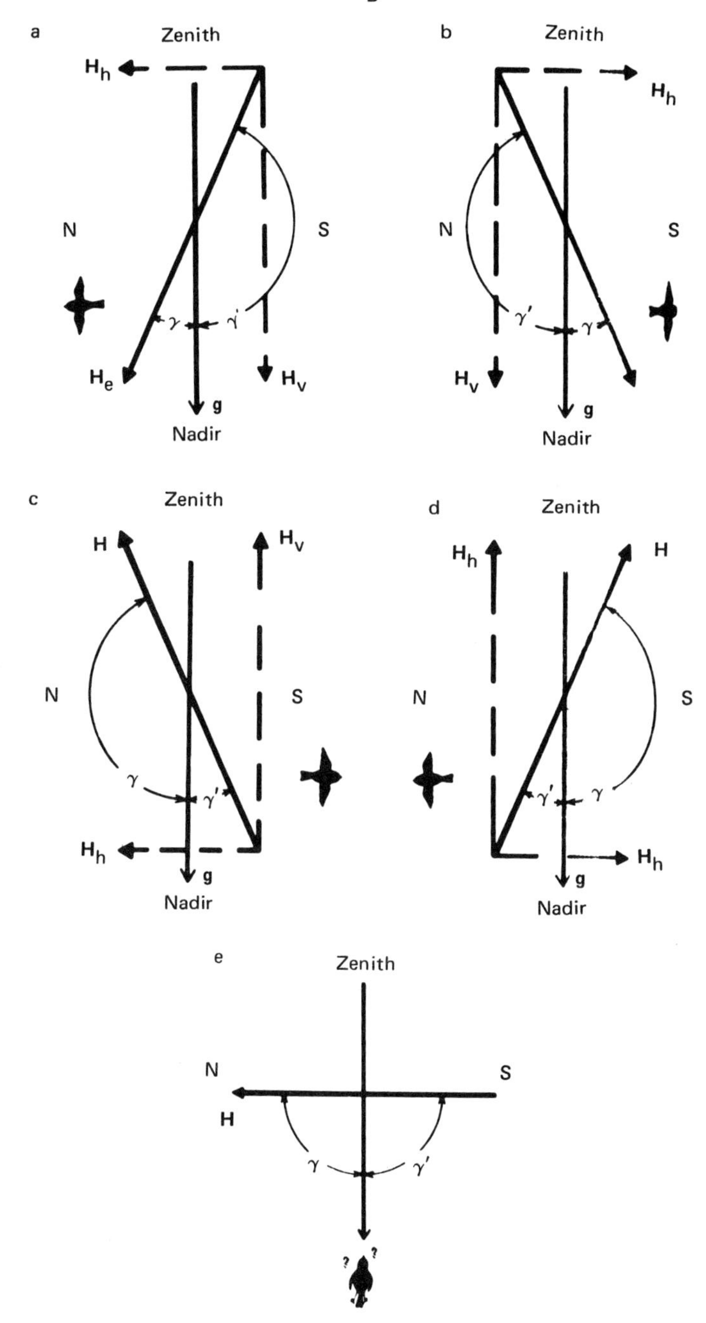

horizontal component; H_v, vertical component; **H**, experimental magnetic field vector; γ, angle between magnetic field vector and gravity vector; γ', supplement of angle γ. (a) GMF in Frankfurt. (a, b, c, d, e) as in 69A. From (1221).

vertical component, and hence there was no dip of the magnetic field and the angles were equal, the birds were unable to distinguish the polarity in the axial magnetic flux of field lines. This artificially created magnetic field destroyed their orientation, and hence the choice of direction was completely random (Fig. 69A,e).

Ornithologists have certainly taken only the first steps in the investigation of this complex and difficult problem. It will be necessary in the future to account not only for the operation of the biocompass system of migrant birds when they cross the equator, where the dip has zero value and the total GMF intensity drops to 0.3 G, but also for the data of various researchers who indicate that birds are capable of using stars as reference points. Even the choice of route to the migration destination, which is almost always determined and remains the same from year to year for each species of migrant bird, is also a very puzzling phenomenon. It is quite possible, in our opinion, that to account for the many puzzling phenomena in the problem of long-range migration of birds (and possibly of fish too) researchers will have to take into account the "skeleton network" of the earth, which is an actual system of specific geophysical fields of still undetermined nature (883, 934).

Thus, the magnetic dip and gravitational force could be factors that provide a fixed system of coordinates, since the number of points with two equal parameters does not exceed four—the points of intersection of the isoclines and the geographical parallels. There are several new hypotheses then, whose experimental verification in the near future can help to resolve this complex problem. It is worth noting that some investigators think that the problem of bird orientation cannot be solved without consideration of the possible effect of the GMF (102, 103).

The above account indicates that further intensive research will be required to reveal the role of the GMF in bird navigation. The problem of orientation of birds, their navigational abilities, and their migration behavior and flights is extremely complex and involves evolutionally consolidated instincts, phylogenetic developmental features, and the complex interaction of internal endogenous hormonal processes, and various external physical factors, whose role is still very obscure.

Fish

The response of fish to electric fields, their high sensitivity to electromagnetic fields, and unique ability to generate "living electricity" are widely known (449). For a long time there have been suggestions that the geomagnetic and geoelectric fields in seas and oceans may be important factors for the orientation, navigation, and behavior of fish (346, 881, 1062). Ocean currents, carrying vast masses of salt sea water intersect the GMF lines of force and create electric currents that can be perceived by fish (796, 1093, 1101, 1195). Geomagnetic disturbances associated with solar activity can also give rise to telluric currents in seas and oceans (452). Another suggestion is that fish can sense electric currents and fields generated in their own bodies as they move through the GMF (62, 63, 961, 1169). Thus electrical phenomena in seas and oceans are attracting more and more interest from scientists of different specialties. Of interest in this connection is the hypothesis that marine telluric currents play a key role in fish migration (346) and data indicating a correlation between fish catches and geomagnetic and solar activity. A test of these data (452) failed to reveal a definite correlation between magnetic activity and fish catches, but the authors nevertheless noted the presence of a direct relation between catches and magnetic activity in autumn and winter. In the opinion of these authors the presence of a correlation between magnetic activity and the behavior and distribution of fish in particular areas of water merits the very close attention of scientists. This is definitely another example of an important role of the GMF in working life, in this case for persons engaged in commercial fishing in seas and oceans.

In addition to theoretical studies there has been an ever-increasing number of investigations in which direct reception of the GMF by fish has been tested and detected (170–174, 231, 233, 235, 1094, 1095, 1159, 1228). Careful experiments have shown that the main role in fish orientation is played by the GMF horizontal component (233) and, in particular, the north part of the GMF horizontal component, i.e., the X component (1158). Tesch (1158) found that the choice of direction by fish was greatly altered if the north part of the GMF horizontal component were compensated by Helmholtz coils (Fig. 70). He found that the water in which the investigations on the fish (eels) were made—fresh water or sea water—was of great importance. An analysis of the existing studies of the effect of the GMF on fish orientation shows a very complex and very contradictory picture since, in fact, only the first steps in the study of this question have been taken so far. For instance, one investigation showed that artificial reversal of the north horizontal component did not alter the behavior in comparison with control fish, nor did compensation of the GMF vertical component have any effect (233). Another very careful investigation by another author, however, showed that reversal of the horizontal component, particularly in experiments in fresh water, led to differences in the choice of direction by fish (1159). Another study (1228) revealed no effect of the GMF on fish orientation. It is apparent that long-term and comprehensive experiments will be required to reveal the reasons for these contradictions in the work of different investigators.

These features and the ability of fish to orient themselves in weak magnetic fields led to hypotheses attributing the orientative ability of migrating fish to the GMF. These hypotheses were based on the production of induction currents in the fish's body (489) or changes in current configuration in the total field of the shoal formed by fish during migration (449). There have been few experimental

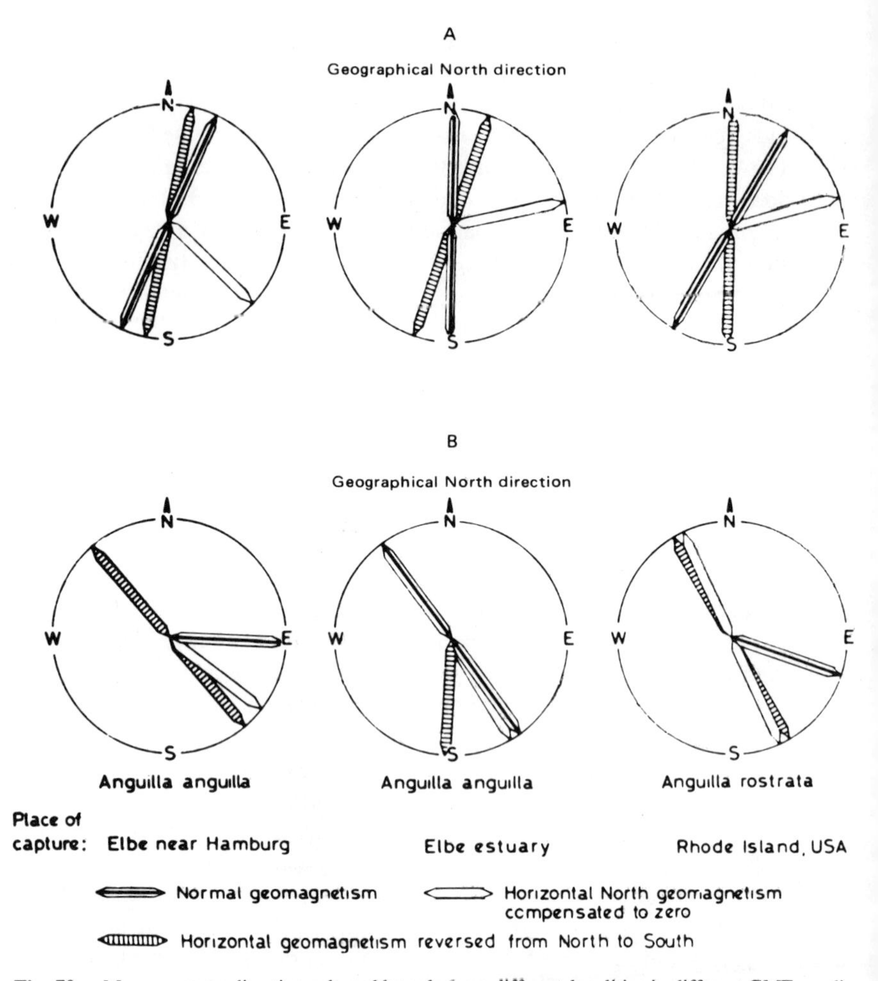

Fig. 70. Mean compass direction selected by eels from different localities in different GMF conditions. (A) Experiments in sea water; (B) experiments in fresh water. From (1159).

studies, however, in which the response of fish to the GMF has been directly investigated. Yet ichthyologists indicate that fish can navigate in the open sea. They can set a course in long migrations and they show the same features that we discussed in the description of bird migration. For instance, fish undertake annual migrations over distances of up to several thousand kilometers. Fish are capable of homing and this ability is not appreciably affected by light intensity, temperature, or salinity. In addition, in a circular body of water they correctly determine direction without any astronomical or hydrological reference points (175, 407, 991, 1157, 1158).

There is no doubt that the lack of data on the actual route of fish during migration and its relation to specific ecological conditions is impeding the study of fish orientation. Even the few experimental investigations, however, clearly show the great importance of the GMF for fish orientation. In one experiment a long-range method of observing migrating fish was used (430, 431). The movement of fish tagged with light floats attached by a Silon line to the dorsal fin was followed from dinghies and a ship. The exact position of the float was determined repeatedly (several times an hour) by sextant and bearings, the actual path of the fish was plotted on a chart, and its speed was calculated. An experiment with one fish lasted several hours to several days, and in toto there were 607 daily and 117 nightly observations of 93 individual fish. The fish used in the experiment were large bream (1–1.5 kg in weight) of the Volga local shoals in Rybinsk Reservoir. The author of this unique investigation noted that the local fish after release from the container swam initially in a circle, then they selected the required direction, and subsequently continued in this direction without deviating significantly to either side. The fish transported from other regions and released in an unfamiliar environment often altered direction after release, and then begin to swim with or against the current according to their direction of motion relative to the current in their own region when they were caught. The direction taken by fish transferred to an unfamiliar region coincided with the direction of the magnetic meridian in 87.5% of the cases, and in 50% for local individuals. This orientation relative to the GMF occurred in a part of the migration route where the fish were traveling through water with uniform hydrophysical indices and was manifested more frequently, the greater the number of stable factors. In motion along the magnetic meridian the speed in 90.5% of the cases differed from that in the preceding part of the migration route, and in 60% was greater. By variance analysis of the collected data, Poddubnyi (430, 431) assessed the role of all the environmental factors we have considered in the orientation of migrating fish. In his opinion temperature, electrical conductivity, and depth are the main factors in short-range orientation of bream, but when the fish is moving in a uniform environment these factors are unreliable and are replaced by others, particularly orientation relative to the GMF.

In addition to investigation of the role of the GMF in the orientation of fish in natural conditions, attempts have been made to study the role of the GMF in strictly controlled environmental conditions (62, 63, 170–175, 292, 595, 693, 961, 1095, 1159, 1228). For this purpose the frequency of selection of three compass directions by fish was determined in a specially designed maze in which the paths were parallel to three axes lying in a plane at an angle of 120° to one another. The setup was essentially as follows (USSR Patent, Author's Certificate No. 300147). The experimental investigation of orientation is usually hindered by the tendency of the animal to move mainly along the walls. The maze, however, was composed of walls alone. Simple Y-shaped mazes, consisting of three branches, have been used in physiology before. The maze under discussion consisted of a system of Y-shaped junctions, in which the fish could wander endlessly and continuously from junction to junction. Each junction ended in a fork, where the animal had to choose its subsequent route: to the left or right. As experiments showed, in one hour of experiment each fish could make more than 1000 choices indicative of its orientative behavior. This gave more than 1000 bits of information for each fish per hour, which ensured high statistical significance of the results and allowed effective use of a computer. The main results of the experiments are given in (171–173, 175, 233, 234, etc.). These results provide a picture of two phenomena investigated: responses to the GMF and orientation relative to the GMF.

The response to the GMF was revealed in the anisotropy of the locomotive behavior of the eel in relation to compass direction. The anisotropy was due to selection of some axes (e.g., the west–east axis) and avoidance of others (usually the north–south axis). This anisotropy depended on the age of the eels (young eels, three–five years old, were used in the experiments), and also on the latitude of the place where the experiments were conducted. It has been impossible so far to link the specific choice of compass direction with any navigational corrections. The response, i.e., the anisotropy of the behavior, disappeared when the GMF in the maze was compensated by Helmholtz coils: The selection of compass direction became isotropic in this case. The eel's response to the GMF could be suppressed by preexposure of the fish to a powerful magnetic field (2700 G) for 1 h. The subsequent behavior of the eels regarding selection of compass direction was isotropic, at least for 40 days after suppression. The experiments also revealed a "gravitational effect." The manifestation of this effect depended on the position of the maze in relation to the magnetic meridian. The maze generally allowed movement along only three axes at angles of 120° to each other in a plane. If one position of the maze led to anisotropic behavior, rotation through 30° (in any direction) led to isotropic behavior. This gravitational effect was manifested in the absence of response to some gradations (angles of application) of the stimulus, which is the direction of the GMF lines of force. In addition, in experiments in which the gravitational effect was detected there was an upper limit to the angular sensitivity of the eel to the direction of the GMF lines of force: In every case it was not more than 30°, but the lower limit is still unknown.

We note also that, according to the authors' data, anisotropic behavior and the gravitational effect are also observed in other species of migratory fish, particularly the Azov anchovy, but a response to the GMF has so far only been found in the European eel. According to the data of the same authors, the adult European eel shows much greater anisotropy of behavior in the selection of compass direction, but proof of its magnetic response will require very special techniques.

The experiments were conducted in mazes in an isolated room at constant temperature and light intensity, with the GMF acting or compensated by Helmholtz coils. Two mazes were oriented with one of their axes at an angle of 0 and 30° to the magnetic meridian, which allowed the investigation of six compass directions. The fish—10- to 15-cm-long elvers—were released one at a time into each maze, where they were kept until they had made 250 directional choices, after which they were replaced by new fish. A total of 11,000 observations on 44 individual eels was made at the latitude of Leningrad. The experiments showed that in normal GMF conditions the eels mainly selected the 30–210° direction (SSW–NNE) and avoided the 90–270° direction (E–W) (reliability of 0.99 by Student's *t* test). When the GMF was compensated the movement was random and the choice of any compass direction was equally probable. Investigations by this method were repeated in Odessa and Kaliningrad and confirmed that eels are receptive to the GMF and can select a particular direction of motion at any geographical point (63). In these investigations the maze was positioned in such a way that one of the three directions of motion of the eels made a right angle with the GMF horizontal component at the site of the investigation. This direction corresponded to the course W–E, and the other directions corresponded to the courses NNW–SSE and NNE–SSW. Twenty-four eels were used in the experiments and each spent 5 min in the maze. Summation of the frequency of selection of each of the three directions in the mazes gave 1460 for W–E, 1176 for NNW–SSE, and 903 for SSW–NNE. The probability of more than 5% deviation from the mathematical expectancy (1180) is about 10^{-7}. In the W–E and SSW–NNE directions, the deviation was 24 and 31%, respectively, which indicates unequal probability of the selected directions and a preference for the W–E direction. It should be noted that the data for Leningrad (63) differed from those cited above: Out of 3851 observations 1264 were W–E, 1181 were SSE–NNW, and 1406 were SSW–NNE. When the mazes were situated inside Helmholtz coils and the GMF was compensated the frequency of appearance in each of the three directions was almost the same—1260, 1228, and 1222, respectively. All three figures lay within the confidence interval corresponding to a significance criterion of 95 or 98%.

A careful investigation with American eels (*Anguilla rostrata*), however, by the above-described method involving an orientation dial and a hexagonal maze showed no effect of the GMF horizontal component on the choice of direction by eels (1228). The authors reported that the position of the horizontal vector—along or across the eel's body—had no effect. In addition, complete compensation of the GMF horizontal component did not affect the choice of direction by the eels, but a weak electric field (10^{-2} $\mu A/cm^2$) had a significant effect. Why this investigation should have given negative results for GMF reception is still not clear, but the authors think that the negative results might be due to age-dependent physiological features.

Thus, for the first time there have been experiments that distinctly show that fish can use the GMF for orientation in space. We can infer that this ability is developed at present in forms of living organisms that are more sensitive to external electromagnetic fields, are more closely related to these external environmental factors, and undertake long migrations. Local shoals of fish, which move over short distances, gradually lose these powers or use them extremely rarely.

Various Representatives of the Animal World

Very carefully conducted investigations, like those of Brown (751, 752), on the role of the GMF in the vital activity of all kinds of organisms are so numerous and adverse (with regard to the processes investigated, groups of animals studied, and methods used) that it is impossible within the scope of this book to give a full account of the results. Since these investigations have been described in detail in the foreign and Russian literature (238, 242, 446, 735, 736, 750–752), we will confine ourselves here to a brief summary of their results. Magnetobiologists have shown experimentally for the first time the role of the GMF in the rhythmicity of functional processes and the spatial orientation of animals, and have thus laid the contemporary theoretical and experimental foundations of geomagnetobiology.

Experimental studies have led to the conclusion that the GMF is a very important space–time coordinate for living organisms (748, 750, 752). This conclusion in its axiomatic form, expressed as long ago as the 1930s (295, 493, 868), has been experimentally confirmed by investigations with artificial (15, 16, 22, 46, 249, 250, 258, 259, 312, 313, 317, 322, 363, 364, 399, 560–563, 624, 705, 706, 711, 712, 1028, 1029, 1037–1047, 1074, 1203–1208) and natural magnetic fields (7, 96, 121, 122, 135, 137, 142, 172–174, 222, 243, 276–282, 297, 393, 569, 570, 629–636, 691–693, 705, 706, 738, 739, 811–815, 969, 1006, 1036, 1064, 1065, 1132, 1142, 1148, 1168, 1195, 1216–1219).

Investigations employing various biological objects (potato, carrot, earthworms, *Drosophilia,* snails, crayfish, salamanders, white mice, etc.) reveal that the metabolism of all organisms shows a solar- and lunar-day periodicity (736, 741, 749, 751, 752). Since the investigations were conducted in constant-pressure and -temperature chambers, where the indices of the external environment were constant, the only regulators of the indicated periodicity could be the GMF and gravitation.

A thorough investigation along these lines was carried out on Mongolian gerbils (*Meriones inguiculatus*) (1148). Using an automatic recording device, the author investigated the spontaneous activity of the animals throughout each day from May through October, 1968, with environmental factors constant. A retrospective analysis of the statistical correlations showed that there was a strong positive and highly significant correlation between the general activity of animals between 1500 and 1800 h and the variation of the GMF horizontal component in the locality where the experiments were conducted ($r = 0.38–0.51, p = 0.04–0.01$).

The author concluded that the period between 1500 and 1800 h is the most important for the activity of the animals investigated and their vital activity in this period is determined directly or indirectly by the strength of the GMF horizontal component: The greater the intensity of the H component, the greater the activity of the animals in this period. The association of the most pronounced effect of the GMF on biological objects with a particular time of day is very interesting. Enhanced sensitivity of animals and plants at particular times of day has been reported before in the case of spontaneous activity of animals, respiration of biological objects (735, 751), permeability of roots and plant tissue cells (134), and so on. Our investigations with plants showed that the effect of an artificial steady magnetic field (40 G) on the dynamics of excretion of organic substances by plant roots was greatest in the period between 1000 and 1700 h.

What we have said above suggests that in the natural geophysical complex, which affects the rhythmicity of the vital activity of animals and plants, there are periods when the effect of a particular factor of the geophysical complex is reduced or enhanced, or another geophysical factor begins to act or becomes more active (134). As we indicated previously, we have the combined effect of the GMF, gravitation, and other factors. The effect of gravitation is very important: At some times of the day the GMF and gravitation act in the same direction and then the object becomes sensitive, while at other times their effects are oppositely directed and the object becomes less sensitive, since they act in opposite ways on the permeability and polarization of biomembranes (134).

This natural complex, however, contains other important geophysical and cosmic factors that affect the living organism. For instance, it has been suggested in several investigations that an important agent affecting the state of living objects and inorganic substances is some unknown kind of radiation (92, 729, 999, 1000, 1098, 1153). These studies merit great attention since they have been carried out very carefully and regularly for many years (10–15 years in succession) and involve a vast mass of experimental material. Hence, methodological or psychological errors due to delusion, which is known in science, are ruled out. A recent study in this direction was concerned with the circadian rhythm of the physicochemical properties of nickel-coated slides covered with a layer of antigen, which are used for the conduction of immunological reactions at a liquid–solid interface (1098). The author found that if such slides were subjected to the action of the north pole of a steady magnetic field of several thousand gauss with the field lines perpendicular to the metal-coated surface, the thickness of the next absorbed layer of antibody was almost doubled (from 40 to 80 Å; the author called such slides "activated") in comparison with the results obtained when the magnetic lines of force were parallel to the slide (nonactivated slides). The investigation revealed that activation decreased during the day, from morning to 1100 h (40 Å). Then, from 1800 h onward it increased, reaching a maximum by 2000 h (70–80 Å), after which it remained constant throughout the night, and in the morning the cycle began again. Similar results were obtained with the slides when the experiments were carried out 25 miles north of New York, outside the walls of the Rockefeller Institute. It was found that the slides could be kept active during the day by placing them inside a metal shield consisting of a steel tube with 0.15-cm-thick walls and 3.5 cm of lead. On the basis of experiments with gamma radiation the author concluded that activation of the slides by night and inactivation during the day were due to penetrating radiation emanating from the sun (the soft cosmic-ray component).

We think that variation of the GMF in the course of the day plays a significant role in the described activation and inactivation effects. The above-described effects and the conclusions regarding them are very similar to the observations and conclusions of the well-known Japanese scientist Moriyama, who has for decades carefully investigated the effects of an unknown cosmic radiation (X agent) acting on biological objects and inorganic matter and physicochemical processes (999, 1000). He believes that the X agent is electromagnetic radiation (gamma rays, X rays, ultraviolet, radiowaves, light and heat rays) with unusual properties, like an abnormal photon. The photons become abnormal when two photons—a positive one (posiphoton) and a negative one (negaphoton)—unite to form a pair. The photon pair (X agent) is absorbed by any substance, particularly by colloidal solutions and dyes, e.g., neutral red. Most normal photons fail to reach the earth's surface, but the abnormal photons easily penetrate and cause their effect. The photon pair is annihilated when it is absorbed by matter and gives rise to an electron pair. The electron pair (a positron and electron) is broken up in turn by heat, giving rise to free electrons, and thus all substances are subjected to the action of free electrons. Photon pairs are produced on the sun and are emitted by it, but stars also emit a similar radiation (999).

We have analyzed the data of scientists who have reported unknown radiations acting on biological and inorganic processes (128) and found that a large proportion of the effects ascribed to unknown radiations can be attributed to the diurnal variation of the GMF. In particular, the sharp changes in the state of biological and inorganic reactions during the day (from 0600 to 1800–2000 h) and the constancy of these reactions during the night (from 2000 to 0400–0500 h) can be attributed to the different state of GMF activity in these periods: In the first period the GMF elements undergo sharper changes than in the second period, when changes in the GMF elements are insignificant. In addition, as evidence in favor of our views we compared in world time the diurnal rhythm of respiration of potato tubers, which was studied by Brown in automatic respirometers in Evanston, Illinois, in 1956–1966, and the diurnal course of hydrolysis of bismuth chloride, investigated in Florence in 1967–1968 by Piccardi. We found that the diurnal courses of these two completely different reactions were almost completely identical (except at 0700 h, when complete counterphase was observed), although it was day according to local time in one locality and late evening in another, and hence the hypothetical active emission did not come directly from the sun. Since the biological and physicochemical reactions investigated at different geographic points (Figs. 9 and 21) have a synchronous course in world time, we can infer that their changes are subject to the effect of a global, all-pervasive factor having unitary world variations, which could be the GMF and gravitation. The synchronous variation of biological reactions and physicochemical processes indicates an identical response of inorganic and biological colloids to the action of the GMF, the geoelectric field, atmospheric electricity, and gravitation. The reason for this could be a change in the properties of water, which is a component common to all these reactions, in response to natural electromagnetic and gravitational fields (139, 260, 420, 815, 1049).

In this case the unknown forms of cosmic radiation postulated by scientists become understandable and obtain a real physical basis. In fact, if as Moriyama (999, 1000) claims all these unknown forms of radiation are included, according to their properties, in the X agent, then we on our part can attribute a large number of the properties of the X agent to the action of individual elements of the natural geophysical complex. A short list of the properties of the X agent clearly confirms our view: (1) the continuous variation of activity—diurnal, seasonal, annual; (2) the activation of substances of any kind—from aqueous solutions to glass, from kaolin to erythrocytes; (3) the effect of compass orientation on the variation of activity; (4) the specific enhanced sensitivity of colloidal solutions to the action of the agent; (5) the relation between the agent and atmospheric pressure; (6) the presence of two components—a hard and a soft one—in the unknown radiation (the soft one can be stopped by thin foil or even paper, while the hard one requires an iron shield and a thick lead layer), and so on.

It is apparent from this short and far from complete survey that some effects caused by the X agent and other unknown kinds of radiation can be largely attributed to the action of individual components of the magnetic and electric fields of the earth and the atmosphere, low-frequency and high-frequency radiowaves, and even sound waves in different ranges. Resonance effects in all these cases can play a decisive role in the interrelation of the natural geophysical complex and biological and inorganic reactions.

We do not reject, however, the existence of still unknown fields and radiations, particularly since in one of our own works (815) we have suggested that one form of polymorphic action on living and nonliving matter could be the gravitational emission of the sun and other cosmic objects. As evidence in favor of this view we analyzed the data on gravitational emission given in the papers of the well-known American physicist Joseph Weber, which he kindly communicated to us, and the data of Moriyama on the effect of the X agent on the optical density of cooled quartz glass (999). In a number of cases the two scientists, who investigated effects of a different nature, noted the occurrence of the emissions they investigated on the same dates, e.g., February 5 and March 21, 1969. Thus our suggestion of a possible gravitational emission acting on biological and inorganic substances receives some confirmation. Gravitational emission interacting with various substances on earth is quantized in various kinds of known electromagnetic radiations, and hence investigators studying unknown kinds of emissions find that their properties recall the action of gamma, x, and ultraviolet rays, etc., since this is the result of quantization of the gravitational emission after action on matter.

It is apparent from the above account that the problem of circadian response of living organisms to the GMF and other environmental physical factors is by no means as simple as appears at first sight. The treatment of this problem will first require determination of the role of the GMF and gravitation and then an investigation of the role of unknown kinds of radiation, but this, as shown above, will require consideration of many aspects of quantum and relativistic physics, cosmology, and several other areas of contemporary physics. This is why the problem of circadian response of living organisms is still far from settled, but the GMF is apparently one of the key factors in this intricate heliogeophysical complex. For instance, geomagnetic activity affects a very important functional index of the reproductive organs: the dynamics of excretion of the luteinizing and follicle-stimulating hormones of the pituitary in the sable (416). These relations are certainly a feature of all animals and, accordingly, the fecundity and numbers of animals are affected.

Subsequent experiments also showed that organisms were sensitive not only to weak magnetic fields (0.5–1 G), but also to electric fields, and that various organisms could orient themselves in compass directions in the complete absence of any external reference points (714, 855, 1031, 1075).

The GMF affects the rhythmicity of all processes occurring in the living organism and the biosphere as a whole. The controlling effect of the GMF on rhythmicity in plants and animals is clearly revealed by the method of direct comparison (53, 122, 124, 270, 324, 325, 648, 729, 740). It is an interesting fact that the diurnal rhythm of radiosensitivity, i.e., variation of the biological effectiveness of ionizing radiation over a period of 24 h, an effect reported in many studies by radiobiologists, depends on the variation of the GMF (Fig. 71).

Thus, the data discussed above indicate the great importance of the GMF for animal life. The GMF by its very nature can give living organisms all the required space–time information. The GMF determines orientation in space, the rhythmicity of processes, and has a strong effect on animal physiology and biochemistry.

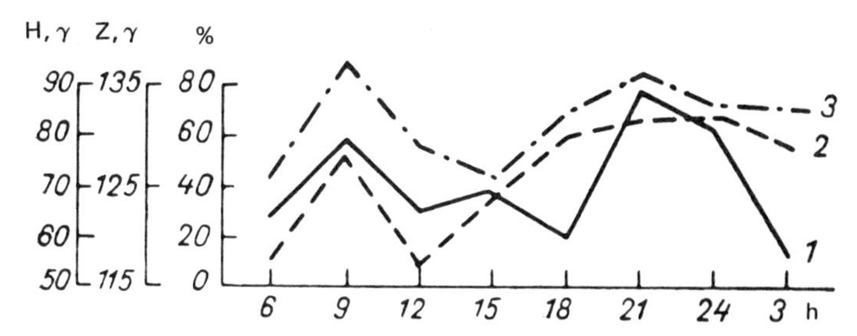

Fig. 71. Diurnal rhythm of radiosensitivity of mice and variation of GMF, September 10, 1964 (127). (1) Mortality of animals (%); GMF components: (2) horizontal (H); (3) vertical (Z).

Plants

Previous sections of the book have shown that it is very difficult to determine the role of the GMF in the vital activity of animals. The complexity of hormonal relations in the organism, evolutionally consolidated instincts, the special features of population interrelations, the almost unlimited movement in space, the continuous interchange of ecologically important environmental factors, and many other factors hinder the experimental investigation of the role of the GMF in animal life. These features have less effect in the case of plants, which makes it easier to determine the effect of the GMF on them. Plants clearly reveal the diverse effects of the GMF on living organisms.

Magnetotropism and Orientation. Studies of the physiological responses of plants to the GMF and artificial magnetic fields have led to the discovery of geomagnetotropism (295–297, 393, 673, 674, 1074, 1168). When plants grow freely in space, parts of them have a particular orientation relative to the geomagnetic poles. In particular, the roots grow mainly in the direction of the south magnetic pole and, accordingly, the properties of plants are altered by orientation in this way. A similar effect was discovered earlier when an artificial magnet was used (1074), the only difference being that the south pole inhibited root growth.

The discovery of magnetotropism led to studies of the effect of spatial orientation of seed embryos on the functional and biochemical properties of plants growing from them (29, 30, 94, 116, 184, 293, 294, 368, 369, 376–378, 394–396, 1055–1066). It was found that the rates of germination and growth were greatest when the seeds were oriented in the direction of the earth's south magnetic pole. This is illustrated by the data in Table 9, which were obtained in experiments with wheat seeds (556).

Geomagnetic orientation of seeds affects not only the growth, but also very diverse functional and biochemical characteristics. This effect is ultimately reflected in crop yield. The discovered relation is very clearly revealed by

Table 9. Effect of Orientation of Seeds Relative to Geomagnetic Poles on Growth of "Krasnozernaya" Wheat Seedlings

Orientation of embryo radicles relative to earth's magnetic poles	Mean length of seedlings, mm		Value of t between M_S and M_N		Significance level p	
	Stem	Root	Stem	Root	Stem	Root
Toward north pole (M_N)	65.8 ± 2.9	236.4 ± 9.5	3.5	5.4	0.001	<0.001
Toward south pole (M_S)	81.5 ± 3.3	312.5 ± 10.2				

geomagnetic orientation of seeds of rye, cucumber, beet, sunflower, pea, melon, barley, oats, wheat, spruce, fir, and other crop plants (13, 29, 30, 184, 188, 399, 441, 567, 570).

Conifer seeds sown with their embryo radicles oriented toward the south germinated 4–5 days earlier than seeds oriented toward the north (441). The time of sowing of the seeds in relation to the phase of the moon was also of great importance. When seeds were sown with their embryo radicles oriented toward the south at full moon the germination rate was increased, but at new moon this effect was less pronounced.

Investigations also revealed that the response of plant seeds to orientation relative to the geomagnetic poles depended on the type of dissymmetry (right- or left-handed) of these seeds (368, 369, 376–381, 399, 512, 544, 550). Left-handed seeds oriented with the tip of the embryo radicle toward the earth's south magnetic pole produced plants with higher growth rate, greater enzymic activity, higher respiration rate, higher chlorophyll content, and accordingly greater yield (13–52%). At the same time, right-handed fruits with the embryo directed toward the earth's north magnetic pole produced plants with higher rates of growth and other physiological processes (7–32%). In addition, the response of left- and right-handed forms of plants to geomagnetic orientation was very complex and also depended largely on the investigated index (107). This indicates that the interrelation of living organisms with the GMF depends on both internal and external factors. Investigators have encountered a similar situation in studies of orientation of plant roots.

The earliest investigations of the orientation of the root system of plants revealed that the side roots in free conditions lay in a north–south direction. Experiments with radioactive isotopes confirmed this and showed that the uptake of labeled phosphorus from the soil was greater on the north and south sides than on the east and west (1055, 1058, 1059, 1063). It was also found, however, that in some of the plants growing in the same field the direction of growth of the roots was different (1055).

Table 10. Effect of Direction of Sowing on Root Orientation of Winter Wheat Grown in Field Conditions, Lethbridge, Alberta, Canada

Direction of sowing	Direction of root growth	Number of plants with root system oriented in a particular direction				
		1958	1959	1960	1961	Mean
North–south	North–south	72	80	72	69	73.4
	East–west	0.0	4.5	8.0	7.8	5.1
	Various ways	28.0	15.5	20.0	22.4	21.5
East–west	North–south	63.8	81.0	71.0	68.6	71.1
	East–west	7.6	5.5	8.0	6.8	7.1
	Various ways	28.3	13.5	21.0	24.6	21.8

Table 10 shows that in field conditions the root orientation of the plants was predominantly north–south, but there were a certain number of plants with different root orientation. Subsequent investigations with wild plants in laboratory and field conditions confirmed these facts. In particular, it was found that in 175 wild oats plants grown in four geographical regions in the central part of North America the root system lay predominantly in the north–south direction. In contrast to this, the roots of rye (*Secale cereale*) sown in the field lay in different directions. However, despite the discovered complexities in the manifestation of magnetotropism in plants, the actual fact of an important role of the GMF has been very clearly demonstrated (29, 30, 393–395, 398, 399, 550, 551, 1055–1066). Various experiments—rotation in a horizontal plane, rotation on a clinostat, subjection to artificial magnetic fields (10–3000 Oe), cultivation in metal and polythylene containers—lead to the conclusion that the roots orient themselves approximately parallel to the horizontal GMF component (1055–1066). This conclusion was made for winter wheat grown in field conditions at different geographic points in Canada, where the magnetic declination was 20°30′ (Lethbridge) and 28°30′ W (St. John). At these points the roots grew approximately in the same direction relative to the GMF horizontal component: they deviated eastward from true north in Lethbridge and westward in St. John. The coefficient *D*, which indicated the direction of root growth, was 88 and 85, respectively, for these points (for complete coincidence with the meridian it was 100). It should be noted that at both points the direction of root growth was different when the containers were rotated daily through 90°. The data illustrating the variation of the coefficient *D* of oat plants grown in different regions of North America (1061) are of interest (Table 11). For instance, plants grown east of the zero line of magnetic declination had a coefficient *D* that differed a little from the

Table 11. Variation of Root Orientation of Wild Oats Plants Grown Randomly in Different Regions of North America

Geographic region	Magnetic declination		Mean value of D^a	
			Theoretically calculated	Actual plants
Pacific Ocean	23°E	20°E	76	73.3 ± 0.7
Midwest	20°E	5°E	86	84.7 ± 1.1
Great Lakes	5°E	0°	97	91.3 ± 2.2
East	0°	15°W	107	95.6 ± 1.6

[a] Calculated from $D = 100 - 5(d_N - d_S)/9$, where d_N and d_S are the declination in degrees (±) from true north and south.

theoretically calculated value. The author believes that the reason for this could be local anomalies in industrial regions east of the zero line, which could disturb the magnetotropic response of the plants.

Of interest in this connection are studies of growth (565–570) and root orientation of root growth in natural anomalous regions (398, 399). In the conditions, for instance, of the Kursk magnetic anomaly there was much greater dispersion of orientation of the root creases of sugar beet (Table 12).

The orientation of other parts of the plant in space is of interest. Some authors (468) conclude from careful experimental studies and a survey of published data that plants exhibit diverse orientation, which depends on many factors: specific and varietal characters, nature of ontogenic development, adaptative features, etc. Yet several studies indicate a particular spatial orientation of parts of plants. For instance, sunflower heads are constantly oriented in a particular direction. This orientation is already fixed before bud formation, i.e., before opening of the marginal ligulate florets, and is retained in the flowering heads. The bulk of the head is turned in a fairly broad fan toward the sun—to the south, southeast, and southwest. The potato has a larger number of leaves in the southeast direction (494, 495). There is evidence (1) that the skeletal branches of the Chinese arbor-vitae, which project east and west from the trunk, fork in vertical planes mainly (67–90% cases) in the north–south and east–west directions (30–50%).

It has been suggested, and in some cases supported by direct and indirect evidence, that the GMF can affect various other properties of plants: polarity, sex determination, dissymmetry (1, 3, 106, 107, 125, 126, 295, 348, 550–552, 586, 674, 794).

We must also mention the careful investigations, using very sophisticated techniques, in which the authors failed to detect an effect of the GMF on germi-

Table 12. Direction of Root Creases of "Ramonskaya" Sugar Beet at Some Points in Kursk Magnetic Anomaly in 1969

Investigated point	Angle between direction of magnetic meridian and furrow, deg	Declination of magnetic needle, deg	Total number of plants	Proportion of oriented root creases, %				
				N–S	W–E	NE–SW	NW–SE	Abnormal roots
Yakovlevo	6	26	1134	48.8	30.8	3.8	9.3	7.3
Gostishchevo	0	13	1797	50.6	29.6	9.3	7.2	3.3
Teterevino	22	38	2158	38.7	27.6	10.6	18.2	4.9
Shoshino[a]	60	48	1086	39.3	26.9	7.6	8.3	17.8

[a]The variety "Yaltushkovskaya" was sown.

nation, rate and amount of growth, change in biomass, or direction of root growth of wheat seedlings when the seeds were oriented relative to the geomagnetic poles and/or the GMF was altered by Helmholtz coils (900, 901). The authors reported, however, that alteration of the GMF by Helmholtz coils led to some teratological and pathological disorders in the seedlings and that this effect on the seedlings could cause acceleration or inhibition of growth (901). The reasons for the general absence of a biological effect of the GMF in these studies were not determined.

Sex Determination. The problem of the sex ratio in progeny is one of the most important in biology. This also applies to plant forms, although here there is a certain specificity. More than 90% of botanical species are hermaphroditic and have flowers with female and male organs, or are diecious, i.e., some plants have flowers with male organs—anthers, and other plants have flowers with female organs—pistils. Hence, it is difficult to determine the effect of the GMF on sex determination in such plants. However, there is a third type of plant: the monecious type, where each individual plant has male and female flowers. Experiments with monecious plants, e.g., cucumbers, have shown that orientation of the seeds affect sex determination. If the embryonic radicle of such plants is oriented toward the north, a greater number of female flowers is formed than in the case of seeds oriented toward the south (1, 3, 11). Since cucumber fruits are produced from the female flowers, northward orientation of the seed radicle will lead, of course, to a greater yield per plant. Although the data show some variation, all the investigators who have studied the effect of the GMF on sex determination in plants have come to the common view that orientation of the seeds with the embryonic radicle (particularly of corn) toward the north promotes femaleness, while southern orientation promotes maleness (106, 107). The question of the effect of the GMF on the sex ratio is by no means conclusively solved, however. According to some data, orientation in the GMF leads to a change in sex ratio in heterosexual monecious plants (cucumber, squash), where sex differentiation of the plants is not very strictly controlled by genetic factors (521). The GMF, however, has no effect on the ratio of male and female individuals in strictly diecious plants (hemp).

Polarity. One of the most important characteristics of all living organisms is their polarity, which is manifested in a specific spatial orientation of activity of the entire organism and its individual parts due to differences in the physicochemical and functional characteristics of opposite sides. Since the polarity is related to spatial orientation of the organism, and this orientation is determined by the GMF, there are grounds for inferring that polarity is related in its origin to the GMF. It has been suggested that plant polarity is due to the magnetic properties of the substance under the action of electromagnetic, gravitational, and other factors (295, 348, 349).

Experimental investigations have consolidated the view that the GMF can affect plant polarity. Different geomagnetic orientation of corn seeds alters the polar characteristics of plants, and also the polarity coefficient, i.e., the ratio of the weight of the aerial and underground parts of plants (106, 107). Southward orientation of the embryonic radicle of seeds leads to greater development of the aerial part, while northward orientation leads to greater development of the root system. In the opinion of Derevenko *et al.* (107) and other researchers (1066), the alteration of the leaf, root, and grain mass of plants by oriented planting in the GMF can be used to improve crop yield.

Dissymmetry. Since the polarity of plants is closely related to their dissymmetric features, the GMF should affect this important characteristic of the plant organism.

Plants, like other organisms, can be divided into right-handed, left-handed, and symmetric (D-, L-, S-) enantiomorphs. This division is based on the morphological and functional characteristics of the plant, which are embodied in the form of dissymmetric objects, altered by mirror reflection to the very opposite form in some respects and hence not coinciding with their mirror images (126, 135, 551, 552, 583–585). The modification to which the plants belong is determined by the characteristic arrangement of the corolla, the bract scales, and the seed embryo, and from the rhythmicity of functional processes in time. For instance, dissymmetric forms are revealed by the clockwise or counterclockwise rotation of the petals, the inclination of the rhythm curves to the left or right, and so on. It is of interest that, in spite of the apparent stability of dissymmetric characters, they are not constant, but vary continuously. These variations in the predominance of a particular dissymmetric modification are cyclic, with a period of about 1 and 11 years (552). Since solar activity is manifested on earth by a change in geomagnetic activity, there are grounds for assuming that variation in the level of dissymmetric forms of plants is also related to the GMF (125, 552, 794). This view is corroborated by the following facts: the correspondence between the change in dissymmetry of the flowers of the same species of plants and the change in dip of the GMF at different geographic points (794), the different numbers of left- and right-handed modifications of plants in different phases of the solar cycle (552), the change in symmetric characteristics of D- and L-forms of plants when they are oriented relative to the geomagnetic poles (106, 379), and the spatial orientation of lateral roots due to the GMF (398, 1055, 1056, 1058, 1063). Hence we can conclude that the GMF has a great effect on the dissymmetry of biological objects and on their many functional characteristics. This is also revealed by a study of rhythmic processes in biological objects.

Rhythmicity of Processes. The GMF determines another important characteristic of plants—the rhythmicity of functional dynamic processes. It is remarkable that the dip and declination of the GMF play a decisive role in this case (121, 122, 125, 136). A change in the GMF elements is accompanied by a change in rhythms with the most varied periods (diurnal, seasonal, and annual) of biological objects. Experiments have shown that the universal relation between the rhythmicity of functional processes and the GMF is based on modification of the permeability of biological membranes by this field (121–123). One should avoid a simplified view of the relation between rhythmicity and the GMF. The fact is that the relation between rhythmicity and the GMF is not definite, but variable. On one day the main factor is the dip of the GMF, and on other days it is the declination. In a number of cases this close relation between one GMF element and a rhythmic functional process cannot be followed completely, but only for 10–16 h, after which there is a relation with another GMF element for 8–14 h. At present there are no experimental data that indicate if this is due to the effect of external factors, e.g., the diurnal variation of gravitation, or to internal factors, e.g., change in bioelectric activity, current systems within cells, electromagnetic fields of the biological objects themselves, the free-radical level.

One of the main pieces of evidence that indicate an important role of the GMF in the rhythmicity of functional processes occurring in living organisms is the long (15 years) continuous investigations of the respiration of potato seedlings (751, 752) or dry onion seeds (755) in a chamber with completely constant temperature, humidity, light intensity, and atmosphere. The conclusion that the GMF affects plant physiological processes is confirmed by studies not only in the strictly controlled conditions of a phytotron, but also in the field, where the external environmental factors vary continuously (202, 230, 375, 500). A comparative analysis shows that in such conditions the GMF greatly affects the periodicity of physiological and biochemical processes (123, 124, 809).

We call the data cited here "indirect" statistical evidence, since they are obtained not by direct experiment, but by comparison of the periodicity of physiological processes with the variation of the GMF in a specific period of time at the place where the investigations were conducted. This does not diminish their value in any way, since our numerous discoveries of exceptional coincidence of curves representing the rhythmicity of the most diverse biological processes and the variation of GMF elements cannot be purely accidental, because according to the estimates of specialists the probability of chance agreement in the vast majority of cases that we cite is exceedingly low (10^{-8}–10^{-9}). For our analysis we only used experimental data of publications in which the authors indicated the exact time (hour, day, month, year) and place of the investigation.

Using the method of direct comparison we found that in approximately 30% of the cases there was a very close correlation ($r = +0.95$) in the course of the day between the variation of a GMF element and a physiological index. In other cases the close correlation between a GMF element and the functional process did not occur throughout the whole day, but only for a period of 10–16 h, and in the next 8–14 h there was a relation with another GMF element. For convenience of exposition we consider separately circadian rhythmicity of physiological processes and rhythmicity (cyclicity) with a period greater than one day.

Circadian Rhythmicity. An analysis shows that the circadian rhythmicity of physiological processes is under the direct control of the GMF. Convincing evidence of this is provided by, for instance, the rhythmicity of the uptake and excretion of substances by plant roots (Fig. 72), respiration (Figs. 73–75), and rate of photosynthesis (Fig. 76) of plants, particularly if the process or index is measured at hourly intervals. We must stress that our analysis, which revealed the leading role of the GMF in the circadian rhythmicity of plants, related to experiments on respiration and excretory activity of plants in a phytotron, i.e., the physiological processes were investigated in optimum light, temperature, and moisture conditions.

A similar close relation with the GMF, however, was also noted when the investigations were conducted in natural plain (Fig. 75) or mountain (Fig. 76) conditions, where the external environmental physical factors vary considerably in the course of a day.

Seasonal Rhythm. Our conclusion that the GMF affects plant physiological processes is also confirmed by studies not only in strictly controlled phytotron conditions, but also in field conditions in the course of a season, when weather factors vary continuously. In such conditions investigators have analyzed plants

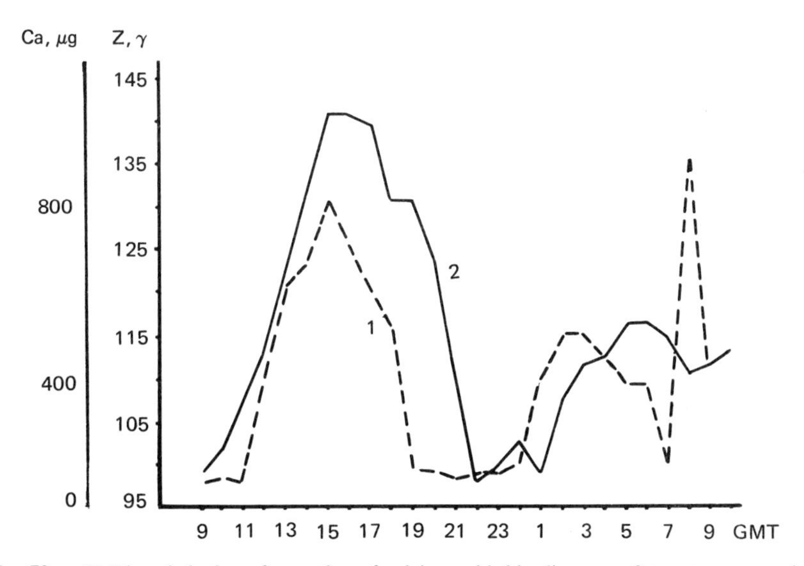

Fig. 72. (1) Diurnal rhythm of excretion of calcium with bleeding sap of tomatoes grown in a phytotron in Moscow and (2) variation of GMF vertical component (Z) on January 10–11, 1957. Krasnaya Pakhra Observatory, Moscow, USSR. From (124). Analysis of data in I. I. Gunar *et al.*, *Izvestiya Timiryazevskoi Sel'skokhozyaistvennoi Akademii,* No. 4, 201 (1957).

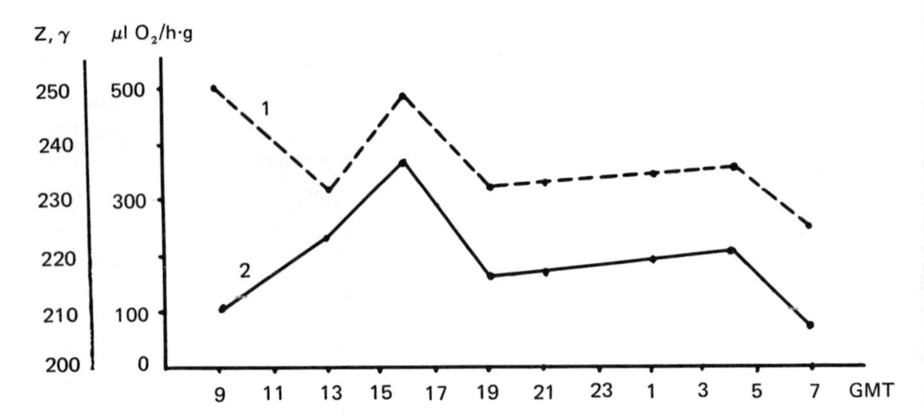

Fig. 73. Diurnal rhythm of respiration of plants and role of GMF. Diurnal rhythm (1) of leaf respiration of "Sil'vestris" tobacco plants grown in short-day conditions in a phytotron in Moscow on July 28–29, 1959, and variation of GMF vertical component (2) in same period.

at intervals of time (days, weeks, months) that have been convenient for various reasons or that have suited the experiment, where factors such as phases of development of the plants, watering times, or dressing with fertilizers have had to be considered. Hence, the periods between individual measurements has varied considerably, but the relation with the GMF can still be traced, although it is more difficult to discover, since there is frequently a phase difference between the biological rhythmicity and variation of the GMF.

As an example we give data relating to allelopathy in different tree plantations (Fig. 77) and seasonal variation of cell-sap concentration in leaves of

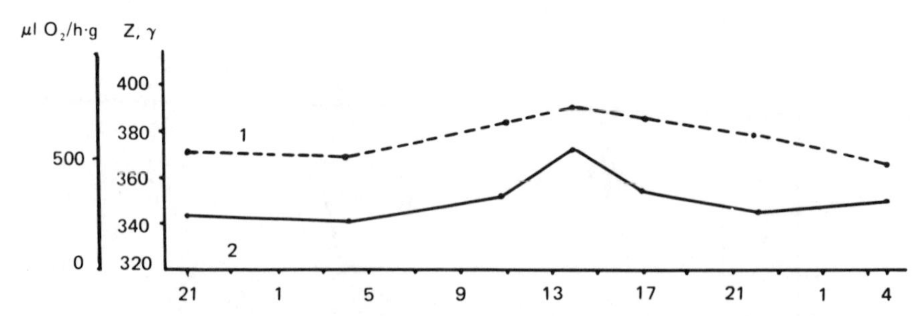

Fig. 74. (1)Diurnal rhythm of respiration of millet plants grown in long-day conditions in a phytotron on August 13–14, 1957; (2) variation of GMF vertical component (Z) in same period. Magnetic data obtained from Krasnaya Pakhra Observatory, Moscow. From (124). Analysis of data published in N. P. Aksenova, Dissertation for Degree of Doctor of Biological Sciences: *Plant Respiration in Relation to Photoperiod,* Inst. of Plant Physiology, Acad. Sci. USSR, Moscow (1962), Figs. 5 and 11.

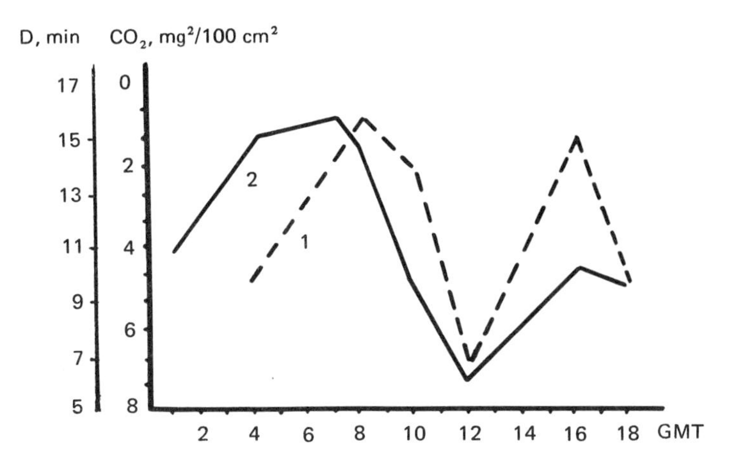

Fig. 75. (1) Diurnal rhythm of respiration of beet leaves in field conditions and (2) declination (*D*) of GMF on July 1, 1933. Zaimishche Observatory, Kazan, USSR. From (124). Analysis of data published in A. L. Kursanov and P. Ugryumov, *Byull. Mosk. Obshch. Isp. Prir.* **43,** 162 (1934).

various citrus plants (Fig. 78). As the graphs show, in both cases the variation of the physiological indices and the GMF are completely synchronous for the specific data of the investigations, although the cited examples relate to quite different processes: In one case (Fig. 77) there is complete synchronism in the variation of the amounts of physiologically active substances excreted by the root systems of pine and oak plantations of different ages, while in the other case (Fig. 78) the cell-sap concentration in different plants was investigated, and quite inexplicable reductions and increases in the investigated concentration index (e.g., an increase in concentration in December) were observed. We would like to stress that the nonaccidental agreement of the physiological processes analyzed and the GMF is confirmed by the discovered universality of the response in plants of different species and ages (beet, tomato, tobacco, pine, oak, lemon, etc.) growing in different geographical localities and in experiments conducted at very different times (1933, 1957, 1965, and other years). We have already mentioned, and we stress again, that in all the examples cited the probability of chance coincidence of the curves representing the variation of different, apparently independent, processes is very low ($p = 10^{-8}$–10^{-9}).

We must, however, point out two important points for investigators who undertake the study of this problem. First, a correct analysis of the changes in physiological functions due to the GMF requires a careful analysis and consideration of the variation of all the GMF elements, and not only the predominant one during the particular period. Second, we must bear in mind that seasonal variations of the GMF elements are obtained in magnetic observatories by averaging a large number of hourly GMF values for all days of the month, while the

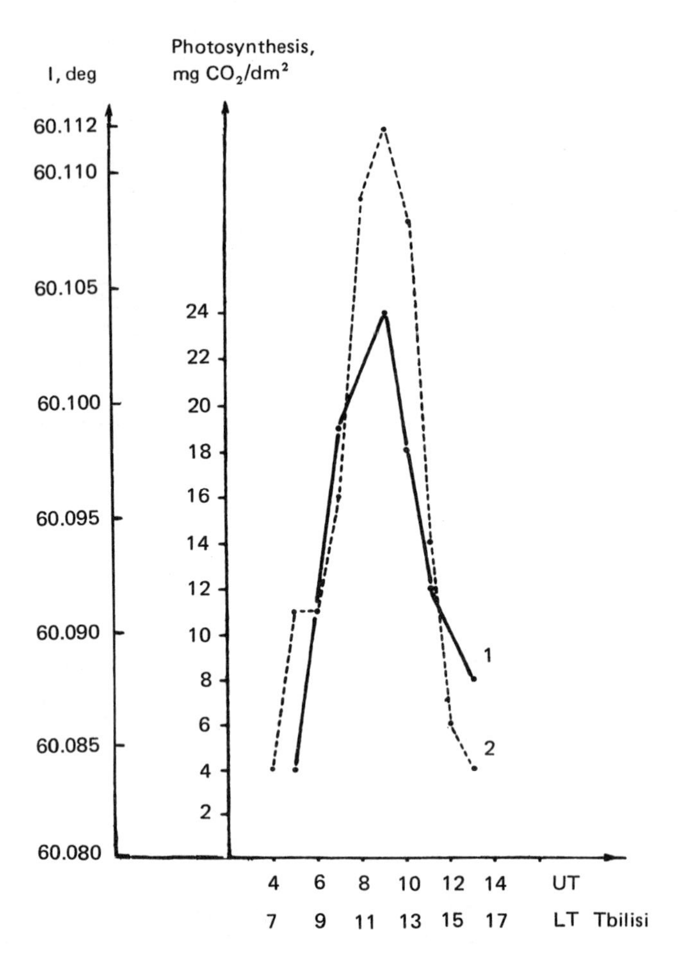

Fig. 76. (1) Diurnal rhythm of photosynthesis of *Scabiosa caucasica* in natural mountain conditions in Georgia and (2) variation of GMF dip (*I*) on August 11, 1965. Dusheti Observatory, Georgian SSR. From (139). Analysis of data published in L. D. Khetsuriani, *Trudy Inst. Botaniki AN Gruz. SSR* **25**(2), 105 (1967), Fig. 6.

experimental data in biology in most cases, as in the examples cited above, are obtained from investigations carried out only on particular dates of the corresponding months of the year in single experiments during a short period of the day. If the physiological data cited by investigators were based on automatic recording of the hourly values of the investigated indices, as was the case in some studies by more advanced techniques (751, 752, 871, 1151), the correlation between the changes in physiological indices and the GMF would be even closer.

Comparison of the variation of physiological processes with the usual

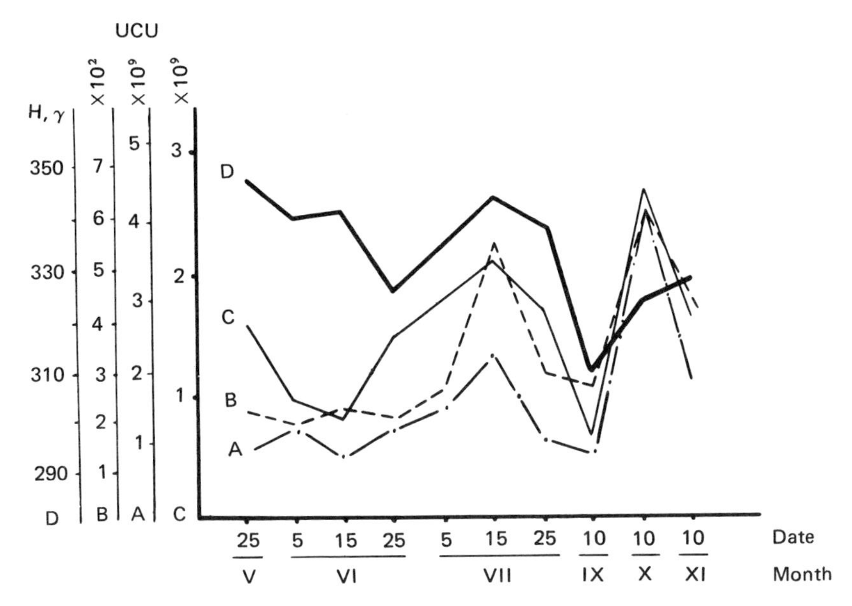

Fig. 77. Dynamics of accumulation of root excretions by different trees during growing season, 1966. Seven-year-old pine stand on moderately fresh soil (A); 24-year-old closed pine stand on dryish sandy soil (B); 30-year-old closed oak stand on fresh sandy loam (C), and variation of GMF horizontal component on these dates (D). UCU, universal commarin units. Dnepropetrovsk Region, Ukr. SSR. Magnetic data obtained from Odesskaya Observatory, Ukr. SSR. From (124). Analysis of data in N. M. Matveev, *Nauchn, Dokl. Vyssh. Shkoly, Biol. Nauk* **8**(68), 84 (1969).

edaphic factors of the external environment (light, temperature, humidity, etc.) have been made by almost all of those who have studied the periodicity of plant function in natural and artificial conditions (877, 1151), but none of these investigators took into account the key role of the GMF in these processes. The universality of the relations that we have discovered is additional evidence that the acting factor is the GMF, which can penetrate to any point on the earth. The material presented shows that no matter what physiological processes of plants have been investigated, their changes show great similarity with variations of GMF elements. Hence we can conclude that the physiological homeostasis of plants, as of animals, depends on the state and variation of natural magnetic and electric fields. The reason for such universal action of the geomagnetic field on different processes in plants (animals) is the controlling effect of the GMF on the permeability of biological membranes, which lies at the base of any physiological homeostasis (120, 122, 809). Since this fundamental property is under the influence of the GMF, all the associated functions of the living organism have a close and very specific relationship with the GMF.

We made this discovery after we had devised and applied a special method

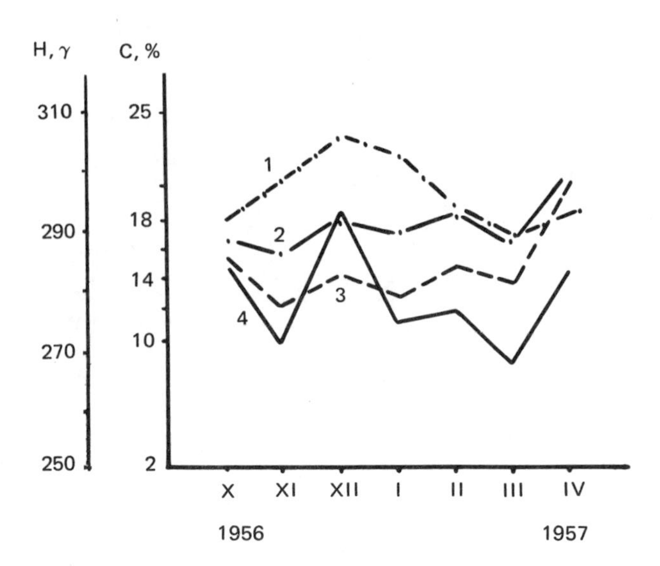

Fig. 78. Seasonal variation of cell-sap concentration (*C*) in citrus plant leaves (as percentage of dry matter) during 1956–1957: (1) Shiva-mikan; (2) "Duncan" grapefruit; (3) "Novogruzinskii" lemon; (4) variation of GMF horizontal component in this period. The work was conducted in Sochi. Magnetic data obtained from Odesskaya Observatory. From (124). Analysis of data in L. I. Surkova, *Fiziol. Rast.* **9,** 607 (1962).

of investigating the rhythmicity of excretion and uptake by roots of living plants—a method that we call the spectrophotometric or SP method (121, 129, 142, 143). The SP method is based on a well-known physical effect—the absorption of ultraviolet rays when they pass through solutions containing organic substances such as amino acids, amino sugars, or nucleic acids and their derivatives (120). Our method is simple, yet very sensitive and accurate, since a high-quality ultraviolet spectrophotometer is used. The principle of the SP method is as follows. Plants excrete various substances of organic origin through the roots into the surrounding medium. The total amount of these root excretions can be determined by measuring the optical density in the ultraviolet (300–400 nm) of the solution in which the plants have been kept (Fig. 79a). In measuring the excretion of plant roots we are really investigating the permeability of biological membranes of the root cells for organic substances of endogenous origin. The measurements can be made hourly, or an automatic system, such as an automatic fraction collector with provision for ultraviolet measurement of the optical density (LKB Products, Sweden), can be used (Fig. 79b).

From investigations by the SP method of the dynamics of excretion of organic substances by the roots of plants in the constant conditions of a phytotron, in darkness, with the GMF compensated by Barenbek coils, with various

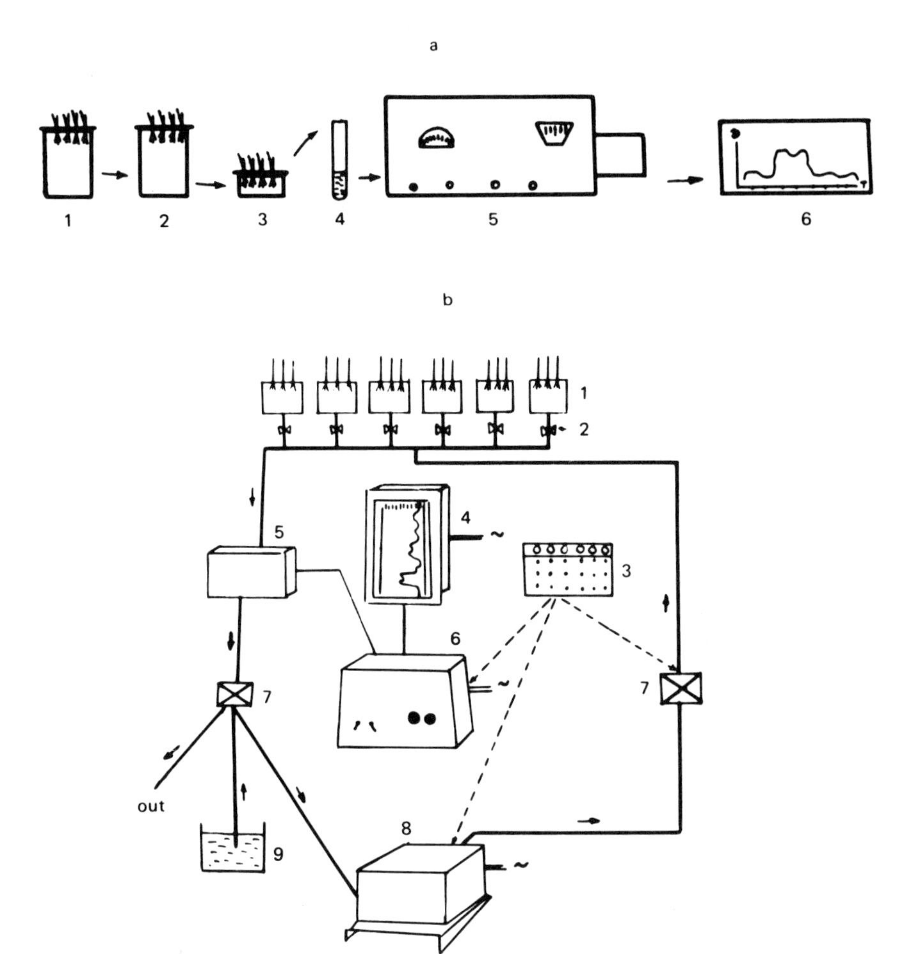

Fig. 79. (a) Setup for spectrophotometric method of determining rhythmicity of plant root excretions (129): (1) main vessel with nutrient solution and experimental plants; (2) washing vessel with pure water; (3) dish with pure water into which plants are put for 10 min hourly for sampling of root excretions; (4) test tube for hourly 2-ml sample of solution containing root excretions; (5) spectrophotometer for measuring ultraviolet absorption of solution; (6) graph of results of hourly measurements of rhythm of root excretions. (b) Block diagram of apparatus incorporating the Swedish LKB-4900A-12 Avtorecichrome automatic fraction collector for automatic recording of rhythm of root excretions by the spectrophotometric method: (1) plant containers with automatic supply of fresh nutrient solution; (2) automatic valves; (3) program (command) device for periodic delivery of fresh solution to plants and withdrawal of solution containing root excretion; (4) recording potentiometer; (5) uv optical cell; (6) uv solution analyzer; (7) automatic switch; (8) pump; (9) vessel containing nutrient solution. From (129).

kinds of shielding, at different points on the earth and at different depths under the earth, and from many other experiments (134), we came to the fundamental conclusion that the external synchronizer of plant rhythmicity is the GMF and the whole intricate complex of heliogeophysical factors that alter its activity—solar activity, the sectoral structure of the interplanetary magnetic field, gravitation, ionospheric processes, and so on (119–124, 133, 134, 143, 809). This effect of cosmic factors is discussed in a very clear and easily understandable manner in a book by American scientists, who examine the two existing viewpoints on biological rhythmicity (746).

As an example we give the results of measurement by the SP method of the root excretions of oat plants in a phytotron, where light, temperature, and humidity were all kept constant, on different dates. The most interesting feature is the differences in circadian rhythm on different days: The curve sometimes has one peak and sometimes two (Fig. 80b), and sometimes has a distinct dip at noon (Fig. 80a). The reason for this is that the plant rhythmicity is affected by different GMF elements—the dip, declination, and other components, and their interrelation with the diurnal variation of gravitation on each specific day (134, 136, 143). It is the gravitation effect that leads to the direct relation of the circadian rhythm in plants with the GMF or to the characteristic phase difference of 5–6 h.

This phase difference and the role of inductive effects are very clearly revealed by a comparison of the circadian rhythm of variation of root permeability with the hourly variations of the geoelectric field, the so-called telluric currents (Fig. 81). The similarity of these curves indicates an important role of inductive current phenomena in the living organism due to diurnal variations of the GMF elements and, in particular, in the mechanism of the effect of the GMF on the permeability of biological membranes. The GMF can apparently alter the electrical current systems responsible for the functioning of the transport mechanism of biomembranes.

Our conclusion is supported by the studies of the well-known American scientist Burr (758, 759), who reported a relation between the electric potential of trees and geomagnetic disturbance, and those of Markson (968), who investigated many years' records of tree potentials given to him by Burr. Although nothing was known at that time about the relation between the circadian rhythm of biological objects and the diurnal variations of the GMF, Markson made a correct inference regarding the essential nature of these relations. He writes: "... both the sun and moon apparently affect tree potential, and the sun obviously acts through an electromagnetic mechanism, while the moon acts through a gravitational or gravitational–electrical mechanism." As the data that we have presented above show, in both cases the effect is based on the membrane mechanism of permeability of living cells and modification of the electrical processes of ionic transport in living cells, whose characteristics can be very unusual from the viewpoint of physical concepts, as we showed above.

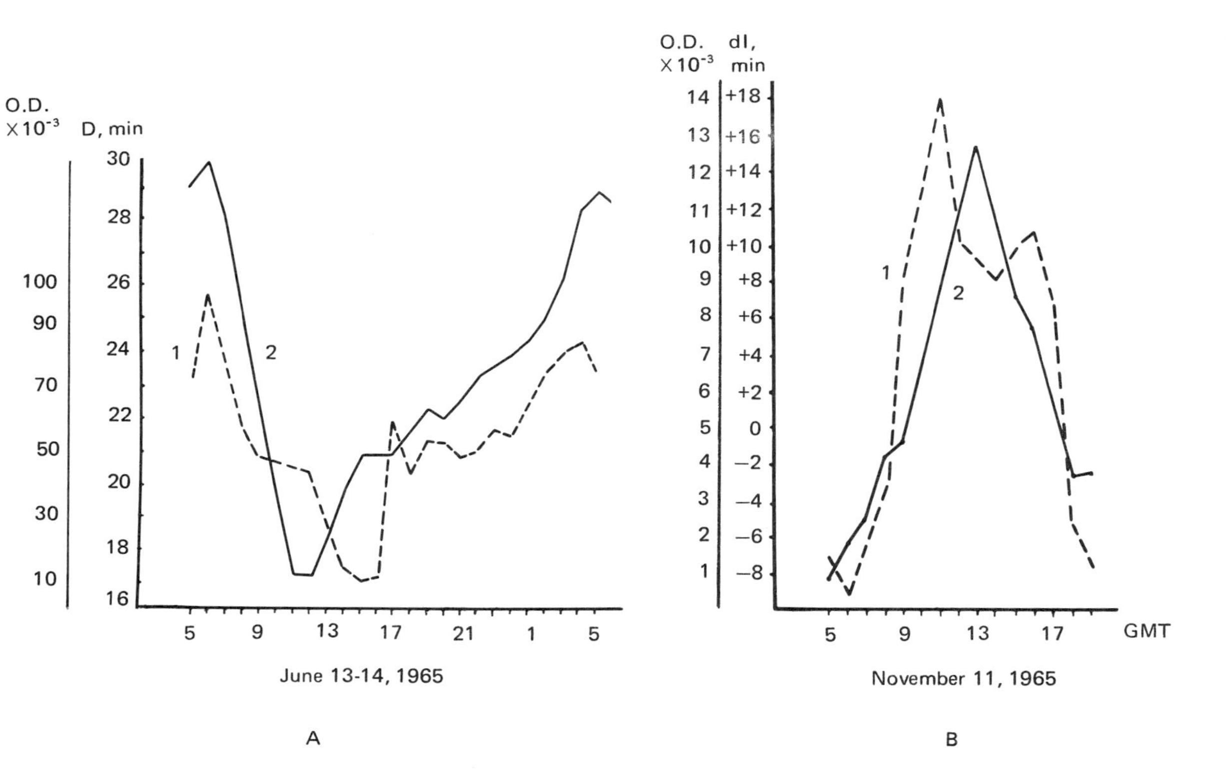

Fig. 80. Circadian rhythm of root excretions of 10-day-old barley seedlings grown in a phytotron in Moscow, June 13 (A) and August 11 (B), 1965, and variation of GMF on same dates. Root excretions measured by SP method. (1) Root excretions, measured in optical density units; (2,A) declination of GMF, minutes; (2,B) dip in GMF, gradient *I*, fractions of a minute. Krasnaya Pakhra Observatory, Moscow. Greenwich time given. From (121).

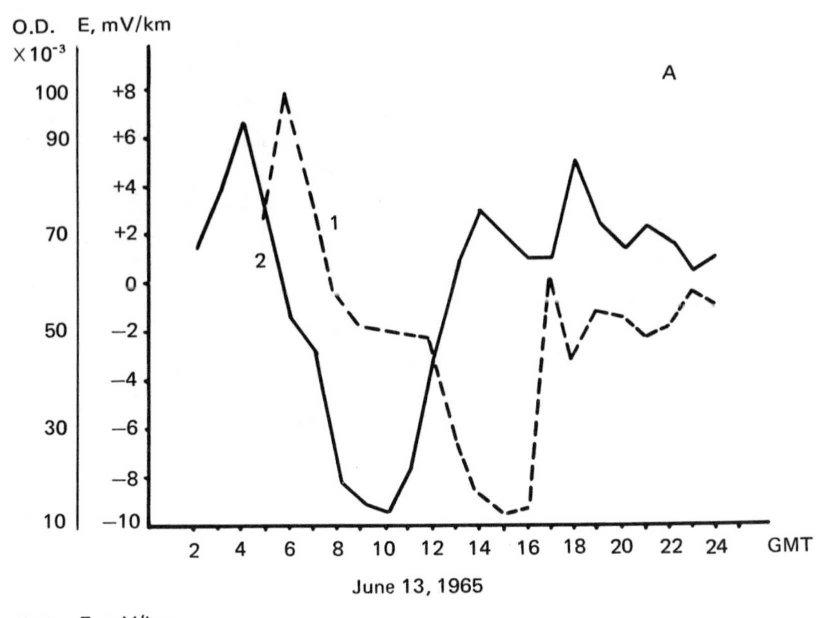

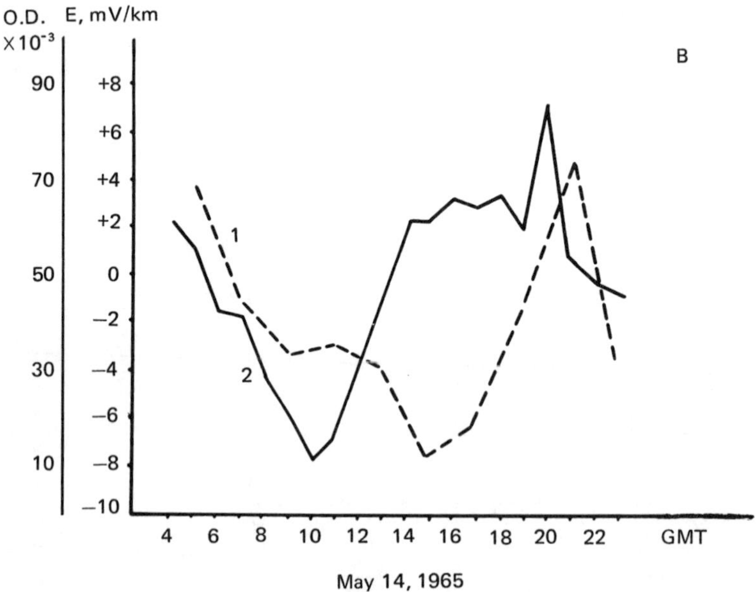

Fig. 81. Circadian rhythm of root excretions (1) of 10-day-old barley seedlings in phytotron in Moscow and diurnal variations of electrical currents of the earth (2) on June 13 (A) and May 14 (B), 1965. (1) Root excretions, measured in optical density units; (2) hourly changes in telluric currents (north–south component) in mV/km. Pleshchenitsy Observatory, Minsk, BSSR. From (121).

At present, however, the effect of the GMF on the rhythmicity of biological processes is a topic of controversy, since most investigators hold the view that these rhythms are of internal (endogenous) origin. Careful experiments in space will probably throw light on this question (871).

Thus, from all the information presented above we can conclude that the GMF probably has a direct effect on such paramount characteristics of living organisms as polarity, symmetry, rhythmicity, differentiation, and genetic properties. This suggests that the GMF is an important evolutionary factor for living organisms on earth and merits the closest attention.

CHAPTER 5

Possible Mechanisms of Biological Effect of the Geomagnetic Field

Having established the existence of a definite relationship between the most diverse properties of living organisms and the GMF we note that the GMF is an important factor affecting such fundamental properties of evolutionary development of all living organisms as heredity and variation, which are responsible for the level and course of natural mutagenesis in nature. Hence, the GMF is a decisive factor in the manifestation of the most basic properties of living organisms, and water molecules play a significant role in this.

Role of Water Molecules

An obvious question that arises is why the GMF should affect living organisms of vastly different levels of organization. Although no answer to this question has yet been obtained, we can postulate that the main substance responsible for the universality of the action of the GMF on living organisms is the water contained in the structure of biological systems and the processes occurring in the water molecule. There is already an adequate basis for this hypothesis. In particular, such a conclusion can be drawn from the results of many years of research on the chemical reactions occurring in an aqueous medium. One such reaction—the hydrolysis of bismuth chloride—has been investigated continuously and thoroughly for 20 years (420, 1049). The high sensitivity of the water molecule to magnetic and electric fields and the modification of its structure by their action have also been investigated (68), and it has been found that these phase transitions do not require large amounts of energy (145, 146, 253, 255–261, 307, 579).

This hypothesis alone provides a key to the understanding of why the GMF has such a universal effect on biological and inorganic systems. It should be noted, however, that there is still no complete and precise explanation of this wide and universal link of living organisms with the GMF. Ideas on this matter have still not got beyond the stage of hypothesis. They are clearly expounded and their physical basis indicated in several fundamental works on magnetobiology, involving a study of the effect of artificial magnetic fields on living organisms (113–115, 677, 678, 782, 816, 1141, 1180), but the corresponding hypotheses for the GMF have not been considered in detail and are still at their earliest stage of development (701).

Role of Biological Membranes

Experimental work on geomagnetobiology has revealed an important mechanism of perception of the GMF by living organisms—the change in the permeability of biological membranes (121–131, 207, 208, 571, 572, 568, 809). The biological effect of alternating electromagnetic fields in the range 0.2–100 kHz is also mediated by a change in the permeability of cell membranes (531, 1081).

The discovery of a connection between the organism and the GMF through the membrane permeability mechanism opens up wide prospects for the understanding and explanation of the biological effects due to the GMF.* At the same time, this would logically account for the universality of a GMF effect in which water molecules play an important role.

In fact, biological membranes, which are structural elements of any cell, are mainly responsible for maintaining the function and for the fine regulation of all organs of the living organism. The coordinated work of the membrane permeability mechanism is responsible for the precise homeostasis of the living organism and the self-regulating ability of all its parts—from submicroscopic particles like microsomes and mitochondria to individual organs.

By altering the permeability of biological membranes, the GMF can affect the whole organism, causing the entire range of changes in the human organism, animals, and plants discussed in previous chapters (1054, 1111, 1197).

We shall show this only for two examples, on the basis of an analysis of the state of the healthy and sick human organism. If we pursue the hypothesis of a universal connection of the permeability of biological membranes with the GMF, we can logically accept the possibility of a direct effect of the GMF on the central nervous system. At present we have evidence of a direct effect of permanent

*At present there are no convincing data indicating that the GMF affects living organisms only through a change in permeability of the cell membranes. In addition, it should be noted that the so-called membrane mechanism does not account for all aspects of the pathogenesis of diseases. [B.R.]

magnetic fields on individual neurons, nervous tissue, and the higher divisions of the animal brain (238, 240, 241).

It is known that active functioning of the brain involves various kinds of mediators and neurohormones, which control both the transmission of individual nerve impulses and the whole work of the central nervous system. The physiological and biochemical processes in the brain are very closely associated with bioelectric activity, whose basis is the membrane mechanism of asymmetry of sodium and potassium ions (262, 645, 1067).

It becomes clear in such a case that the action of the GMF on the membrane permeability mechanism has a pronounced effect on the entire organism as a whole and gives rise to the extensive chain of events that are observed in the most diverse manifestations of neuropsychic activity.

It should be borne in mind that the direct effect of the GMF on the membrane permeability mechanism operates at the same time as a large number of diverse adaptive and regulatory responses in the organism, which strive to maintain the organism in an optimum state in its interrelations with the environment. The indisputable effect of the GMF on the permeability of all membranes of the living organism leads to a change in its functional-dynamic indices.

In the sick organism, however, where the homeostatic reactions that maintain the dynamic equilibrium of the organism with the environment fail to occur, the interrelations with the GMF are different. This is well illustrated by the example of diseases of the cardiovascular system, where the reactions to changes in the GMF are most pronounced.

According to modern ideas on the pathogenesis of diseases of the cardiovascular system, disturbances of the tonus and permeability of the vascular wall followed by morphological changes in it are of great importance. Changes in the permeability of cell membranes and the associated changes in electrolyte balance occurring in response to the action of the GMF can thus be regarded as one of the components of a complex pathogenic mechanism in diseases of the circulatory organs. At the same time, the effect of the GMF extends also to the central nervous system, thus altering the membrane permeability mechanism and the bioelectric activity of the brain cells. Thus the GMF can affect the organism by directly acting on the vascular wall, and also indirectly, through the central nervous system, i.e., by altering the regulatory function. We note, incidently, that the question of the role of the endocrine and vegetative nervous system in relation to geomagnetic disturbances and the course of cardiovascular diseases was raised a considerable time ago (1069).

It should be especially noted that all that we have said above about the direct effect of the GMF on the membrane permeability mechanism, and through it on the bioelectric activity of the brain, applies in equal measure to another vitally important organ—the heart (299). The disturbance of the fine membrane permeability mechanism in the heart muscle and changes in its electric and magnetic properties, particularly when pathological changes are already present, can have serious consequences, leading to disturbances of function. In particular, this may be manifested in various disturbances of the heart rhythm.

It is clear from the above account how well the postulated membrane mechanism of the biological effect of the GMF accounts for the various effects in the sick and healthy organism and many fundamental biological effects.

However, in addition to the important and unique role of biological membranes in securing a relation between the GMF and the living organism, there is another very important property that can account for this connection—the presence of a biomagnetic field in living organisms.

Role of Biomagnetic Fields

In speaking of the connection between the GMF and the homeostasis of biological objects, we must take into account the intrinsic magnetic field of the living organism, although very little is known about it. It is probably a result of the complex interaction of the intrinsic magnetic fields at all levels of organization of living matter, from the subatomic level upward (249, 250, 501, 756, 759, 795, 836, 1005, 1013, 1121, 1193).

Of particular importance for the organism as a whole are the magnetic properties of the blood and its constituent elements. One of the first persons to draw attention to this was Chizhevskii (91), who discovered radial–annular structures in moving blood and investigated theoretically the magnetic interaction of rotating erythrocytes (48). The changes occurring in the blood when the GMF is disturbed can be better understood if one postulates that in addition to its action through the vegetative nervous system the GMF has a direct effect on the magnetic properties of blood elements. In particular, it has been suggested that the erythrocytes in the blood of patients with cardiovascular diseases have anomalous magnetic properties and their ability to be ''magnetized'' leads to a change in the viscosity of the blood, the structure of the blood flow, and altered hemodynamics (179, 182).

Magnetobiological investigations have revealed that living organisms have their own magnetic field, but electric effects, which are closely connected with biomagnetic activity, have also been the subject of close and thorough investigation for a long time. The scientific literature on this question is immense (189–193, 227, 228, 262, 564, 593, 645, 1067) and hence these investigations will be mentioned here only if they have some connection with biomagnetic fields. It should be noted that although electric potentials and currents in living organisms were discovered a considerable time ago, and have been carefully measured and investigated, the mechanism responsible for their appearance is still unknown. One of the main hypotheses ascribes great importance to active sorption and ion transfer by high-energy compounds. It is quite possible, however, that the GMF and gravitation play a special role in the generation of biopotentials in living organisms, since these two factors can produce asymmetry in the distribution of action currents.

An effect of the GMF on the asymmetry of the electromagnetic fields of living objects has been reported by several authors (295, 296, 493). Experimental tests have confirmed that these factors are implicated in the induction of bioelectrical activity (730, 1224) and in the alteration of the value of the biopotentials in plants (568, 571, 572).

In the 1960s, as a result of the successful development of the technique of measuring ultraweak low-frequency electromagnetic fields, experimenters managed to detect the intrinsic magnetic fields of biological objects, which confirmed earlier hypotheses of their existence (445). We now have experimental data indicating that there is a magnetic field associated with heart activity, the passage of a nerve impulse along a fiber (771, 849, 1123, 1145), and the electrical activity of the nerve and the brain (484, 485, 681, 731, 770, 772, 1122).

The detection of such weak magnetic fields involved several techniques: shielding of the subject with a permalloy screen, the use of special detector coils with 2000–2,000,000 turns of calibrated wire, and the use of low-noise amplifiers. This setup reduced interference of various kinds to a minimum, reduced the effect of the GMF, and made it possible to detect the ultraweak magnetic biological fields.

Variations of the intrinsic magnetic field of the human brain have been found in normal conditions without permalloy shielding—by the use of a special superconducting quantum interference device (731). Using this device the authors could not only determine the strength of the brain's magnetic field but also the characteristic features of its configuration around the human head. In particular, it was found that the strongest field signal was detected at a distance of 5–7.5 cm from the inion and on a 3-cm length along the midline of the scalp. The magnetic field is located exactly in the region of the brain visual regions, and the magnetic field, induced by a light flash of 50 msec and equal in absolute magnitude to 5×10^{-9} G, changes phase at a point situated between 5 and 6 cm on the side of the midline of the head above the inion. This study opens up broad perspectives for the further comprehensive investigation of the intrinsic magnetic field of living organisms.

The stength of the alternating magnetic field of the heart is 10^{-7}–10^{-8} G, while that around the head is 10^{-9} G. These magnetic fields are believed to be derivatives of the ionic electric current in the brain and, accordingly, in the muscular groups of the heart. The magnetic signal of the heart and brain has been identified as a true **B** vector produced by ionic currents within the corresponding parts of the body (Fig. 82). In the case of the human head the decisive current is the alpha rhythm, which consists of electrical oscillations in the range 8–12 Hz.

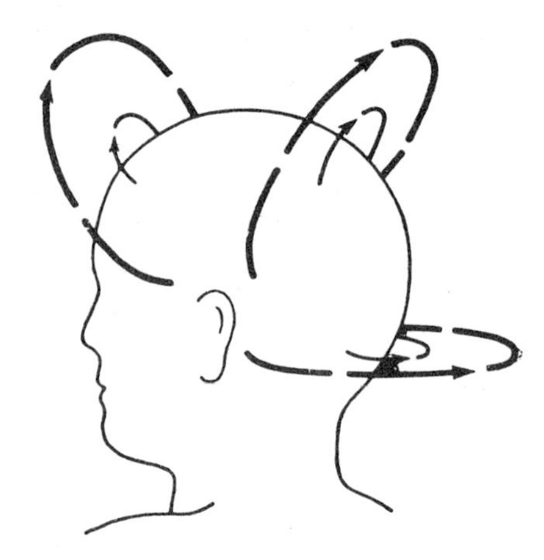

Fig. 82. General scheme of distribution around hear of measured magnetic **B** vector due to alpha-rhythm currents (771).

The decisive role in the case of the heart is due to currents arising in the heart muscle and reflecting the propagation of excitation through the heart musculature with characteristic QRST peaks. As a consequence, the magnetocardiogram also has these characteristic spikes.

It should be noted that the magnetic lines of force of the human head are directed from the left half to the right half, i.e., have left–right symmetry. This is very remarkable, since it provides an explanation of a series of hypnosis experiments described a long time ago (226). In these experiments the presence of a horseshoe magnet near the head of the hypnotist or the subject altered (inhibited or, on the contrary, assisted) the conduction of the hypnotic influence when the magnet poles were in a particular position relative to the head.

Information on the magnetic field of other organisms is inadequate at present. According to the theoretical and experimental data of some investigators, the magnetic field of the insect is too small to account for its orientation in the GMF (924). Other investigators, however, who found that even dead insects suspended on a fine thread, or mounted on a float resting in water, assumed a particular position in the GMF (694), believe that the insect's body is probably a magnetic dipole (147).

There are data (248, 251, 319) indicating that the intrinsic magnetic field of wheat, barley, and rye seeds is several gammas, and that the south magnetic pole is at the germ end and the north pole at the opposite end of the seeds. The position of the magnetic poles in seeds is probably not so simple. This is indicated by data indicating functional dissymmetry (135) and a different reaction of left- and right-handed seeds of the same plants when they are oriented toward the north or south magnetic pole of the earth (379, 380, 544, 550, 552, etc.).

The first experimental evidence of the existence of a magnetic field in living organisms revealed ways of elucidating the mechanism of the biological effect of natural long- and short-period magnetic fields. The discovery of the biological effectiveness of short-period oscillations of the GMF confirms the hypotheses of an effect of magnetic fields of a wide frequency range [from 7–12 Hz (696–698) to 0.029–0.031 Hz (891)] on living organisms. The first frequency interval, as is known, corresponds to the frequency of the alpha-rhythm electrical activity, while the second corresponds to the ultraslow oscillations of the brain potentials (39). New discoveries relating to biomagnetic fields will probably help to provide a complete picture of their spatial disposition, and the mechanisms responsible for their appearance and variation. This in turn will help toward a correct explanation of such well-known effects as the biological action of the peculiar "terrestrial radiation" (683, 684, 889, 934), the change in electric potential of trees during geomagnetic storms (758) or the induction of desynchronization in the case of rapid travel of people (926, 1126).

Magnetoecology

Another interesting feature in addition to the universality of the action of the GMF, which is a unique space–time coordinate for living organisms, is the diversity of the reactions of biological systems to the GMF. This indicates the complexity of the relationship between living organisms and the GMF and the need to take into account the numerous operating factors, both internal, dependent on the properties and features of the biological objects, and external, of terrestrial and cosmic origin (130). We have repeatedly indicated the need to take these two groups of factors into account in the study of the relationship of living organisms with the GMF, e.g., the role of the functional dissymmetry of biological objects and the specific effect of gravitation, electric potential of the atmosphere, and so on.

The diversity of the responses of living organisms also reflects the complex relationship between living organisms and the extremely diverse magnetic and gravitational environment of the habitat (587). The relationship between living organisms and local and global natural electric and magnetic fields will be the subject of study of a new discipline—electromagnetic ecology. This scientific discipline—a branch of biogeophysics—will probably have to answer such complex questions as the induction of desynchronization when living organisms travel rapidly over great distances and the orientation and homing of living organisms. The first steps have been taken along these lines. First, polar regions with special geomagnetic conditions, where the specific action of cosmic and geophysical factors is possible, have been clearly distinguished (516).

It should be noted that the specific nature of the natural electromagnetic environment distinguishes the region of the equator and the adjacent regions (159, 972), where, as is known, powerful equatorial current streams flow.

Geomagnetobiology has now taken its first steady steps, confirming by experiment and observation that "life came into being and has evolved in the presence of the geomagnetic field" (749). One can hope that the numerous and many-sided investigations will lead shortly to a deep understanding of the special significance that the GMF has for life on earth, and to the development of a geomagnetic theory of development of the biosphere. Geomagnetobiology has laid firm new bases for a study of biological rhythmicity and opens up great opportunities for the further development of natural scientific disciplines; in this lies its importance as a science of the future.

Appendix

To facilitate the analysis and comparison of the biological experimental data with the variation of heliogeophysical parameters we provide the following tabular data. This material is taken from the journals *Kosmicheskie Dannye* for 1964–1971, the *IAGA Bulletin,* No. 12, 1970–1971, and other publications.

Table A1. Number and Date of First Days of Solar Revolutions

Year	Number	Date of first day		Year	Number	Date of first day		Year	Number	Date of first day	
		Month	Day			Month	Day			Month	Day
1964	1886	I	22	1967	1826	I	6	1970	1867	I	17
	87	II	18		27	II	2		68	II	13
	88	III	16		28	III	1		69	III	12
	89	IV	12		29	III	28		70	IV	8
	90	V	9		30	IV	24		71	V	5
	91	VI	5		31	V	21		72	VI	1
	92	VII	2		32	VI	17		73	VI	28
	93	VII	29		33	VII	14		74	VII	25
	94	VIII	23		34	VIII	10		75	VIII	21
	95	IX	21		35	IX	6		76	IX	17
	96	X	18		36	X	3		77	X	14
	97	XI	14		37	X	30		78	XI	10
	98	XII	11		38	XI	26		79	XII	7
1965	1799	I	7		39	XII	23	1971	1880	I	3
	1800	II	3	1968	1840	I	19		81	I	30
	01	III	2		41	II	15		82	II	26
	02	III	29		42	III	13		83	III	25
	03	IV	25		43	IV	9		84	IV	21
	04	V	22		44	V	6		85	V	18
	05	VI	18		45	VI	2		86	VI	14

	06	VII	15		46	VI	29		87	VII	11
	07	VIII	11		47	VII	26		88	VIII	7
	08	IX	7		48	VIII	22		89	IX	3
	09	X	4		49	IX	18		90	IX	30
	10	X	31		50	X	15		91	X	27
	11	XI	27		51	XI	11		92	XI	23
	12	XII	24		52	XII	8		93	XII	20
1966	1813	I	20	1969	1853	I	4	1972	1894	I	16
	14	II	16		54	I	31		95	II	12
	15	III	15		55	II	27		96	III	10
	16	IV	11		56	III	26		97	IV	6
	17	V	8		57	IV	22		98	V	3
	18	VI	4		58	V	19		99	V	30
	19	VII	1		59	VI	15		1900	VI	26
	20	VII	28		60	VII	12		01	VII	23
	21	VIII	24		61	VIII	8		02	VIII	19
	22	IX	20		62	IX	4		03	IX	15
	23	X	17		63	X	1		04	X	12
	24	XI	13		64	X	28		05	XI	8
	25	XII	10		65	XI	24		1906	XII	5
					66	XII	21				

Table A2. Dates of Selected Five Magnetically Quiet and Disturbed Days

Month	Quiet days					Disturbed days				
	1964									
I	14	15	21	22	27	2	3	10	16	31
II	3	11	16	18	19	6	8	13	25	26
III	2	18	19	28	31	4	5	22	23	30
IV	12	14	22	23	24	1	2	19	27	28
V	7	8	9	12	20	1	14	15	24	25
VI	3	5	6	16	30	10	11	12	20	21
VII	14	15	24	27	28	3	7	8	17	18
VIII	10	15	24	28	30	4	5	11	12	31
IX	13	14	15	19	20	7	8	22	28	30
X	11	22	23	30	31	4	5	19	21	26
XI	7	14	19	24	25	1	2	9	15	23
XII	5	12	27	30	31	7	13	16	17	19
	1965									
I	6	11	16	24	25	2	8	12	13	22
II	2	12	13	17	19	7	8	21	23	25
III	8	10	11	81	30	3	4	23	25	26
IV	2	3	8	21	28	9	17	18	19	20
V	2	11	13	14	19	5	8	9	10	16
VI	10	13	20	21	24	9	15	16	17	30
VII	4	5	11	17	31	6	8	10	23	28
VIII	5	6	10	13	28	18	19	20	24	25
IX	8	9	10	14	30	16	17	19	27	28
X	4	6	15	16	21	2	8	23	24	28
XI	3	10	16	23	28	5	6	19	20	30
XII	14	15	16	17	21	1	18	25	26	28
	1966									
I	1	12	13	16	31	20	21	22	23	24
II	1	9	14	26	28	5	19	20	23	24
III	1	2	7	24	31	14	19	23	26	28
IV	11	19	25	26	27	1	2	13	22	30
V	10	14	15	23	24	2	4	11	26	31
VI	9	10	11	18	22	1	2	23	24	25
VII	7	13	14	18	25	8	9	10	12	21
VIII	2	16	17	22	28	19	23	24	30	31
IX	11	12	13	18	22	1	3	4	6	8
X	2	11	21	22	23	4	5	6	16	31
XI	9	14	22	23	25	1	3	28	29	30
XII	3	9	11	12	31	5	13	14	26	27
	1967									
I	4	12	24	30	31	1	7	8	13	14
II	2	3	10	12	13	7	8	16	17	25
III	8	11	12	15	16	9	18	19	20	27
IV	13	14	26	27	28	1	19	22	23	24
V	8	9	16	20	22	3	25	26	28	29
VI	1	18	20	23	24	5	6	25	26	27

Table A2. (Continued)

Month	Quiet days					Disturbed days				
VII	3	9	10	22	31	1	5	11	23	30
VIII	1	2	3	22	23	10	11	17	18	25
IX	5	10	11	12	23	20	21	28	29	30
X	20	21	24	25	26	9	10	12	28	29
XI	7	17	18	19	20	3	8	12	13	24
XII	11	13	14	26	28	1	8	19	20	31
		1968								
I	4	8	9	10	25	1	2	6	20	29
II	6	7	14	25	26	10	11	15	20	28
III	7	8	9	13	22	14	15	16	24	30
IV	8	9	19	20	21	1	5	6	14	26
V	4	5	6	26	27	7	9	12	20	21
VI	5	6	21	24	25	10	11	12	13	14
VII	9	20	24	29	31	3	10	13	14	22
VIII	1	2	28	29	30	14	15	16	17	24
IX	18	24	25	26	27	8	13	14	15	23
X	5	11	21	22	23	2	12	13	29	31
XI	12	14	15	29	30	1	2	3	4	9
XII	7	14	17	20	26	3	4	5	23	25
		1969								
I	3	6	10	13	29	15	17	18	25	26
II	1	9	17	18	22	2	3	11	15	27
III	3	4	10	27	28	12	17	23	24	25
IV	10	18	21	23	26	1	13	17	28	30
V	7	11	26	27	29	2	13	14	15	16
VI	6	18	22	28	29	12	13	14	16	17
VII	4	5	19	20	29	1	13	14	26	27
VIII	1	11	25	29	30	3	12	19	26	27
IX	1	2	13	22	26	6	18	28	29	30
X	8	14	15	26	30	1	2	3	6	10
XI	1	6	14	15	21	3	9	10	27	30
XII	13	20	21	30	31	5	6	9	16	23
		1970								
I	4	11	13	25	26	2	9	16	17	30
II	7	8	11	21	22	2	4	14	24	28
III	16	21	22	24	25	6	7	8	9	31
IV	10	13	14	15	28	6	17	20	21	22
V	8	9	10	11	26	12	20	27	28	29
VI	6	11	12	22	23	1	18	20	21	27
VII	7	5	18	19	28	9	10	21	25	29
VIII	1	3	45	24		8	9	17	18	26
IX	6	9	11	12	29	1	13	14	21	27
X	7	8	9	15	21	4	16	17	18	23
XI	1	15	20	29	30	7	11	21	22	23
XII	1	10	11	17	31	8	14	15	24	28

Table A2. (Continued)

Month	Quiet days					Disturbed days				
		1971								
I	7	8	9	12	26	3	20	27	28	30
II	3	4	5	13	22	14	15	16	25	26
III	21	22	23	28	29	8	13	14	15	31
IV	8	20	24	25	26	4	9	10	14	15
V	12	13	27	28	31	6	7	17	18	30
VI	7	9	12	19	20	1	2	3	25	29
VII	7	10	17	20	25	1	2	21	26	31
VIII	3	6	14	20	27	2	11	18	23	31
IX	2	3	21	22	23	7	18	25	27	30
X	17	18	19	26	27	1	2	8	9	29
XI	3	14	15	16	17	22	23	24	25	26
XII	6	7	8	10	14	17	18	22	29	30
		1972								
I	6	7	8	13	14	16	22	23	26	28
II	9	12	22	27	29	13	14	17	24	25
III	10	12	14	19	21	7	16	24	29	30
IV	3	9	24	25	26	4	18	21	28	29
V	7	8	19	20	21	1	2	15	16	28
VI	9	10	11	12	30	4	17	18	19	27
VII	5	13	14	21	29	7	23	24	25	26
VIII	3	13	23	24	25	4	5	6	9	28
IX	1	5	20	21	22	13	14	15	17	29
X	3	5	6	8	17	13	14	19	29	31
XI	5	10	13	14	30	1	2	15	16	20
XII	5	6	10	21	27	13	15	16	23	30

Table A3. Dates of Large (1) and Very Large (2) Magnetic Storms in the Period 1961–1970

	Start			End			
Year	Month	Day	Hour	Month	Day	Hour	Characteristic
1964	I	1	02	I	5	00	1
1965				No large storms			
1966	IV	25	23–27	IV	27	03	1
	VIII	29	13	VIII	31	24	1
	IX	3	08	IX	5	18	2
1967	I	7	07	I	8	23	1
	I	13	12	I	14	14	1
	II	15	23–47	II	18	13	1
	V	1	19	V	4	05	1
	V	24	17–27	V	27	01	2
	V	30	14	V	31	22	1
	VI	4	11	VI	7	07	1
	IX	18	09	IX	22	06	1
	XII	30	16–30	XII	1	24	1
1968	VI	9	19	VI	14	24	1
	X	30	09	XI	4	23	2
1969	II	2	15	II	4	04	1
	II	10	21	II	12	02	1
	III	23	16	III	24	19	1
	V	14	19–30	V	16	15	2
1970	III	5	08–06	III	10	02	2
	IV	20	11–23	IV	22	15	1
	VII	8	23–18	VII	11	6	1
	VII	24	23–50	VII	26	16	1
	VIII	16	22	VIII	19	19	1
	X	16	09–18	X	18	23	1
1971	IV	14	12–44	IV	16	06	1
	V	16	23	V	18	23	1
	XII	16	19–06	XII	18	22	2
1972	VI	17	06–28	VI	17	17	2
	VIII	4	01–20	VIII	7	18	2
	IX	13	12–40	IX	15	19	1
	X	31	16–56	XI	3	05	2

Table A4. International Characteristic Numbers C_i of Magnetic Disturbance

Year	I	II	III	IV	V	VI	VII	VIII	IX	X	XI	XII	Mean for year
1964	0.6	0.7	0.6	0.7	0.6	0.5	0.5	0.4	0.6	0.5	0.4	0.3	0.53
1965	0.4	0.6	0.5	0.4	0.4	0.5	0.5	0.5	0.6	0.4	0.4	0.4	0.45
1966	0.4	0.4	0.6	0.4	0.4	0.4	0.5	0.6	0.9	0.5	0.5	0.6	0.52
1967	0.5	0.5	0.4	0.5	0.8	0.7	0.5	0.5	0.7	0.6	0.6	0.7	0.58
1968	0.6	0.8	0.7	0.6	0.7	0.7	0.6	0.6	0.6	0.6	0.6	0.6	0.65
1969	0.4	0.6	0.8	0.7	0.6	0.5	0.4	0.5	0.6	0.5	0.5	0.4	0.54
1970	0.3	0.3	0.6	0.6	0.4	0.5	0.7	0.5	0.5	0.5	0.6	0.4	0.52
1971	0.6	0.5	0.5	0.6	0.6	0.5	0.5	0.5	0.6	0.6	0.4	0.4	0.58

Table A5. Polarity of Sectors of Interplanetary Magnetic Field (328), 1971–1975[a]

	I	II	III	IV	V	VI	VII	VIII	IX	X	XI	XII
						1971						
1	+	+	+!	−	+!	−	−	−	−	+!	+!	±
2	±	+!	−!	+!	±	−	−!	∓	−!	+!	+	−
3	+!	+!	+	+	+	−!	−	−	±	+	+!	±
4	+	±	−!	+	+	−!	−!	−	−	+	∓	−!
5	+	−	−!	+	±	−	−	−!	+	+	−!	+
6	+	−	−	+	−	±	+	−	+	−!	±	−
7	+	∓	−!	+	−	+	+!	−!	+	−!	+	−
8	+	+!	+!	+	−	∓	−!	+	+	+	−	−!
9	+	−!	+	−!	−	−	−	+!	+	+!	+	−
10	∓	+!	+!	−	−!	−	−	+	+!	+	±	−
11	−!	+	+!	−	∓	−	−!	∓	+!	+	−!	−!
12	−!	+	∓	−!	−!	+	+!	+	+	−	−	∓
13	∓	+!	−	−	−!	−!	+!	+!	−	+	+	+
14	+!	−!	−	−!	−!	−	+!	+	∓	−	−!	+
15	+!	−!	−	+	−!	−	+!	−	+	−	∓	−!
16	+!	−!	−	−	−	+!	−	−	+	−!	−	−
17	+	−	−!	−	+	+	−	∓	±	+	−	+!
18	±	−	−!	−	+!	+	−	+	+	−	−	∓
19	−	−	−	+	±	+!	−!	+	−	−	∓	+!
20	−	∓	−	+!	−	+!	−	+!	−	−	−	+
21	−	−	−	+	+	∓	−	+!	−	−	∓	+
22	−	−	−!	±	+!	−	∓	+	−!	+	∓	+!
23	−	−	+!	+	±	+	+	−	+	−	+!	+!
24	−	+	+!	+!	−	±	+	−!	+	×	+	+!
25	+	+!	+!	+	−	+	+	−	±	×	+!	±
26	−!	−	−	−!	−!	±	±	+	−	−	±	+!
27	+!	−	+	−!	−	+!	−!	−	−	−	+!	+!
28	±	+!	+!	−!	−	−	−	−	−	+	+!	−

Table A5. (Continued)

	I	II	III	IV	V	VI	VII	VIII	IX	X	XI	XII
29	+		+	±	+	−!	+	−!	∓	+	+!	+!
30	+		−	+	+	−!	−!	−!	∓	±	+	+
31	+		−		+		−	−!		+		+
						1972						
1	+	−	+!	−	∓	+!	+!	−!	+	+	+	+!
2	−	+!	−!	−	+!	+!	+	+!	+!	−	+	±
3	−	∓	−	∓	+	±	−!	+	+	−	±	±
4	∓	±	−	+!	+!	−!	∓	∓	−!	−	−!	−!
5	−	−	−	+!	+	−	−	−	−	−!	−	−!
6	−!	−	−	+	−!	+	−	±	−	±	−	−
7	±	−	+	−	+	−!	−	−	−!	−	+	∓
8	−	−!	+!	−	−	−!	+!	−!	∓	−!	+	−
9	−	−!	+!	+	−!	∓	−	−!	+!	±	−	+
10	−	∓	+!	±	−!	+!	−	−	∓	∓	∓	−
11	−!	+!	±	−	−!	+	+!	−!	+!	+!	±	−
12	−!	+!	+	−	−!	+	+	−	+	−!	∓	±
13	−	+!	+	−	−	∓	−	−	∓	+	−	±
14	−!	+	+	∓	−!	+	∓	−!	+	+	∓	∓
15	−!	+!	+	−!	−	+!	+!	−	+!	+!	∓	∓
16	+	+!	−!	−	+!	−!	−	−!	∓	+	+!	+!
17	∓	+	−!	−	−	−!	−!	+	+	±	+!	+
18	+!	−!	−	−!	+	+!	−	∓	+	+	+!	+
19	−	−	−!	+!	+	+!	−!	+	+	+!	+	+!
20	+!	+!	−	−!	±	−!	−!	+	±	+!	+	−!
21	+	+	−!	−	+	−!	−	+	+	+!	+!	+!
22	−!	−	∓	±	±	−	∓	+	+	+	∓	−!
23	−!	−	+	+!	±	−	∓	+!	−!	+	−	−!
24	−!	±	−	+	−	∓	∓	+	±	±	±	−
25	±	−	+	+!	+	+	−	+!	+	±	−	+!
26	+	−!	−	+	−	+!	−	∓	+	±	∓	+
27	−	−!	+	−	−	+	+!	+!	+!	−	+	−
28	−!	−	+!	+!	+!	+	+!	+!	+!	−	±	−!
29	−	−	−	+	−	∓	+	+	+	+	+	−!
30	−!		−	−	+	∓	−	+	+	+	∓	−
31	−		−		+		∓	−		−!		−
						1973						
1	−	±	−	−!	±	+	+	+	−	−	−	+!
2	−!	−	±	+!	+!	+!	+	−	−	+!	+!	+!
3	−	−	±	+	±	+!	+	−!	−!	+	+	−
4	−!	+	−	+	+!	+!	−	−!	+	+	−	−
5	∓	+!	−	+!	+	+!	−	−	+	−!	∓	−
6	−	+!	+	+	−	∓	+!	+!	+	±	−!	−!
7	+	+!	+!	−	+!	+	+	+	+	±	±	−
8	±	+	+	+	−	+	−	−	+	±	−	−!

Table A5. (Continued)

	I	II	III	IV	V	VI	VII	VIII	IX	X	XI	XII
9	+!	+!	+!	+	−	+	−!	+	−	−	−	−
10	+!	−	∓	+	+	−!	−	+	−	−!	−!	∓
11	+!	+!	±	±	+	−	+	−	−	−	−!	+!
12	±	+	+	−	−!	∓	−	−	−	−!	−!	−!
13	+!	+	±	−	±	−	−	±	+	−	−	−!
14	+	±	+	+	−	−	±	−	+!	−!	+	−!
15	+!	+	−	−	+	−!	−	−!	−	±	+	+!
16	+!	+	±	−	−!	∓	+	+	−!	∓	±	+!
17	+	−	+	±	−	+!	+	−!	−	+!	±	−
18	+!	−	±	−	−	−!	+	+	−	+	+!	−
19	−	±	−	−	−	−	+	+!	−	+!	+!	±
20	+!	−	−	−	±	−	+	−	−	+!	+	+
21	−	−	−	−	−	−	−	−	+!	+	+	+!
22	−!	−!	−	+	−	−	+!	+	−!	±	−!	+
23	−	−!	−	−	−!	±	−	+!	+!	±	−	+!
24	∓	±	−!	±	+	+!	−	+	+	−!	+	+!
25	−!	±	−	+!	±	+!	−	+!	+	−!	+!	+
26	−!	±	−	±	−	∓	+!	+	∓	−!	±	+
27	−!	±	−	−	±	±	+	+!	+	+	+	±
28	−!	−	−	−	+!	−!	+	+!	+	+!	+!	−!
29	+		−	+!	+	+	+	−	∓	+!	+	−
30	−		−!	+!	+!	+!	+	+	−	+	+!	−
31	∓		±		+		+	+!		−		−
						1974						
1	−	±	−	∓	−	+	+!	+	−	−	−	∓
2	−	±	−	∓	∓	+	+!	−	±	−!	−	+!
3	−	∓	−!	∓	+	+	+!	−	−!	−!	−	−!
4	+	−	−	+	+	+!	+!	−	−	−!	−	−
5	±	−!	−	+	±	+	∓	−	±	−!	∓	∓
6	−	−	−	+	+!	∓	∓	−	−	−	−	−
7	−	−	+	+	±	+!	−!	−	∓	∓	∓	∓
8	−	−!	+!	∓	+	+	−	±	±	−	−	+
9	±	∓	+!	+	∓	−!	−	±	−	+!	+!	+
10	−	+	+!	+!	+!	−!	−!	−	−	−	+!	+!
11	−!	+!	+!	+	∓	∓	−!	∓	−!	−!	+	±
12	+!	+!	+!	+	+	−!	−	−	−!	∓	+	+
13	±	+!	+!	+	±	−	−	−	+	+	+	+!
14	−	+	+!	+!	−!	−!	−	−	±	+	+	+
15	+!	+	−!	+	−!	−!	−	−	−!	+!	+	+
16	×	+!	∓	−!	−!	−	−!	−	±	+!	+	+
17	−	∓	+	∓	−!	−	+	−	+!	+	±	−
18	+	+	+	∓	∓	−	±	±	+!	+	∓	−
19	+	+!	+!	−	−!	−!	−	+	+!	+	−	−
20	+	−	−	−	−	−	−!	+	+!	+	−	−
21	+!	−	−	−	−	−	−	+	∓	∓	−	−

Table A5. (Continued)

	I	II	III	IV	V	VI	VII	VIII	IX	X	XI	XII
22	×	−!	−	−	−	−	−	+	±	+	−!	−
23	+	−!	−	−	∓	±	+!	+	+	−	−	−!
24	±	−	+	−	−	−	+!	+	±	±	−!	−
25	−!	−!	+	−	∓	∓	±	+	−	−	−	−!
26	−!	−!	−	−	−!	+!	+!	±	+!	±	−	+
27	−	−	−	∓	+	+	+!	∓	∓	−	−	+!
28	−	−	−	±	−	+	+!	−	−!	−	−	+
29	−!		−	+	−!	+	+	−	−!	−!	−	+
30	−		−	+	+	+!	+!	−!	−!	−	−	±
31	−		−		+!		+	−!		−!		−

[a]The main direction on each Greenwich day is given. Symbols: + or −, the IMF is directed away from or toward the sun, respectively; ± or ∓, the IMF was directed away from the sun in the first part and away from the sun in the second part, and vice versa; !, the most reliable determination; ?, a difficult case; ×, the absence of data.

References

1. V. V. Abros'kin, Some results of the action of the earth's magnetic field on plants, *in: Conf. Effect of Magnetic Fields on Biological Objects,* Abstracts of Papers, Scientific Council on the Complex Problem "Cybernetics," Moscow (1966), p. 11.
2. V. V. Abros'kin, Position of mammalian and human embryo in the uterus, Abstracts of Sci. Conf. Results of Research in 1965, Zoological Faculty of Voronezh Agricultural Institute (1966), pp. 29–30.
3. V. V. Abros'kin, Relation between orientation of germinating seeds and developing plants and their sex determination (as exemplified by hemp and cucumber), *Fiziol. Rast.* **15**(1), 167 (1968).
4. V. V. Abros'kin, A possible effect of geomagnetism and solar activity on some characters of *Drosophila, in: Proc. Sci. Conf., 1969,* K. D. Glinka Voronezh Agricultural Institute, Veterinary Medicine, No. 2, Voronezh (1969), p. 69.
5. V. V. Abros'kin, Effect of orientation of eggs in incubator on sex of incubated chicks, *Zap. Voronezhskogo S/kh. Inst. im. K. D. Glinki* **38,** 219 (1970).
6. V. V. Abros'kin, The magnetic field of the earth (MFE) and sex determination of the human embryo, *18th Sci. Conf. Physiologists of the South RSFSR,* Vol. 2, Voronezh (1971), p. 8.
7. V. V. Abros'kin, Sexual features of animals and terrestrial magnetism, *Zap. Voronezhskogo S/kh. Inst. im. K. D. Glinki* **46,** 111–119 (1971).
8. V. V. Abros'kin, Effect of solar activity and magnetic field of the earth (MFE) on vital activity and sexual characters of horses in ontogenesis, *in: The Sun, Electricity, and Life, Izd. MGU, Moscow* (1972), p. 80.
9. V. V. Abros'kin, Effect of the earth's magnetic field on early ontogenesis, *in: Physicomathematical and Biological Problems of Effect of Electromagnetic Fields and Ionization of Air,* Vol. 2, Nauka, Moscow (1975), pp. 78–81.
10. V. V. Abros'kin, A. V. Grebenyuk, A. I. Lakomkin, and É. A. Mel'nikov, Some data on the effect of a steady magnetic field and the geomagnetic field on seed germination rate, *in: Materials of Second All-Union Conference on Effect of Magnetic Fields on Biological Objects,* Moscow (1969), p. 18.
11. V. V. Abros'kin and P. G. Zadonskii, Effect of orientation of cucumber seedlings in the earth's magnetic field, *Zap. Voronezhskogo S/kh. Inst. im. K. D. Glinki* **34**(2), 86 (1968).
12. V. V. Abros'kin, L. M. Il'in, I. G. Kondratenko, and A. M. Ponomareva, Embryogenesis and fecundity of animals in relation to solar activity cycles and planetary index of magnetic disturbance, *Proc. 3rd All-Union Symp. Effect of Magnetic Fields on Biological Objects,* Kaliningrad State University, Kaliningrad (1975), p. 43.

13. V. V. Abros'kin and A. P. Salei, Effect of magnetic fields and different concentrations of iron in nutrient media on seed germination rate, *in: Reaction of biological Systems to Weak Magnetic Fields,* Moscow (1971), p. 84.
14. V. V. Abros'kin, A. P. Salei, and Yu. A. Dmitrieva, Possible role of iron-containing solutions in effects observed in orientation of biological objects in a magnetic field, *in: Reaction of Biological Systems to Weak Magnetic Fields,* Moscow (1971), pp. 92–94.
15. Yu. N. Achkasova, Metabolism and rate of multiplication of microorganisms developing in conditions of shielding from electric and magnetic fields, *Trudy Krymskogo Meditsinskogo Inst.* **53,** 51–56 (1973).
16. Yu. N. Achkasova, N. I. Bryzgunova, T. A. Sarachan, and K. D. Pyatkin, Effect of electromagnetic fields similar to natural fields on vital activity of microorganisms, *in: Physicomathematical and Biological Problems of Effect of Electromagnetic Fields and Ionization of Air,* Vol. 2, Nauka, Moscow (1975), pp. 131–132.
17. Yu. N. Achkasova and B. M. Vladimirskii, Effect of low-frequency electromagnetic fields on microbiological objects, *in: Adaptation of Organism to Physical Factors,* NII Éksperim. i Klin. Med., Vilnius (1969), p. 250.
18. Yu. N. Achkasova and B. M. Vladimirskii, Reaction of microorganisms to a magnetic field with micropulsations of the pc2 type, *in: Effect of Natural and Weak Artificial Magnetic Fields on Biological Objects,* Belgorod (1973), pp. 127–128.
19. Yu. N. Achkasova, B. M. Vladimirskii, K. D. Pyatkin, and N. I. Bryzgunova, Sectoral boundaries of interplanetary magnetic field and vital activity of bacteria, *Proc. 3rd All-Union Symp. Effect of Magnetic Fields on Biological Objects,* Kaliningrad State University, Kaliningrad (1975), pp. 71–72.
20. Yu. N. Achkasova, B. M. Vladimirskii, and A. I. Smirnov, Effect of magnetic field with pc1 SPF frequency on microorganisms, *in: Reaction of Biological Systems to Weak Magnetic Fields* Moscow (1971), p. 97.
21. Yu. N. Achkasova, L. V. Monastyrskikh, M. G. Guk, and A. K. Grigor'eva, Change in properties of microorganisms due to shielding from natural magnetic fields, *in: Effect of Natural and Weak Artificial Magnetic Fields on Biological Objects,* Belgorod (1973), pp. 129–130.
22. Yu. N. Achkasova, T. A. Sarachan, and A. K. Grigor'eva, Effect of low-strength magnetic fields on transmission of chromosome markers and episomal determinants, *Trudy Krymskogo Meditsinskogo Inst.* **55,** 43–45 (1974).
23. Yu. N. Achkasova and L. V. Monastyrskikh, Effect of ultralow-frequency electromagnetic fields on white mice, *in: Effect of Electromagnetic Fields on Biological Objects,* Kharkov State Medical Institute, Kharkov (1973), pp. 46–47.
24. A. S. Adamchik, Effect of solar activity on blood system, *in: Abstracts of Papers of Sci. Conf. Young Scientists and Specialists of the Sci.-Res. Inst. of Medical Radiology*, Obninsk (1972), p. 3.
25. A. S. Adamchik, Seasonal changes in indices of homeostasis of healthy people and geomagnetic disturbances, *Proc. 3rd Sci. Conf. Young Scientists of Kuban Medical Inst.,* Krasnodar (1974), pp. 3–4.
26. A. S. Adamchik, Effect of seasonal and heliogeophysical factors on composition of peripheral blood and blood clotting of healthy people in northern Urals, Author's abstract of candidate's dissertation, Institute of Hematology and Blood Transfusion, Ministry of Health of the USSR, Moscow (1974).
27. V. G. Adamenko, Physical conditions leading to cold emission of living organisms, *in: Some Questions of the Biodynamics and Bioenergetics of the Organism in the Normal and Diseased State. Biostimulation by Laser Radiation,* Part 2, Kazakh State University, Alma-Ata (1972), pp. 32–34.
28. V. I. Afanas'eva, Families of geomagnetic storms in 1957–1964. II. 27-Day periodicity of

geomagnetic storms with sudden and gradual commencements, *Geomagnetizm i Aeronomiya* **9**(3), 697 (1969).
29. V. A. Afanas'ev and E. P. Kollegov, Some results of oriented sowing of tree seeds in Kamchatka, *in: Effect of Natural and Weak Artificial Magnetic Fields on Biological Objects*, Belogorod (1973), pp. 90–91.
30. V. A. Afanas'ev and E. P. Kollegov, Effect of magnetic orientation of sowing of pine and larch seeds on survival of shoots in the Petropavlovsk-Kamchatka Region, *Proc. 3rd All-Union Symp. Effect of Magnetic Fields on Biological Objects,* Kaliningrad State University, Kaliningrad (1975), p. 192.
31. V. M. Afonina, V. B. Chernyshev, and S. A. Yarovenko, Effect of shielding from electromagnetic fields on life span of *Drosophila, in: Effect of Natural and Weak Artificial Magnetic Fields on Biological Objects,* Belgorod (1973), pp. 83–84.
32. L. M. Aitmukhanova and G. K. Bogdanovskaya, Frequency of vascular accidents and blood clotting in patients with coronary atherosclerosis and hypertensive disease in relation to geomagnetic disturbances, *in: Abstracts of Papers of 39th Concluding Sci. Conf. Alma-Ata Med. Inst.* (1967), p. 367.
33. V. M. Akhutin, E. Ya. Voitinskii, M. E. Livshits, *et al.*, Experimental investigation on real time scale of effect of weak magnetic field on power spectrum of electrical activity of brain, *in: Biological and Medical Cybernetics. Part 3. Neurocybernetics,* Nauka, Moscow-Leningrad (1973), pp. 9–12.
34. V. M. Akhutin, N. I. Muzalevskaya, and G. B. Kuznetsova, Effect of a weak alternating (infralow-frequency) magnetic field on sorption of neutral red by cells of intact and regenerating rat liver, *in: Effect of Natural and Weak Artificial Magnetic Fields on Biological Objects,* Belogorod (1973), p. 124.
35. V. M. Akhutin and N. I. Muzalevskaya, Effect of weak alternating infralow-frequency magnetic field on ESR in experiments *in vivo* and *in vitro, in: Effect of Natural and Weak Artificial Magnetic Fields on Biological Objects,* Belgorod (1973), pp. 57–58.
36. V. N. Akimov and A. E. Evgen'ev, Correlation between some otorhinolaryngological diseases and several meteorological and geocosmic factors, *Vestnik Otorinolar.*, No. 1, 74–75 (1975).
37. Yu. I. Alabovskii and A. N. Babenko, Heliogeophysical factors, meteorological conditions, and complications of cardiovascular diseases, *Solnechnye Dannye,* No. 7, 104 (1968).
38. Yu. I. Alabovskii and A. N. Babenko, Mortality from vascular brain diseases in years with different levels of magnetic activity, *in: Effect of Solar Activity on Earth's Atmosphere and Biosphere,* Nauka, Moscow (1971), p. 189.
39. N. A. Aladzhalova, Slow electric processes in the brain, *Izd. AN SSSR,* Moscow (1962).
40. L. M. Anan'ev and E. N. Tarvid, Some features of the design of biotelemetric equipment for the study of animals in a magnetic field, *in: Technique of Biological Telemetry and Its Application in Biology and Medicine,* Moscow (1972), pp. 87–88.
41. T. I. Andronova, Effect of heliogeophysical and meteorological factors on cardiovascular system of healthy people in northern conditions, *in: Adaptation and Health of Man in the Far North,* Krasnoyarsk (1971), pp. 15–20.
42. T. I. Andronova, Meteotropic reactions of the healthy human organism in North Europe, Author's abstract of doctoral dissertation, Novosibirsk State Med. Inst., Novosibirsk (1975).
43. T. I. Andronova and L. M. Andronov, Complex of statistical characteristics for investigation of the biological effect of the environment, *in: Physicomathematical and Biological Problems of Effect of Electromagnetic Fields and Ionization of Air,* Nauka, Moscow (1975), pp. 242–245.
44. T. N. Anisimova (compiler), *Animal Orientation and Navigation. A Bibliographic List of Russian and Foreign Literature,* Chaps. I and II, Moscow (1969).
45. A. I. Appenyanskii and V. Ya. Pechurin, Analysis of the ability of animals to orient them-

selves in relation to the geomagnetic field, *in: Reaction of Biological Systems to Weak Magnetic Fields,* Moscow (1971), p. 23.

46. V. A. Artishchenko, S. A. Vinogradov, and V. G. Perederii, Effect of weak low-frequency electromagnetic fields on myocardial morphology, *Trudy Krymskogo Meditsinskogo Inst.* **53,** 42–45 (1973).
47. Yu. A. Azhitskii, Some features of the dynamics of negative responses of patients with cerebral atherosclerosis to a change in the components of the pulsed electromagnetic field of the atmosphere, *in: Climato-Medical Problems and Questions of Medical Geography in Siberia,* Vol. 1, Tomsk (1974), pp. 165–182.
48. M. K. Babunashvili, On A. L. Chizhevskii's "Electrical and Magnetic Functions of Erythrocytes," *in: The Sun, Electricity, and Life,* Izd. MGU, Moscow (1972), p. 112.
49. R. P. Baranova, A characterization of the biological activity of some soils of forest-steppe Transuralia, *Trudy Sverdlovskogo S/kh Inst.* **19,** 38 (1970).
50. S. S. Barats, E. D. Rozhdestvenskaya, and I. M. Kheinonen, Variations of geomagnetic activity as a factor affecting the results of hormone research, *in: Current Problems of Clinical and Laboratory Diagnostics,* Moscow (1975), pp. 16–18.
51. S. S. Barats, E. D. Rozhdestvenskaya, and I. M. Kheinonen, Functional state of sympatho-adrenal system in patients with ischemic heart disease in periods of different strength of the earth's magnetic field, *in: Humoral Disturbances in Ischemic Heart Disease,* (S. S. Barats, ed.), Sverdlovsk (1975), pp. 38–46.
52. I. L. Baumgol'ts, *Effect of Earth's Magnetic Field on the Human Organism,* Sevkavkazgiz, Pyatigorsk (1936).
53. S. S. Belokrysenko, M. M. Gorshkov, and M. A. Davydova, Level of spontaneous phage production as a test of solar activity, *in: The Sun, Electricity, and Life,* Izd. MGU, Moscow (1972), p. 88.
54. N. P. Ben'kova, Solar activity, disturbances of the earth's electromagnetic field, and their possible effect on the human organism, *in: 2nd Sci. Conf. Questions of Climatology of Cardiovascular Diseases,* Moscow (1962), pp. 7–8.
55. N. P. Ben'kova, Solar activity, disturbances of the earth's electromagnetic field, and their possible effects on the human organism, *in: Climate and Cardiovascular Pathology,* Meditsina, Moscow (1965), p. 246.
56. N. P. Ben'kova, The earth's magnetic field and its variations, *in: Physicomathematical and Biological Problems of Effect of Electromagnetic Fields and Ionization of Air,* Vol. 1, Nauka, Moscow (1975), pp. 13–23.
57. R. L. Berg and S. N. Davidenkov, *Heredity and Hereditary Human Diseases,* Nauka, Leningrad (1971).
58. A. L. Bil'dyukevich, L. S. Valeeva, and U. Sh. Akhmerov, Possible role of liquid-crystal components of organism in perception of external magnetic influences, *in: Materials of 2nd All-Union Conf. Effect of Magnetic Fields on Biological Objects,* Moscow (1969), p. 42.
59. E. N. Biryukov and I. G. Krasnykh, Change in optical density of bone tissue and calcium metabolism in the cosmonauts A. G. Nikolaev and V. I. Sevast'yanov, *Kosmicheskaya Biologiya i Meditsina,* No. 6, 42 (1970).
60. L. A. Blyumenfel'd, The problem of biomagnetism, *Nauki i Zhizn',* No. 7, 89 (1961).
61. L. T. Boguslavskii and A. B. Vannikov, *Organic Semiconductors and Biopolymers,* Nauka, Moscow (1968) [*English transl.:* Plenum Press, New York (1970)].
62. G. G. Branover, A. S. Vasil'ev, S. I. Gleizer, and A. B. Tsinober, Assessment of the role of magnetohydrodynamic effects in the orientation of migrating fish, *Magnitn. Gidrodin.* **4**(1), 76 (1970).
63. G. G. Branover, A. S. Vasil'ev, and S. I. Gleizer, An investigation of the behavior of the eel in natural and artificial magnetic fields, and an analysis of the mechanism of its receptions, *Voprosy Ikhtiologii* **11**(4), 720 (1971).

64. S. E. Bresler and V. M. Bresler, The liquid-crystalline structure of biological membranes, *Dokl. Akad. Nauk SSSR* **214**(4), 936–939 (1974).
65. Z. I. Brodovskaya, N. A. Temur'yants, and O. G. Tishkin, Morphology of puppy bone marrow after the action of a weak infralow-frequency electromagnetic field, *Trudy Krymskogo Meditsinskogo Inst.* **53,** 31–33 (1973).
66. G. R. Broun, Yu. N. Andrianov, and O. B. Il'inskii, Ability of electroreceptive system of Black Sea rays to perceive a magnetic field, *Dokl. Akad. Nauk SSSR* **216**(1), 232 (1974).
67. G. R. Broun, O. B. Il'inskii, and N. K. Volkova, Study of some features of electroreceptive structures of lateral line of Black Sea rays, *Fiziol. Zh. SSSR* **58**(10), 1499 (1972).
68. A. L. Buchachenko, *Chemical Polarization of Nuclei and Electrons,* Nauka, Moscow (1974).
69. V. M. Bukalov and A. A. Narusbaev, *Design of Atomic Submarines (from Foreign Press Material),* Sudostroenie, Leningrad (1968).
70. S. Burlatskaya, *Archeomagnetism,* Nauka, Moscow (1965).
71. L. V. Chastokolenko and E. N. Nemirovich-Danchenko, Effect of seed orientation in earth's magnetic field on dynamics of germination and growth of radicles of onion, *in: Effect of Natural and Weak Artificial Magnetic Fields on Biological Objects,* Belgorod (1973), pp. 98–99.
72. L. I. Chavrenko, Effect of geomagnetic disturbances of different intensity on some indices of the blood clotting system of healthy people and rheumatic patients as a special manifestation of the reactivity of the organism, *Proc. 1st Sci. Conf. Young Scientists of Stavropol Med. Inst.,* Stavropol (1974), pp. 9–10.
73. V. I. Chechernikov, *Magnetic Measurements,* Vysshaya Shkola, Moscow (1969).
74. A. Ya. Chegodar', Effect of an electromagnetic field of low frequency and strength on frog heart activity, *18th Sci. Conf. Physiologists of the South of the RSFSR,* Vol. 1, Voronezh (1971), p. 161.
75. A. Ya. Chegodar', Analysis of effect of electromagnetic field of low frequency and different strengths on cardiovascular system of animals in a long exposure, *in: Effect of Electromagnetic Fields on Biological Objects,* Kharkov (1973), pp. 18–21.
76. A. Ya. Chegodar', Author's abstract of dissertation: *Effect of Electromagnetic Fields of Low Frequency and Different Strengths on Cardiovascular System of Animals,* Crimean State Med. Inst., Simferopol (1972).
77. A. M. Chernukh, L. I. Vinogradova, B. M. Gekht, M. N. Gnevyshev, and K. F. Novikova, Effect of solar activity (geomagnetic disturbances) on patients with defective hypothalamic brain structures, *in: Physicomathematical and Biological Problems of Effects of Electromagnetic Fields and Ionization of Air,* Vol. 2, Nauka, Moscow (1975), pp. 57–59.
78. V. B. Chernyshev, Effect of disturbances of earth's magnetic field on insect activity, *in: Conf. Effect of Magnetic Fields on Biological Objects,* Abstracts of Papers, Scientific Council on the Complex Problem "Cybernetics," Moscow (1966), p. 80.
79. V. B. Chernyshev, Psychological errors in determination of orientation of living objects, *Zhur. Obshch. Biol.* **31**(6), 42 (1970).
80. V. B. Chernyshev, Disturbances of earth's magnetic field and biological rhythmicity of the beetle *Trogoderma, Zhur. Obshch. Biol.* **29**(6), 719 (1968).
81. V. B. Chernyshev, Insect behavior and electromagnetic fields, *in: Materials of 2nd All-Union Conf. Effect of Magnetic Fields on Biological Objects,* Moscow (1969), p. 248.
82. V. B. Chernyshev, Disturbance of earth's magnetic field and insect motor activity, *in: Effect of Solar Activity on Earth's Atmosphere and Biosphere,* Nauka, Moscow (1971), p. 215.
83. V. B. Chernyshev, Solar activity, disturbances of geomagnetic field and insect behavior, *in: The Sun, Electricity, and Life,* Izd. MGU, Moscow (1972), p. 87.
84. V. B. Chernyshev, Behavior of insects in natural and artificial fields, *Dokl. Moskovskogo Ob. Ispytat. Prirody,* 26–27 (1972).
85. V. M. Chervinets, Variation of *Escherichia* due to experimentally simulated fluctuations of

the geomagnetic field, *Proc. 3rd All-Union Symp. Effect of Magnetic Fields on Biological Objects,* Kaliningrad State University, Kaliningrad (1975), p. 54.

86. D. N. Chetaev and V. A. Yudovich, Directional analysis of magnetotelluric observations, *Izv. Akad. Nauk SSSR, Fiz. Zemli,* No. 12, 61 (1970).
87. D. N. Chetaev, V. A. Morgunov, V. A. Troitskaya, and K. Yu. Zybin, The vertical electric component of the geomagnetic pulsation field, *Dokl. Akad. Nauk SSSR* **218**(4), 828–829 (1974).
88. V. M. Chibrikin, A. M. Kogan, A. I. Appenyanskii, and L. A. Piruzyan, Effect of an electromagnetic field on the living organism, *Proc. 3rd All-Union Symp. Effect of Magnetic Fields on Biological Objects,* Kaliningrad State University, Kaliningrad (1975), p. 39.
89. V. A. Chigirinskii, Possible use of dark adaptation to investigate the effect of a magnetic field on the human organism, *in: Materials of 2nd All-Union Conf. Effect of Magnetic Fields on Biological Objects,* Moscow (1969), p. 251.
90. A. L. Chizhevskii, *Epidemic Catastrophes and Periodic Activity of Sun,* Izd. VOVG, Moscow (1930).
91. A. L. Chizhevskii, *Structural Analysis of Moving Blood,* AN SSSR, Moscow (1959).
92. A. L. Chizhevskii, One form of specifically bioactive or Z emission of the sun, *in: The Earth in the Universe,* Mysl', Moscow (1964), p. 342.
93. P. P. Chuvaev, Effect of plant growth stimulators and inhibitors and phases of moon on seeds oriented differently in the earth's magnetic field, *in: Conf. Effect of Magnetic Fields on Biological Objects,* Scientific Council on the Complex Problem "Cybernetics," Moscow (1966), p. 82.
94. P. P. Chuvaev, Effect of compass orientation on speed of germination and nature of growth of seedlings, *Fiziol. Rast.* **14**(3), 540 (1967).
95. P. P. Chuvaev, Effect of an extremely weak steady magnetic field on seedling root tissues and on some microorganisms, *in: Materials of 2nd All-Union Conf. Effect of Magnetic Field on Biological Objects,* Moscow (1969), p. 252.
96. V. I. Danilov, Possible mechanism of action of electromagnetic fields in the earth's biosphere, *in: Physicomathematical and Biological Problems of Effect of Electromagnetic Fields and Ionization of Air,* Vol. 1, Nauka, Moscow (1975), pp. 168–174.
97. V. I. Danilov, G. G. Demirchoglyan, Kh. O. Nagepetyan, and Sh. V. Grigoryan, Possible mechanisms of magnetosensitivity of birds, *in: Materials of 2nd All-Union Conf. on Effect of Magnetic Fields on Biological Objects,* Moscow (1969), p. 70.
98. V. I. Danilov, B. S. Fedorenko, R. D. Govorun, V. N. Gerasimenko, S. V. Vorontsova, L. A. Koshcheeva, and D. Ya. Oparina, Effect of a magnetic field slowly varying in time on the animal organism, *in: The Magnetic Field in Medicine,* Kirgizskii Gos. Med. Inst., Frunze (1974), pp. 102–103.
99. V. I. Danilov, G. G. Demirchoglyan, and Z. A. Avetisyan, Possible mechanisms of magnetosensitivity of birds, *Biol. Zhurnal Armenii* **23**(8), 26 (1970).
100. G. M. Danishevskii, *Acclimatization of Man to the North,* Medgiz, Moscow (1955).
101. I. V. Dardymov, Effect of magnetic field of Kursk magnetic anomaly on incidence of disease in the population, *in: Conf. Effect of Magnetic Fields on Biological Objects,* Abstracts of Papers, Scientific Council on the Complex Problem "Cybernetics", Moscow (1966), p. 23.
102. G. P. Dement'ev, Questions of bionics in ornithological investigations, *in: Migration of Birds and Mammals,* Nauka, Moscow (1965), p. 11.
103. G. P. Dement'ev, The class Aves. General characterization, *in: Animal Life,* Vol. 5, Moscow (1970), p. 44.
104. V. V. Dem'yanenko, Biological processes in an amagnetic space, Abstracts of Papers of Conf. of Young Scientists of the Ivan Franko Med. Inst., Ivano-Frankovsk (1966), pp. 272–273.

105. V. V. Dem'yanenko, Role of cellular and serum components of blood in magnetobiological reactions, *in: Hygienic Assessment of Magnetic Fields,* Izdanie Scientific Council on the Complex Problem "Cybernetics," (1972), pp. 61–66.
106. A. S. Derevenko and G. Kh. Molotkovskii, Possible effect of earth's magnetic field on sex of enantiomorphic forms of corn plants, *Fiziol. Rast.* **17**(6), 1217 (1970).
107. A. S. Derevenko and G. Kh. Molotkovskii, Orientation in the geomagnetic field and sex of corn enantiomorphs, *in: Reaction of Biological Systems to Weak Magnetic Fields,* Moscow (1971), p. 81.
108. N. R. Deryapa, N. P. Neverova, A. P. Solomatin, G. A. Zherebtsov, V. S. Posnyi, L. B. Al'perin, and I. A. Isavina, Effect of geomagnetic field on state of adrenal cortex fraction, *in: Physicomathematical and Biological Problems of Effect of Electromagnetic Fields and Ionization of Air,* Vol. 2, Nauka, Moscow (1975), pp. 77–78.
109. Yu. K. Didyk, Physicomathematical basis of Kainosymmetry, *Zh. Obshch. Khim.* **44**(12), 2601–2605 (1974).
110. Yu.K. Didyk, É. V. Artamonov, and B. K. Vasil'ev, Substantiation of optimal variants of periodic systems and the periodic law, Collection of Scientific Works of Norilsk Evening Industrial Institute, No. 17, Krasnoyarsk (1975), pp. 91–108.
111. I. V. Dobroserdova, Effect of microelements on water regime of some trees, *Fiziol. Rast.* **9**(5), 586 (1962).
112. N. T. Dolinina, Relation between geomagnetic rhythms and incidence of disease, *in: Treatment and Prevention Assistance for Workers in Chelyabinsk Metallurgical Plant,* Chelyabinsk (1974), pp. 76–80.
113. Ya. G. Dorfman, *Magnetic Properties and Structure of Matter,* GITTL, Moscow (1955).
114. Ya. G. Dorfman, *Physical Mechanism of Effect of Static Magnetic Fields on Living Systems,* Izd. VINITI, Moscow (1966).
115. Ya. G. Dorfman, Physical effects induced in living objects by steady magnetic fields, *in: Effect of Magnetic Fields on Biological Objects,* Nauka, Moscow (1971), p. 15.
116. E. A. Draganets and M. P. Travkin, Nature of orientation of root creases of radish in a geomagnetic field, *in: Effect of Natural and Weak Artificial Magnetic Fields on Biological Objects,* Belgorod (1973), p. 161.
117. I. P. Druzhinin and N. V. Kham'yanova, *Solar Activity and Sudden Changes in Course of Natural Processes on Earth,* Nauka, Moscow (1969).
118. I. P. Druzhinin, B. I. Sazonov, and V. N. Yagodinskii, *Space–Earth. Prognoses,* Mysl', Moscow (1974).
119. A. P. Dubrov and E. V. Bulygina, Effect of some geophysical factors and a steady magnetic field on exogenous rhythmicity of physiological processes in plants, *in: Conf. Effect of Magnetic Fields on Biological Objects,* Abstracts of Papers, Nauchn. Sovet po Kompl. Probleme "Kibernetika", Moscow (1966), p. 29.
120. A. P. Dubrov, *Genetic and Physiological Effects of Ultraviolet Radiation on Higher Plants,* Nauka, Moscow (1968).
121. A. P. Dubrov, Effect of heliogeophysical factors on membrane permeability and diurnal rhythm of excretion of organic substances by plant roots, *Dokl. Akad. Nauk SSSR* **187**(6), 1429 (1969).
122. A. P. Dubrov, Effect of natural electric and magnetic fields on permeability of biological membranes, *in: Materials of 2nd All-Union Conf. Effect of Magnetic Fields on Biological Objects,* Moscow (1969), p. 79.
123. A. P. Dubrov, Contemporary heliobiology, *Nauka i Zhizn'* **9,** 97 (1970).
124. A. P. Dubrov, Effect of geomagnetic field on physiological processes in plants, *Fiziol. Rast.* **17**(4), 836 (1970).
125. A. P. Dubrov, Dissymmetry of biological reactions and the geomagnetic field, *in: Reaction of*

Biological Systems to Weak Magnetic Fields, Moscow (1971), p. 9.
126. A. P. Dubrov, Role of dissymmetry of biological systems and their reactions to the effect of heliogeophysical factors, *in: Symmetry in Nature,* Leningrad (1971), p. 365.
127. A. P. Dubrov, Relationship between radiosensitivity of animals and the geomagnetic field, *Radiobiologiya,* **11**(4), 613 (1971).
128. A. P. Dubrov, Global changes in biochemical and physicochemical processes due to the geomagnetic field, *in: Questions of Theory and Practice of Magnetic Treatment of Water and Aqueous Systems,* Tsvetmetinformatsiya, Moscow (1971), p. 302.
129. A. P. Dubrov, Spectrophotometric method of determining excretory substances, *in: Physiological and Biochemical Bases of Interaction of Plants in a Phytocenose,* No. 2, Naukova Dumka, Kiev (1971), pp. 158–162.
130. A. P. Dubrov, Effect of cosmic factors on micro- and macroevolutionary processes in earth's biosphere, *in: Cosmos and Evolution,* Moscow (1974), p. 176.
131. A. P. Dubrov, Heliobiological bases of rhythmicity in earth's biosphere, *in: Rhythmicity of Natural Processes,* Gidrometeoizdat, Leningrad (1971), p. 53.
132. A. P. Dubrov, "Global nonreproducibility of biological and physicochemical reactions, *in: The Sun, Electricity, and Life,* Izd. MGU, Moscow (1972), p. 48.
133. A. P. Dubrov, Effect of heliogeophysical factors on rhythmicity of excretion of organic substances by plant roots, *in: The Sun, Electricity, and Life, Izd.* MGU, Moscow (1972), p. 76.
134. A. P. Dubrov, Heliobiological factors and dynamics of excretion of organic substances by plant roots, *in: Effect of Some Cosmic and Geophysical Factors on Earth's Biosphere,* Nauka, Moscow (1973), p. 67.
135. A. P. Dubrov, Functional symmetry and dissymmetry of biological objects, *Zh. Obshch. Biol.* **34**(3), 440 (1973).
136. A. P. Dubrov, Some aspects of heliobiological responsibility for rhythmicity in elements of biosphere, *in: Lectures in Memory of L. S. Berg,* XV–XIX, Nauka, Leningrad (1973), p. 233.
137. A. P. Dubrov, Effect of cosmic factors on micro- and macroevolutionary processes in the biosphere, *in: The Cosmos and the Evolution of Organisms,* Paleontol. Inst., AN SSSR, Moscow (1974), pp. 156–158.
138. A. P. Dubrov, Effect of geomagnetic field on genetic homeostasis, *in: Investigation of Species Productivity in Range,* Mintis, Vilnius (1975), pp. 168–175.
139. A. P. Dubrov, Global synchronous experiments for investigation of the biological effect of natural geophysical factors, *Proc. 3rd All-Union Symp. Effect of Magnetic Fields on Biological Objects,* Kaliningrad (1975), pp. 3–7.
140. A. P. Dubrov, Artificial and natural electromagnetic fields as an important factor in human ecology, *in: The Biosphere and Man,* Nauka, Moscow (1975), pp. 48–49.
141. A. P. Dubrov, Biological effects of an electric field, *in: Physicomathematical and Biological Problems of Effect of Electromagnetic Fields and Ionization of Air,* Vol. 1, Nauka, Moscow (1975), pp. 147–154.
142. A. P. Dubrov and E. V. Bulygina, Rhythmicity of excretion of organic substances by cereal roots, *Fiziol. Rast.* **14**(2), 257–263 (1967).
143. A. P. Dubrov and E. V. Dubrovina, Effect of geophysical and cosmic factors on endogenous rhythmicity in plants, *in: 4th All-Union Conf. Problems of Planetology,* Vol. 2, Leningrad (1968), pp. 98–100.
144. A. P. Dubrov, A. T. Platonova, T. G. Neeme, S. A. Mamaev, and P. P. Chuvaev, Simultaneous investigation of excretory rhythmicity at different geographic points, *in: All-Union Conf. Interrelations of Plants in Phytocenoses,* Abstracts, Minsk (1969), p. 14.
145. A. G. Dudoladov and K. S. Trincher, Ferroelectric properties of intracellular water and their role in magnetobiology, *in: Materials of 2nd All-Union Conf. Effect of Magnetic Fields on Biological Objects,* Moscow (1969), p. 87.

146. A. G. Dudoladov and K. S. Trincher, Ferroelectric properties of intracellular water and their role in magnetobiology, *Biofizika* **16**(3), 547 (1971).
147. S. I. Dumbadze, M. A. Khvedelidze, M. A. Sokolova, S. N. Zhorzholiani, and É. K. Gabritsidze, The magnetic dipole in insects, *Soobshch. AN Gruz. SSR* **55**(2) p. 285 (1968).
148. H. Durville, *How to Cure Diseases with the Earth's Magnetism and Electricity,* Izd. Z. S. Bisskii, Kiev (1913).
149. H. Durville, *Treatment of Diseases with Magnets,* Izd. Z. S. Bisskii, Kiev (1913).
150. A. V. D'yakov, Use of information on sun's activity in long-range hydrometeorological forecasting (experience of 1941–1972), *in: Abstracts of Papers of 1st All-Union Conf. Solar-Atmospheric Relations in the Theory of Climate and Weather Forecasting,* Moscow (1972), p. 35.
151. B. L. Dzerdzeevskii, Analysis of variation of nature of general circulation of atmosphere and indices of climatic elements on earth's surface over many years, *Geofiz. Bull. Mezhduved. Geofiz. Komiteta pri Prezidiume AN SSSR,* No. 14 (1964).
152. L. V. Egorova, Physical fields—an important ecological factor, *in: The Sun, Electricity and Life,* Izd. MGU (1969), p. 30.
153. M. S. Eigenson, *The Sun, Weather, and Climate,* Gidrometeoizdat, Leningrad (1963).
154. M. S. Eigenson, M. N. Gnevyshev, A. I. Ol', and B. M. Rubashev, *Solar Activity and Its Terrestrial Manifestations,* Part 3, Ogiz, Moscow-Leningrad (1948).
155. M. S. Éigenson, *Essays on Physical Geographic Manifestations of Solar Activity,* Izd. LGU, Lvov (1957).
156. A. L. Él'darov and Yu. A. Kholodov, Effect of a steady magnetic field on motor activity of birds, *Zhur. Obshch. Biol.* **25**(3), 224 (1964).
157. G. T. Ermolaev, Effect of meteorological and heliogeophysical factors on patients with hypertensive disease in a Riga seashore resort, *in: Materials of 4th Sci. Conf. Problem "Climate and Cardiovascular Pathology,"* Moscow (1969), p. 73.
158. G. T. Ermolaev, Effect of geomagnetic disturbances on patients with cardiovascular disease, *in: Physicomathematical and Biological Problems of Effect of Electromagnetic Fields and Ionization of Air,* Vol. 2, Nauka, Moscow (1975), p. 74.
159. M. M. Ermolaev, Some mechanisms of effect of earth's magnetosphere on biological processes, *Proc. 3rd All-Union Symp. Effect of Magnetic Fields on Biological Objects,* Kaliningrad State University, Kaliningrad (1975), pp. 50–51.
160. Ya. I. Fel'dshtein, Variations of magnetic fields in interplanetary space and on earth's surface, *Vestnik AN SSSR,* No. 8, 15–26 (1973).
161. Ya. I. Fel'dshtein, Interplanetary magnetic fields and their relation to variations of the magnetic field on the earth's surface, *in: High-Latitude Geophysical Phenomena,* Nauka, Leningrad (1974), pp. 22–61.
162. E. Z. Gak, Effect of external and local magnetic fields on biological objects, *Trudy Leningradskogo Obshch. Est.* **76**(1), 57–59 (1971).
163. I. E. Ganelina, *The Acute Period of Myocardial Infarct,* Medgiz, Leningrad (1970).
164. I. E. Ganelina and B. A. Ryvkin, Effect of some meteorological and heliogeophysical factors on course of primary acute myocardial infarct, *Kardiologiya* **8,** 21–29 (1973).
165. I. E. Ganelina, S. K. Churina, and N. V. Savoyarov, State of environmental physical factors and incidence of main complications of acute myocardial infarct, *Kardiologiya* **10,** 112–118 (1975).
166. I. E. Ganelina, S. K. Churina, and N. V. Savoyarov, Strength of magnetic field and incidence of myocardial infarct, *in: Physicomathematical and Biological Problems of Effect of Electromagnetic Fields and Ionization of Air,* Vol. 2, Nauka, Moscow (1975), pp. 75–76.
167. M. K. Geikin, Analogs of multipoint electrical conduction in living nature, *in: The Light of Helium-Neon Lasers in Biology and Medicine,* Alma-Ata (1970), p. 69.
168. B. B. Gershevich, Ya. M. Dymshits, and I. V. Tyun'kov, Amounts of some elements in

organs of white mice subjected to a constant magnetic field, *in: The Magnetic Field in Medicine,* Kirgiskii Gos. Med. Inst., Frunze (1974) pp. 130–131.
169. M. S. Gilyarov, Termites in the humid tropics, *Entomol. Ob.* **40**(3), 713 (1961).
170. S. I. Gleizer, A magnetobiological experiment with the European eel, *in: Reaction of Biological Systems to Weak Magnetic Fields,* Moscow (1971), p. 20.
171. S. I. Gleizer, Early ontogenesis of unconditioned responses of the eel to the natural magnetic field of the earth, *in: Questions of Fish Behavior,* AtlantNIRO, Kaliningrad (1971), p. 6.
172. S. I. Gleizer, Candidate's dissertation: *Investigation of Response of European Eel to the Earth's Natural Magnetic Field,* Kaliningrad State University, Kaliningrad (1971).
173. S. I. Gleizer, Some features of the geomagnetic orientation of the young European eel, *in: Animal Behavior,* Nauka, Moscow (1972), p. 112. (1st All-Union Conf. Ecological and Evolutionary Aspects of Animal Behavior, Abstracts of papers.)
174. S. I. Gleizer, A. M. Gorodnitskii, R. M. Demenitskaya, N. N. Trubyatchinskii, V. A. Khodorkovskii, and V. N. Yakovlev, Effect of geomagnetic field on navigation of young eels, *in: Geophysical Survey Methods in the Arctic,* Trudy Nauchno-Issled, Inst. Geologii Arktiki, No. 6, Leningrad (1971), pp. 101–106.
175. S. I. Gleizer and V. A. Khodorkovskii, Experimental determination of geomagnetic reception in the European eel, *Dokl. Akad. Nauk SSSR* **201**(4), 964 (1971).
176. M. N. Gnevyshev and K. F. Novikova, Effect of solar activity on earth's biosphere, *in: The Biosphere and Its Resources,* Nauka, Moscow (1971), p. 237.
177. M. N. Gnevyshev, K. F. Novikova, A. I. Ol', and N. V. Tokareva, Premature death due to cardiovascular diseases and solar activity, *in: Effect of Solar Activity on Earth's Atmosphere and Biosphere,* Nauka, Moscow (1971), p. 179.
178. M. N. Gnevyshev and A. I. Ol' (eds.), *Effect of Solar Activity on Earth's Atmosphere and Biosphere,* Nauka, Moscow (1971).
179. N. K. Golobokii, Physicochemical features of induction of magnetic susceptibility in blood erythrocytes, *in: Biological and Medical Electronics,* No. 3, Sverdlovsk (1972), pp. 58–61.
180. A. S. Golovatskii, S. I. Sikora, and V. I. Boiko, Effect of constant magnetic field of environment on electronic activity of neurons, *Biophysical Aspects of Pollution of the Biosphere,* Moscow (1973), pp. 33–34.
181. B. A. Golubchak and L. V. Vasilik-Parkulab, Investigation of ESR of patients with pulmonary tuberculosis in a space partially shielded from the geomagnetic field, *in: Effect of Natural and Weak Artificial Magnetic Fields on Biological Objects,* Belgorod (1973), p. 71.
182. N. K. Golobokii, Possible mechanisms of effect of geomagnetic fields on blood of patients with cardiovascular diseases, *in: Abstracts of Papers of 21st Ukrainian Republic Scientific-Technical Conf.,* Kiev (1972), p. 68.
183. B. I. Gorokhovskii, A. A. Dmitriev, P. L. Lokshina, and G. A. Remizov, The environment, its effect on incidence of myocardial infarct and consequent mortality, *Sov. Med.* **4,** 105 (1971).
184. V. S. Gorya, Effect of orientation of seeds in soil in relation to geomagnetic poles on growth and development of corn, *in: Brief Survey of Work of Moldavian Scientific-Research Institute of Selection, Seed-Raising, and Agrotechnics of Field Corps,* Kishinev (1969), p. 103.
185. R. M. Granovskaya, On the electromagnetic field of the brain, *Trudy Len. Ob. Est.* **72**(1), 111 (1961).
186. A. V. Gul'el'mi and V. A. Troitskaya, *Geomagnetic Pulsations and Diagnostics of the Magnetosphere,* Nauka, Moscow (1973).
187. L. V. Gubskii and V. I. Gubskii, A characterization of some circadian rhythms in patients with diencephalic pathology, *Zh. Nevropatol. i Psikhiatr.* **68**(3), 353–358 (1968).
188. A. V. Grebenyuk, Experimental data on the effect of a steady magnetic field on biological systems, *Materials of Rev. Sci. Conf. Biology and Soil Faculty of Voronezh State University,* No. 4, Voronezh (1969), p. 104.

189. P. I. Gulyaev, V. I. Zabotin, and N. Ya. Shlippenbakh, Various questions of aurotronics, *in: The Nervous System,* Izd. Leningradskogo Gos. Univ. Nos. 9–15 (1967–1974).
190. P. I. Gulyaev, V. I. Zabotin, and N. Ya. Shlippenbakh, Electroaurogram of frog nerve, muscle, and heart and human heart and muscle, *Dokl. Akad. Nauk SSSR* **180**(6), 1504 (1968).
191. P. I. Gulyaev, V. I. Zabotin, and N. Ya. Shlippenbakh, Electromagnetic fields associated with the movement of insects, birds, and beasts, and their possible biological significance, *in: Problems of Bionics,* Nauka, Moscow (1973), pp. 188–191.
192. P. I. Gulyaev, V. I. Zabotin, and N. Ya. Shlippenbakh, Aurogram of single axon of stretch receptor of the crayfish *Astacus astacus, Dokl. Akad. Nauk SSSR* **207**(3), 750–752 (1972).
193. P. I. Gulyaev, V. I. Zabotin, and N. Ya. Shlippenbakh, Aural electric field of isolated nerve and muscle, *Biofizika,* **19,** No. 2, 290–294 (1974).
194. P. I. Gulyaev, V. I. Zabotin, and N. Ya. Shlippenbakh, Atmospheric electromagnetic fields of biological origin, *in: Physicomathematical and Biological Problems of Effect of Electromagnetic Fields and Ionization of Air,* Vol. 1, Nauka, Moscow (1975), pp. 68–69.
195. P. I. Gulyaev, V. I. Zabotin, and N. Ya. Shlippenbakh, Electromagnetic fields associated with the movement of insects, birds, and beasts and their possible biological significance, *in: Problems of Bionics,* Nauka, Moscow (1973), p. 188.
196. N. G. Gulyuk, Effect of solar activity, variation of terrestrial magnetism, and other factors of cosmic and geophysical origin on labor rhythm and cyclicity of menstruation in women, *in: Topical Questions of Midwifery and Gynecology,* Uzhgorod (1965), p. 295.
197. L. I. Il'ina, N. A. Kostyukhina, and M. I. Mel', Solar and magnetic activity as factors inducing hypertensive crises, *2nd Sci Conf. Problems of Medical Geography,* No. 1, Leningrad (1965), p. 80.
198. L. I. Il'ina, N. A. Kostyukhina, and M. I. Mel', Climate and cardiovascular pathology, *Proc. 2nd Sci Conf. Inst. of Therapy, Academy of Med. Sci. of the USSR on Questions of Climatology of Cardiovascular Diseases,* Meditsina, Leningrad (1965), p. 83.
199. L. I. Il'ina, N. A. Kostyukhina, and M. I. Mel', Frequency of occurrence of hypertensive crises and geomagnetic activity, *in: Materials of 4th Sci. Conf. Problem ''Climate and Cardiovascular Pathology,''* Moscow (1969), p. 78.
200. E. N. Indeikin and V. P. Zhokhov, Materials on relation between variation of the earth's magnetic field and frequency of acute disturbances of retinal blood circulation, *in: Materials of Conf. All-Union Sci. Med. Soc. Ophthalmologists,* Ordzhonikidze (1970), p. 50.
201. V. M. Inyushin and P. R. Chekurov, *Biostimulation by a Laser Beam and Bioplasm,* Izd. Kazakhstan, Alma-Ata (1975).
202. V. M. Inyushin, V. A. Semykin, A. Il'in, and I. B. Beklemishev, Short-period variations in fluctuations of the optical activity of biological objects, *Proc. 3rd All-Union Symp. Effect of Magnetic Fields on Biological Objects,* Kaliningrad State University, Kaliningrad (1975), pp. 40–41.
203. V. P. Iskhakov, Problem of effect of solar activity on psychic diseases, *in: The Sun, Electricity, and Life,* MOIP, Moscow (1972), pp. 70–71.
204. V. P. Iskhakov, Periodicity in schizophrenia and its possible causation by solar activity, *Abstracts of Papers of 21st Ukrainian Republic Scientific-Technical Conf. Dedicated to 50th Anniversary of the USSR (5th Ukrainian Conf. Bionics),* Kiev (1972), pp. 53–54.
205. V. P. Iskhakov, Correlation between solar activity and incidence of schizophrenia, *in: Abstracts of 6th Sci. Morphological-Clinical Conf. Andizhan Section of All-Union Sci. Assoc. of Anatomists, Histologists, and Embryologists,* Andizhan (1972), pp. 166–167.
206. V. P. Iskhakov, Question of possible effect of solar activity on increase in number of schizophrenic patients in population, *Trudy Molodykh Uch.-Med. Uzb.* **3,** 106–107, Tashkent (1973).
207. I. V. Kachevanskaya, Effect of heliogeophysical factors on decompensation of glaucomatous

process and vascular permeability in patients with primary glaucoma, *Oftal'mol. Zh.* **1**, 42–46 (1975).

208. I. V. Kachevanskaya, Author's abstract of candidate's dissertation: *Role of Disturbances of General Vascular Permeability and Metabolic Processes in Pathogenesis of Primary Glaucoma and Effect of Heliogeophysical Factors on Its Course,* Leningrad Medical Institute of Sanitation and Hygiene (1975).
209. G. Kaden, *Electromagnetic Shields in High-Frequency and Electrocommunication Technology,* (transl.) Gosenergoizdat, Moscow–Leningrad (1957).
210. M. S. Kaibyshev, Effect of earth's magnetic field on human heart beat, *in: Abstracts of Papers of Concluding Sci. Conf. Professional and Teaching Staff of Turkmen State Med. Inst. Dedicated to 50th Anniversary of Soviet Power,* Ashkhabad (1967), p. 9.
211. M. S. Kaibyshev, Disturbances of geomagnetic field and heart rhythms, *Solnechnye Dannye,* No. 11, 96 (1968).
212. M. S. Kaibyshev, Change in minute heart rate in the course of a geomagnetic disturbance, *in: Abstracts of Papers of 21st Ukrainian Republic Scientific-Technical Conf.,* Kiev (1972), p. 55.
213. M. S. Kaibyshev, Effect of geomagnetic field components on minute heart rate, *in: Effect of Natural and Weak Artificial Magnetic Fields on Biological Objects,* Belgorod (1973), pp. 66–67.
214. M. S. Kaibyshev, Effect of changes in geomagnetic field on work capacity of a flight crew, *in: Questions of Medico-Biological Research,* Moscow (1974), pp. 63–65.
215. M. S. Kaibyshev, Author's abstract of Candidate's dissertation: *An Investigation of Preflight Medical Inspection in Civil Aviation with a View to Finding Ways of Improving It,* Inst. Medico-Biological Problems, Ministry of Health, USSR, Moscow (1975).
216. M. S. Kaibyshev, V. G. Morozova, E. N. Moskalyanova, and M. P. Travkin, Synchronous reactions of tryptophan and its metabolites at different latitudes, *in: Physicomathematical and Biological Problems of Effect of Electromagnetic Fields and Ionization of Air* (1975), pp. 110–113.
217. A. G. Kalashnikov, *Terrestrial Magnetism and Its Practical Application,* Znaniye, Moscow (1952).
218. A. G. Kalashnikov, History of the geomagnetic field, *Izv. Akad. Nauk SSSR, Ser. Geofiz.,* No. 91, 1243 (1961).
219. T. N. Karapina, A. F. Skugarevskii, and V. G. Vier, Relation between geomagnetic activity and course of epilepsy, *Proc. 5th All-Union Congr. Neuropathologists and Psychiatrists,* Vol. 2, Moscow (1969), pp. 146–147.
220. A. F. Karavai, Question of effect of geomagnetic factors on some blood clotting components in patients with chronic coronary insufficiency, *in: Abstracts of Papers of Conf. Physiotherapists and Health-Resort Specialists of Central Kazakhstan,* Karaganda (1970), p. 151.
221. A. F. Karavai and M. Sh. Tastambekova, Geomagnetic factors and some blood clotting components in patients with coronary atherosclerosis, *in: Abstracts of Materials of 1st Republican Conf. Kazakhstan Therapists,* Alma-Ata (1971), p. 57.
222. A. A. Karpenko, Some features of the behavior of the planarian *Dugesia tigrina* in the earth's and artificial magnetic fields, *Vestnik LGU Biologiya* **15**(3) 5–11 (1974).
223. T. V. Karsaevskaya, *Social and Biological Factors Leading to Changes in Human Physical Development,* Meditsina, Moscow (1970).
224. Yu. I. Kats, Study of brachiopods in light of problem of planetary periodism, *Abstracts of Papers of 2nd All-Union Conf. Mesozoic and Cainozoic Brachiopods,* Kharkov (1971), pp. 23–25.
225. Yu. I. Kats and A. I. Bereznyakov, Geomagnetic reversals: rotational causation and correlation with geological processes and the evolution of organisms, *in: The Cosmos and the*

Evolution of Organisms, Paleontologicheskii Inst. AN SSSR, Moscow (1974), pp. 199–216.
226. B. B. Kazhinskii, *Biological Radio Communication,* Izd. AN USSR, Kiev (1962).
227. V. P. Kaznacheev and S. P. Shurin, Spontaneous electromagnetic radiation in extreme cell states and problem of adaptive behavior, *in: Adaptation and the Problem of General Pathology,* Novosibirsk (1974), pp. 2–14.
228. V. P. Kaznacheev, L. P. Mikhailova, and S. P. Shurin, Informational interaction in biological systems due to electromagnetic emission in optical range, *in: Progress in Biological and Medical Cybernetics,* Moscow (1974), pp. 314–328.
229. V. P. Kaznacheev, A. P. Solomatin, and E. F. Vasilenko, Geomagnetic disturbances and strokes in Novosibirsk, *in: Some Questions of the Medical Geography in Siberia,* Nauka, Novosibirsk (1975), pp. 14–15.
230. P. P. Kazymov, Motion of kidney-bean leaves in very weak electromagnetic fields, *Fiziol. Rast.* **20**(5), 915–920 (1973).
231. V. A. Khodorkovskii, Methodological approach to study of behavior of animals in magnetic fields, *in: Materials of All-Union Symposium Effect of Artificial Magnetic Fields on Living Organisms,''* Baku (1972), p. 32.
232. V. A. Khodorkovskii, Behavior of young eels in weak nonuniform magnetic fields, *in: Materials of All-Union Symposium ''Effect of Artificial Magnetic Fields on Living Organisms,* Baku (1972), p. 37.
233. V. A. Khodorkovskii, Author's abstract of candidate's dissertation: *An Experimental Investigation of the Orientation of Young European Eels in Magnetic Fields,* Institute of Cybernetics, Kiev (1975).
234. V. A. Khodorkovskii and R. I. Polonnikov, Study of extremely weak magnetic reception in fishes, *in: Questions of Fish Behavior,* Kaliningrad (1971), p. 72.
235. V. A. Khodorkovskii and S. I. Gleizer, Effect of a uniform magnetic field on orientation of young eels in a maze, *in: Materials of All-Union Symposium ''Effect of Artificial Magnetic Fields on Living Organisms,''* Baku (1972), p. 34.
236. Yu. A. Kholodov, Author's abstract of dissertation: *Physiological Analysis of Effect of Magnetic Field on Animals,* Moscow State University (1959).
237. Yu. A. Kholodov, *Effect of Electromagnetic and Magnetic Fields on Central Nervous System,* Nauka, Moscow (1966).
238. Yu. A. Kholodov, *Magnetism in Biology,* Nauka, Moscow (1970).
239. Yu. A. Kholodov, Electromagnetic fields—New stimulants, *The Future of Science,* No. 4, Znaniye, Moscow (1971).
240. Yu. A. Kholodov, Reactions of nervous system to magnetic fields, *Issledovaniya po Geomagnetizmu, Aeronomii, i Fizike Solntsa,* No. 17, 55 (1971).
241. Yu. A. Kholodov (ed.), *Effect of Magnetic Fields on Biological Objects,* Nauka, Moscow (1971).
242. Yu. A. Kholodov, *Man in the Magnetic Web,* Znaniye, Moscow (1972).
243. Yu. A. Kholodov, *Response of Nervous System to Electromagnetic Fields,* Nauka, Moscow (1975).
244. A. N. Khramov, G. N. Petrova, A. G. Komarov, and V. V. Kochegura, Technique of paleomagnetic investigations, *Trudy VNIGRI,* No. 161, 56 (1961).
245. A. N. Khramov, *Paleomagnetic Correlation of Sedimentary Strata,* Gostoptekhizdat, Leningrad (1958).
246. A. N. Khramov and L. E. Sholpo, *Paleomagnetism,* Nedra, Leningrad (1967).
247. V. V. Khromenko, Study of daily course of photosynthesis in differently illuminated parts of crown of apple tree, *Fiziol. Rast.* **19**(2), 445–447 (1972).
248. M. A. Khvedelidze, S. I. Dumbadze, and M. Sh. Lomsadze, Controllable magnetic properties of plant seeds, *Soobshch. AN Gruz. SSR* **49,**(2), 287 (1968).
249. M. A. Khvedelidze, S. I. Dumbadze, M. Sh. Lomsadze, and N. A. Detabashvili, An

investigation of the orientation of plant seeds in a constant magnetic field before the start of germination, *Élektronnaya Obrabotka Materialov,* No. 1, 58–66 (1968).
250. M. A. Khvedelidze, S. I. Dumbadze, M. Sh. Lomsadze, N. A. Datebashvili, V. G. Zhorzholiani, and M. A. Sokolova, Bionic aspects of magnetobiological effects, *in: Problems of Bionics,* Nauka, Moscow (1973), pp. 196–201.
251. M. A. Khvedelidze, S. I. Dumbadze, M. Sh. Lomsadze, N. A. Datebashvili, V. G. Zhorzholiani, and M. A. Sokolova, Bionic aspects of magnetoelectric effects, *in: Problems of Bionics,* Nauka, Moscow (1973), p. 196.
252. S. S. Kiryushkin and A. S. Kim, Biological effect of a weak steady field, *in: Materials of 2nd All-Union Conf. Effect of Magnetic Fields on Biological Objects,* Moscow (1969), p. 112.
253. L. D. Kislovskii, Possible molecular mechanism of effect of solar activity on processes in the biosphere, *in: Effect of Solar Activity on Earth's Atmosphere and Biosphere,* Nauka, Moscow (1971), p. 147.
254. L. D. Kislovskii and B. M. Vladimirskii, Inversion of geomagnetic field and evolution of biosphere, *in: Reaction of Biological Systems to Weak Magnetic Fields,* Moscow (1971), p. 7.
255. L. D. Kislovskii and V. V. Puchkov, Metastable structures in aqueous solutions, *in: Questions of Theory and Practice of Magnetic Treatment of Water and Aqueous Systems,* Tsvetmetinformatsiya, Moscow (1971), p. 25.
256. L. D. Kislovskii and V. V. Puchkov, Possible role of water in mechanism of "direct" effect of solar activity on biological processes, *in: Adaptation of the Organism to Physical Factors,* NII Eksperim. i Klin. Med., Vilnius (1969), p. 275.
257. L. D. Kislovskii and B. M. Vladimirskii, Possible mechanism for effect of geomagnetic field reversals on evolution of biosphere, *in: The Cosmos and the Evolution of Organisms,* Paleontologicheskii Inst., AN SSSR (1974), pp. 159–164.
258. A. B. Kistyakovskii, An investigation of bird orientation and navigation in Kiev University, *in: Analyzer Systems and Orientational Behavior of Birds,* Izd. Mosk. Gos. Univ., Moscow (1971), p. 32.
259. A. B. Kistyakovskii, Duplicated systems of bird orientation and navigation, *in: Orientation and Territorial Associations of Bird Populations,* Zinatne, Riga (1973), p. 62.
260. V. I. Klassen, Status of work on the effect of magnetic fields on water, *in: Application of New Physical and Physicochemical Effects for Mineral Enrichment,* Nauka, Moscow (1967), p. 5.
261. V. I. Klassen, Magnetic treatment of water and aqueous systems, *in: Questions of Theory and Practice of Magnetic Treatment of Water and Aqueous Systems,* Tsvetmetinformatsiya, Moscow (1971).
262. A. B. Kogan, *Electrophysiology,* Vysshaya Shkola, Moscow (1969).
263. A. F. Kolchanov and M. P. Travkin, The anomalous geomagnetic field as an ecological factor, *in: Effect of Natural and Weak Artificial Magnetic Fields on Biological Objects,* Belgorod (1973), pp. 145–146.
264. A. M. Kolesnikov and M. P. Travkin, The anomalous geomagnetic field and incidence of disease in the population, *in: Hygienic Assessment of Magnetic Fields (Proc. Symp. May 20–22, 1972),* AN SSSR, Moscow (1972), pp. 156–160.
265. V. P. Kolodchenko, Distribution of incidence of myocardial infarct and geomagnetic disturbances, *Solnechnye Dannye,* No. 6, 112 (1969).
266. V. P. Kolodchenko, Incidence of myocardial infarct in Kiev and geomagnetic disturbances, *Solnechnye Dannye,* No. 12, 107 (1969).
267. V. P. Kolodchenko, Correlation between erythrocyte sedimentation rate and state of disturbance of earth's magnetic field, *in: Materials of 2nd All-Union Conf. Effect of Magnetic Fields on Biological Objects,* Moscow (1969), p. 24.
268. V. P. Kolodchenko, Effect of geomagnetic disturbances on higher nervous activity of elderly

people, *in: Abstracts of Papers of 21st Ukrainian Republic Scientific-Technical Conf.*, Kiev (1972), p.54.

269. V. P. Kolodchenko, Effect of geomagnetic field on dynamics of pulse and arterial pressure in people more than 60 years old, *in: Abstracts of Papers of 21st Ukrainian Republic Scientific-Technical Conf.*, Kiev (1972), p. 56.
270. V. P. Kolodchenko, L. D. Tikhomirov, A. K. Podshibyakin, V. V. Sirotskii, and V. I. Shakhova, Elements of effect of geomagnetic disturbances on man and animals, *in: Materials of 2nd All-Union Conf. Effect of Magnetic Fields on Biological Objects,* Moscow (1969), p. 24.
271. A. S. Kompaneets, *Symmetry in the Microworld,* Znanie, Moscow (1965).
272. V. M. Kondrashenko, Effect of geomagnetic disturbances on induction of hemorrhages from the gastrointestinal tract, *Abstracts of Papers of 7th Congr. of Surgeons in Belorussia,* Minsk (1973), pp. 44–45.
273. N. N. Koroleva, Author's abstract of dissertation: *Effect of Heliogeophysical and Meteorological Factors on Course of Hypertensive Crises in the Irkutsk Climate,* Kaunas Medical Inst. (1968).
274. A. D. Koryak, Effect of geomagnetic field on germination and growth of corn, *in: Conf. on Effect of Magnetic Fields on Biological Objects,* Scientific Council on the Complex Problem "Cybernetics," Moscow (1966), p. 43.
275. A. V. Koval'chuk, Role of geomagnetic field as a factor in changes of reactivity of the organism, *in: Materials of All-Union Symp. Effect of Artificial Magnetic Fields on Living Organisms,* Baku (1972), p. 61.
276. A. V. Koval'chuk, Dynamics of earth's magnetic field as a factor affecting level of physiological processes, *in: Effect of Natural and Weak Artificial Magnetic Fields on Biological Objects,* Belgorod (1973), pp. 37–38.
277. A. V. Koval'chuk, Many-day biorhythms and problems of homeostatic regulation, *in: Biological and Medical Cybernetics,* Part 1, Moscow–Leningrad (1974), pp. 70–73.
278. A. V. Koval'chuk, Cosmically caused many-day rhythms of physiological processes as a factor in the evolution of the animal kingdom, *in: The Cosmos and the Evolution of Organisms (Proc. Conf.),* Paleontologicheskii Inst., AN SSSR, Moscow (1974), pp. 133–149.
279. A. V. Koval'chuk, The problem of the organism–environment relation and prolonged biological rhythms, *in: Cybernetic Aspects of Adaptation of the "Man–Environment" System,* Moscow (1975), pp. 61–66.
280. A. V. Koval'chuk, G. N. Gurlach, and V. B. Perekrest, Dynamics of blood hemoglobin and fluctuations of the geomagnetic field, *in: Effect of Natural and Weak Artificial Magnetic Fields on Biological Objects,* Belgorod (1973), pp. 59–61.
281. A. V. Koval'chuk and P. K. Matveevich, Dynamics of cosmo-geophysical conditions and some questions of mathematical investigation of circadian biorhythms, *in: Circadian Rhythms of Man and Animals,* Izd. ILIM, Frunze (1975), pp. 219–221.
282. A. V. Koval'chuk and M. I. Sukhoviya, Effect of a constant magnetic field on RNA and DNA in solution, *in: Proc. 2nd All-Union Symp. Effect of Artificial Magnetic Fields on Living Organisms,* Baku (1972), pp. 26–28.
283. A. V. Koval'chuk and M. K. Chernyshev, Role of many-day biorhythms in securing the relation between living organisms and environment, *in: Physicomathematical and Biological Problems of the Effect of Electromagnetic Fields and Ionization of Air,* Vol. 2, Nauka, Moscow (1975), pp. 47–50.
284. V. V. Koval'skii and I. A. Pletneva, Diurnal rhythm of blood sugar, *Dokl. Akad. Nauk SSSR* **6**(8), 835–838 (1947).
285. V. V. Koval'skii and I. A. Pletneva, Biological rhythms and diurnal periodism of carbohydrate function of liver, *Trudy Inst. Akusherstva i Ginekologii AMN SSSR* **1,** 88–110 (1948).

286. V. A. Kozlov, A. K. Kozlova, A. Z. Neznanova, and B. I. Naledakov, Effect of solar activity on variation of some somatometric and physiological indices in the acceleration process, *in: Physicomathematical and Biological Problems of Effect of Electromagnetic Fields and Ionization of Air,* Vol. 2, Nauka, Moscow (1975), pp. 67–69.
287. L. N. Kozlova and M. M. Kozlova, Effect of steady field on human ESR, *in: Reaction of Biological Systems to Weak Magnetic Fields,* Moscow (1971), p. 52.
288. L. G. Kozyr', Candidate's dissertation: *Effect of Meteorological and Heliogeophysical Factors on Induction and Course of Myocardial Infarct in the Climatic Conditions of Voroshilovgrad,* Central Scientific-Research Institute of Health-Resort Studies and Physiotherapy, Ministry of Health, USSR, Moscow (1974).
289. N. N. Kramarenko and A. L. Chepalyga, Problem of effect of cosmic factors on the evolution of organisms and paleontology, *in: The Cosmos and the Evolution of Organisms (Proc. Conf.),* Paleontologicheskii Inst., AN SSSR, Moscow (1974), pp. 6–17.
290. V. A. Krasilov, Reconstruction of extinct plants, *Paleontol. Zh.* No. 1, 3–11 (1969).
291. E. V. Krasnov, Correlation between cyclic changes in solar radiation, paleotemperatures, and the organic world in the Cryonogene, *in: The Cosmos and the Evolution of Organisms,* PIN AN SSSR, Moscow (1974), pp. 269–275.
292. V. V. Krasyuk, A comparative characterization of the effect of magnetic and light factors of the environment on the behavior of the migrating achovy, *Proc. 3rd All-Union Symp. Effect of Magnetic Fields on Biological Objects,* Kaliningrad State University, Kaliningrad (1975), pp. 37–38.
293. M. I. Krukover, Experience of practical application of geomagnetotropism, *Proc. 3rd All-Union Symp. Effect of Magnetic Fields on Biological Objects,* Kaliningrad State University, Kaliningrad (1975), pp. 197–198.
294. M. I. Krukover, Some results of a study of the effect of the vertical component of the geomagnetic field on plant development, *Proc. 3rd All-Union Symp. Effect of Magnetic Fields on Biological Objects,* Kaliningrad State University, Kaliningrad (1975), pp. 196–197.
295. A. V. Krylov, Magnetotropism in plants, *Izv. Akad. Nauk SSSR, Ser. Biol.,* No. 2, 221 (1961).
296. A. V. Krylov, Magnetotropism in plants, *in: The Earth in the Universe*, Mysl', Moscow (1964), p. 471.
297. A. V. Krylov and G. A. Tarakanova, Magnetotropism in plants and its nature, *Fiziol. Rast.* **7**(2), 191 (1960).
298. E. V. Kuchis, *Methods of Investigating the Hall Effect,* Sovetskoe Radio, Moscow (1974).
299. B. S. Kulaev, *Reflexogenic Zone of Heart and Disturbance of Blood Circulation,* Nauka, Leningrad (1972).
300. N. Ya. Kunin and N. M. Sardonnikov, Cyclical variation of the magnetic field and the earth's climate in the Phanerozoic, *in: The Cosmos and the Evolution of Organisms,* Paleontologicheskii Inst., AN SSSR, Moscow (1974), pp. 61–82.
301. V. N. Kupetskii, Use of sun–earth relations for long-term forecasting of hydrometeorological events, *in: Sun–Atmosphere Relations in the Theory of Climate and Weather Forecasting* (É. R. Mustel', ed.), Gidrometeoizdat, Leningrad (1974), pp. 452–462.
302. S. N. Kupriyanov and I. V. Gering-Galaktionova, Oncological diseases in Turkmenia in different periods of solar activity, *Zdravookhranenie Turkmenistana,* No. 11, 25 (1967).
303. A. L. Kursanov and P. Ugryumov, Reasons for irregular course of photosynthesis during the day. Observations of daily course of respiration in sugar beet leaves, *Buyll. Mosk. Ob. Ispytatelei Prirody* **43**(1), 159 (1934).
304. N. V. Kursevich and M. P. Travkin, Effect of magnetic fields of different strength on activity of some enzymes in "Val'titskii" barley seedlings, *in: Effect of Natural and Weak Artificial Magnetic Fields on Biological Objects,* Belgorod (1973), p. 102.

305. N. V. Kursevich and M. P. Travkin, Effect of weak magnetic fields on radicle growth and respiration rate of "Val'titskii" barley seedlings, *in: Effect of Natural and Weak Artificial Magnetic Fields on Biological Objects,* Belgorod (1973), pp. 104–105.
306. S. S. Kuznetsova, Diurnal rhythm of variations of radiosensitivity of mammals, *in: Questions of General Radiobiology,* Atomizdat, Moscow (1971), p. 180.
307. L. A. Lapaeva, Mechanism of action of weak electromagnetic fields on living organisms, *Trudy Krymskogo Med. Inst.* **53,** 13–18 (1973).
308. R. Lauterbakh, Biogeophysics at the service of the biosphere, *in: Science of the Future,* Znanie, Moscow (1975), pp. 142–152.
309. L. Z. Lautsevichus, Ya. P. Yushenaite, and S. I. Blinstrubas, Some indices of solar activity, disturbances of geomagnetic field, and cardiovascular accidents, *in: Effect of Solar Activity on Earth's Atmosphere and Biosphere,* Nauka, Moscow (1971), p. 187.
310. M. D. Lebedev, Some features of the reactions of the white blood during travel in a submarine, *Voenno-med. Zh.,* No. 11, 67 (1967).
311. S. F. Leisle and V. V. Nikulin, Effect of low-strength magnetic fields on growth processes of corn, sunflower, and sugar beet, *Zap. Voronezhskogo S/kh Inst. im. K. D. Glinki* **34**(1), 113 (1967).
312. V. S. Levashev, M. M. Gorshkov, S. S. Belokrysenko, and M. G. Davydova, Level of spontaneous phage production in an *E. coli* K_{12} lysogenic system as a test of solar activity, *in: Effect of Some Cosmic and Geophysical Factors on the Earth's Biosphere,* Nauka, Moscow (1973), pp. 189–194.
313. V. S. Levashev, V. I. Danilov, M. G. Davydova *et al.* Effect of magnetic field slowly varying in time on phage production by lysogenic bacteria, *Zh. Mikrobiol. Epidemiol. Immunol.,* No. 2, 20–24 (1974).
314. Ts. A. Levina, *Prophylactic Therapy of Hypertensive Disease and Angina Pectoris,* Kiev (1970).
315. Ts. A. Levina and E. B. Ternovskaya, Relation between heliogeophysical factors and myocardial infarcts in Odessa, *2nd Sci. Conf. Problems of Medical Geography,* No. 1, Leningrad (1965), p. 82.
316. Ts. A. Levina and E. B. Ternovskaya, Relation between heliogeophysical factors and myocardial infarcts in Odessa, *in: Materials of 4th Sci. Conf. on the Problem "Climate and Cardiovascular Pathology,"* Moscow (1969), p. 82.
317. A. I. Likhachev, A magnetic field as a control factor for the central nervous system, *in: Neurobionics,* Kiev (1973), pp. 61–64.
318. N. V. Lipskaya, M. S. Babushnikov, N. P. Vladimirov, N. A. Deniskin, M. K. Kravtsova, Yu. N. Kuznetsov, and N. N. Nikiforova, *Variations of the Natural Electromagnetic Field and Their Relation to the Electrical Conductivity of the Earth's Interior,* Izd. Nauka i Tekhnika, Minsk (1972), pp. 1–21.
319. M. Sh. Lomsadze, Author's abstract of dissertation: *Investigation of Magnetoelectric Properties of Plant Seeds in Relation to Their Vital Activity,* Tbilisi State Pedagogic Institute (1971).
320. N. V. Lovelius, Rhythmic variation of increment of tree, *in: Lectures in Memory of L. S. Berg, XV–XIX,* Nauka, Leningrad (1973), p. 209.
321. O. B. Lutsyuk and G. K. Nazarchuk, Question of possible orientation of birds relative to the geomagnetic field, *Vestnik Zoologii,* No. 3, 35 (1971).
322. O. B. Lutsyuk, An investigation of the orientation of robins in an artificial magnetic field by the round-cage method, *in: Orientation and Territorial Relations of Bird Populations,* Zinatne, Riga (1973), p. 67.
323. E. V. Maksimov, Cosmic glaciation factors, *Izv. Vses. Geogr. Ob.* **102**(4) (1970).
324. D. I. Malikov, Effect of natural variations of geomagnetism on reproductivity capacity of

sheep, *in: Reaction of Biological Systems to Weak Magnetic Fields,* Moscow (1971), p. 168.

325. D. I. Malikov, Relation between changes in quality of ram semen and variations of geomagnetism, *Trudy Vses. Nauchn.-Issled. Inst. Ovtsevodstva i Kozovodstva* **1**(23), 225 (1972).
326. S. M. Mansurov, New evidence of relationships between magnetic field in space and earth, *Geomagnetizm i Aeronomiya* **9,** 622–623 (1969).
327. S. M. Mansurov and L. G. Mansurova, Relation between magnetic fields in space and earth, *Geomagnetizm i Aeronomiya* **2,** 115–118 (1971).
328. S. M. Mansurov, G. S. Mansurov, and L. G. Mansurova, Catalog of determinations of polarity of interplanetary magnetic field sectors during 1957–1974, *in: Antarktika,* No. 15, Nauka (1976).
329. B. P. Manteifel', N. P. Karpov, and V. É. Yakobi, Orientation and navigation in the animal world, *in: Bionics,* Nauka, Moscow (1965), p. 246.
330. V. I. Marchenko, Author's abstract of dissertation: *Effect of Changes in Environmental Factors on Fibrinolysis and Fibrinogenolysis in Healthy People,* Krasnoyarsk State Pedagogic Inst. (1972).
331. Masamura Shiro, Strong effect of solar activity in road accidents, *in: Effect of Solar Activity on Earth's Atmosphere and Biosphere,* Nauka, Moscow (1971), p. 209.
332. N. M. Matveev, Seasonal changes in strength of allelopathic regime in artificial tree plantings in steppe zone, *Nauchnye Dokl. Vysshei Shkoly Biol. Nauki,* No. 8 (68), 84 (1969).
333. V. I. Marchenko, Effect of solar activity on fibrinolysis and fibrinogenolysis, *Issledovaniya po Geomagnetizmu, Aeronomii i Fizike Solntsa,* No. 17, 13 (1971).
334. V. I. Marchenko, Fibrinolysis, and fibrinogenolysis on magnetoactive days, *Issledovaniya po Geomagnetizmu, Aeronomii i Fizike Solntsa,* No. 17, 16 (1971).
335. G. N. Matyushin, An unusual child of the monkey, *Khimiya i Zhizn'* **8,** 38–53 (1940).
336. G. N. Matyushin, Role of ionizing radiation in anthropogenesis, *in: The Cosmos and the Evolution of Organisms,* Paleontologicheskii Inst., AN SSSR, Moscow (1974), pp. 276–292.
337. L. D. Melikadze, É. G. Lekveishvili, and M. N. Tevdorashvili, Polyvariance of photocondensation of alkylphenanthrene hydrocarbons with maleic anhydride, *Soobshch. Akad. Nauk Gruz. SSR* **56**(2), 318–320 (1969).
338. V. P. Mel'nikov and N. A. Kryukov, Equipment for investigation of biological objects in very weak homogeneous magnetic fields, *in: Reaction of Biological Systems to Weak Magnetic Fields,* Moscow (1971), p. 137.
339. V. V. Menner, K. V. Nikiforova, M. A. Pevzner, M. N. Alekseev, Yu. B. Gladkov, G. Z. Gurarii, and V. M. Trubikhin, Paleomagnetism in the detailed stratigraphy of the Upper Cenozoic, *Izv. Akad. Nauk SSSR, Ser. Geol.,* No. 6, 3–17 (1972).
340. V. D. Mikhailova-Lukasheva, A. V. Skrikal', *et al.,* An investigation of the effect of weak electromagnetic field (EMF) differentials on man, *Dokl. AN BSSR* **16**(12), 1147–49 (1972).
341. V. M. Mikhailovskii, M. M. Krasnogorskii, K. S. Voichishin, L. I. Grabar, and V. M. Zhegar', Perception of weak magnetic fields by people, *Dopovidi Akad. Nauk URSR* **10,** Series B, 929 (1969).
342. V. N. Mikhailovskii, K. S. Voichishin, and L. I. Grabar, Perception of infralow-frequency oscillations of magnetic field by some people and means of protection, *in: Reaction of Biological Systems to Weak Magnetic Fields,* Moscow (1971), p. 147.
343. V. N. Mikhailovskii, N. N. Krasnogorskii, K. S. Voichishin, L. I. Grabar, and V. N. Zhegar', Perception of slight variations in strength of magnetic field by people, *in: Problems of Bionics,* Nauka, Moscow (1973), p. 202.
344. A. L. Mikhnev, A. K. Podshibyakin, and K. I. Shesternev, Heliogeomagnetic disturbances and incidence of myocardial infarct in Kiev, *in: The Most Important Questions of Cardiovascular Pathology,* Kiev (1967), p. 141.
345. A. A. Minkh, *Atmospheric Electricity and Medicine,* Meditsina, Moscow (1974).

346. A. T. Mironov, An electric current in the sea and the effect of the current on fish, *Trudy Morskogo Gidrofizi. Inst. AN SSSR* **1,** 56–74 (1948).
347. A. N. Mitropol'skii, Effect of season of year and magnetic storms on peripheral blood of donors, *Problemy Gematol.* **18**(8), 25–28 (1973).
348. G. Kh. Molotkovskii, *Polarity of Plant Development,* Lvov (1961).
349. G. Kh. Molotkovskii, *Polarity of Development and Physiological Genetics of Plants,* Chernovtsy (1968).
350. I. I. Morgunov, Preliminary data on effect of renal colic due to earth's magnetic field and use of "magnetic" water in kidney stone disease, *in: Problems of Clinical Pathology,* Ryazan (1965), p. 105.
351. I. I. Morgunov, Periodicity of occurrence of renal colic in patients and endemicity of renal calculus, *Urologiya i Nevrologiya,* No. 5, 3–6 (1972).
352. V. K. Morozov, Does the flower head of the sunflower show heliotropism?, *Bot. Zh.* **48**(6), 885 (1963).
353. E. N. Moskalyanova, Seasonal and diurnal variations of indicator reaction for L- and DL-tryptophan, *in: Effect of Natural and Weak Artificial Magnetic Fields on Biological Objects,* Belgorod (1973), pp. 26–28.
354. E. N. Moskalyanova and A. P. Salei, Seasonal and diurnal variations of condensation of xanthydrol with tryptophan, *in: Physicomathematical and Biological Problems of Effect of Electromagnetic Fields and Ionization of Air, All-Union Tech. Symp.,* Yalta (1975), pp. 114–118.
355. E. N. Moskalyanova and A. P. Salei, Effect of solar activity and earth's magnetic field on invariance of indicator reaction with L- and DL-tryptophan and its metabolites, *in: Proc. 3rd All-Union Symp. Effect of Magnetic Fields on Biological Objects,* Kaliningrad (1975), p. 205.
356. E. R. Mustel' Method of superposed epochs, *Byull. Nauchnye Informat. Astrono. Soveta AN SSSR,* No. 10, 98 (1968).
357. E. R. Mustel', Solar activity and the troposphere, *in: Effect of Solar Activity on Earth's Atmosphere and Biosphere,* Nauka, Moscow (1971), p. 32.
358. N. I. Muzalevskaya, Biological activity of disturbed geomagnetic field, *in: Adaptation of Organism to Physical Factors,* NII Éksperim. i Klin. Med. Vilnius (1969), p. 272.
359. N. I. Muzalevskaya, Biological activity of disturbed geomagnetic field, *in: Effect of Solar Activity on Earth's Atmosphere and Biosphere,* Nauka, Moscow (1971), p. 119.
360. N. I. Muzalevskaya, Characterization of disturbed geomagnetic field as a stimulant, *in: Effect of Some Cosmic and Geophysical Factors on Earth's Biosphere,* Nauka, Moscow (1973), pp. 123–142.
361. N. I. Muzalevskaya, Inhomogeneity of low-frequency fluctuations of electromagnetic field and its ecological significance, *in: Human Adaptation System and the External Environment,* Leningrad (1975), pp. 113–115.
362. N. I. Muzalevskaya and T. A. Larkina, Change in state of blood clotting and anticlotting system of white rats due to a weak alternating magnetic field in the infralow-frequency range in experiments, *in: Magnetic Field in Medicine,* Kirgizskii Gos. Med. Inst., Frunze (1974), p. 90.
363. N. I. Muzalevskaya and R. N. Pavlova, Nature of change in thermodynamic equilibrium of blood serum proteins due to a weak alternating magnetic field, *Proc. 3rd All-Union Symp. Effect of Magnetic Fields on Biological Objects,* Kaliningrad State University, Kaliningrad (1975), p. 35.
364. N. I. Muzalevskaya and M. G. Spasskaya, Role of set of magnetic field characteristics in formation of responses of biological systems, *in: Hygienic Assessment of Magnetic Fields (Proc. Symp., May 20–22, 1972),* AN SSSR, Moscow (1972), pp. 147–150.
365. N. I. Muzalevskaya, I. A. Okhonskaya, and M. G. Spasskaya, Response of hind lobe of

hypophysis and adrenal cortex in rats to magnetic disturbance, *in: Reaction of Biological Systems to Weak Magnetic Fields,* Moscow (1971), p. 127.

366. V. V. Navrotskii, Reactions of pulmonary tuberculosis patients to passage of atmospheric fronts, and to variation of the electromagnetic field and solar activity, *Abstracts of Symp. Cyclicity of Processes in Patients with Tuberculosis and Other Diseases,* TsNIItuberkuleza, Moscow (1973), pp. 43–45.
367. B. A. Neiman, Possible interaction of magnetic field and biological objects, *in: Bionics,* Nauka, Moscow (1965), p. 381.
368. E. N. Nemirovich-Danchenko and L. V. Chastokolenko, Effect of geomagnetic field on seed germination, *in: Questions of Botany, Zoology, and Soil Science,* No. 1, Tomsk (1973), pp. 55–59.
369. E. N. Nemirovich-Danchenko and L. V. Chastokolenko, Role of the geomagnetic field as an ecologically significant factor in the reaction of enantiomorphic onion seedlings to a constant magnetic field, *Proc. 3rd All-Union Symp. Effect of Magnetic Fields on Biological Objects,* Kaliningrad State University, Kaliningrad (1975), pp. 186–187.
370. N. P. Neverova, Author's abstract of doctoral dissertation: *State of Vegetative functions in Healthy People in the Far North,* Novosibirsk State Med. Inst., Novosibirsk (1972).
371. N. P. Neverova, T. I. Andronova, and M. I. Mochalova, Physiological mechanisms in initial period of acclimitization in the Arctic, *in: Adaptation of Man,* Nauka, Leningrad (1972), pp. 181–186.
372. N. P. Neverova, Changes in blood flow rate and adrenalin excretion in relation to fluctuations of earth's electromagnetic field in the north, *Vrachebnoe Delo,* No. 4, 87–89 (1973).
373. L. M. Nikiforova, A. A. Khadzhimetov, and S. G. Chernomorchenko, Effect of constant magnetic field on dynamics of microelements in the animal organism, *in: The Magnetic Field in Medicine,* Kirgizskii Gos. Med. Inst., Frunze (1974), pp. 138–139.
374. L. F. Nikitina, L. D. Sharygin, M. G. Éngel', *et al.,* Dynamics of cardiovascular reactions in hypertensive patients in relation to state of earth's magnetic field, *in: The Medical Service in the Health Resorts of Chelyabinsk Region,* Chelyabinsk (1974), pp. 96–100.
375. S. N. Nikolaichuk and A. S. Il'in, Some correlations between the K_p index and barley seed germination, *Proc. 3rd All-Union Symp. Effect of Magnetic Fields on Biological Objects,* Kaliningrad State University, Kaliningrad (1975), p. 191.
376. A. V. Nikulin, Orientation of root creases in left- and right-handed isomers of sugar beet in a geomagnetic field, *in: Effect of Natural and Weak Artificial Magnetic Fields on Biological Objects,* Belgorod (1973), pp. 94–95.
377. A. V. Nikulin, Effect of orientation of root creases of right- and left-handed sugar beet roots in the geomagnetic field on their seed yield, *in: Effect of Natural and Weak Artificial Magnetic Fields on Biological Objects,* Belgorod (1973), pp. 96–97.
378. A. V. Nikulin, Nature of orientation of root creases in left- and right-handed isomers of sugar beet in the geomagnetic field, *in: Proc. 3rd All-Union Symp. Effect of Magnetic Fields on Biological Objects,* Kaliningrad (1975), p. 18.
379. A. V. Nikulin, Effect of orientation of left- and right-handed fruits of sugar beet in earth's magnetic field on some physiological processes in plants developing from them, *Izv. Akad. Nauk SSSR, Ser. Biol.,* No. 6, 922 (1966).
380. A. V. Nikulin and V. F. Leisle, Responsiveness of plants grown from left- and right-handed sugar-beet fruits to orientation in earth's magnetic field and forms of nitrogen nutrition, *Fiziol. Rast.* **17**(3), 471 (1970).
381. A. V. Nikulin and V. F. Leisle, Orientation of left- and right-handed sugar beet roots in geomagnetic field, *in: Reaction of Biological Systems to Weak Magnetic Fields,* Moscow (1971), p. 79.
382. I. Novak, Treatment of some diseases with weak magnetic fields, *in: Effect of Solar Activity on Earth's Atmosphere and Biosphere,* Nauka, Moscow (1971), p. 202.

383. V. P. Novikov, Reasons for laying up and limited suitability of submarines in service, *Voenno-med. Zh.*, No. 5, 72 (1966).
384. K. F. Novikova, Author's abstract of dissertation: *Materials on Epidemiology of Myocardial Infarct and Its Early Anticoagulant Therapy from Sverdlovsk,* Sverdlovsk State Med. Inst. (1968).
385. K. F. Novikova, M. N. Gnevyshev, N. V. Tokareva, and A. I. Ol', Effect of solar activity on incidence of myocardial infarct and mortality from it, *Kardiologiya* **8**(4), 109 (1968).
386. K. F. Novikova, T. N. Panov, and A. P. Shushakov, Geomagnetic disturbances and myocardial infarct, *in: Abstracts of the Competition Symp. of Young Scientists "Arterial Thromboses and Embolisms,"* Sverdlovsk (1966).
387. K. F. Novikova, T. N. Panov, and A. P. Shushakov, Geomagnetic disturbances and myocardial infarct, *in: Materials of Scientific Session on the Problem "Climate and Cardiovascular Pathology,"* Moscow (1966), p. 80.
388. K. F. Novikova, T. N. Panov, and A. P. Shushakov, Geomagnetic disturbances and myocardial infarcts, *Solnechnye Dannye,* No. 2, 69 (1966).
389. K. F. Novikova and B. A. Ryvkin, Solar activity and cardiovascular diseases, *in: Effect of Solar Activity on Earth's Atmosphere and Biosphere,* Nauka, Moscow (1971), p. 164.
390. K. F. Novikova and A. P. Teplyakova, Investigation of diurnal fluctuations of heart-cycle phases in rheumatic patients by kineto-cardiography in relation to geomagnetic field, *in: Questions of Health-Resort Treatment of Patients with Diseases of the Circulatory Organs,* Pyatigorsk (1972), pp. 88–100.
391. K. F. Novikova and A. P. Teplyakova, Change in some physiological indices of rheumatic patients in period of adaptation to climate of Kislovodsk foothills in relation to state of geomagnetic field, *in: Questions of Health-Resort Treatment of Patients with Diseases of the Circulatory Organs,* Pyatigorsk (1972), pp. 93–97.
392. K. F. Novikova, N. S. Chubarova, and E. S. Gubina, Effect of geomagnetic disturbances on course of tubercular process, *in: Abstracts of Papers of Symp. Cyclicity of Processes in Patients with Tuberculosis and Other Diseases,* TsNIItuberkuleza, Moscow (1973), pp. 41–42.
393. Yu. I. Noviskii, The magnetic field in plant life, *in: Effect of Some Cosmic and Geophysical Factors on the Earth's Biosphere,* Nauka, Moscow (1973), pp. 104–108.
394. Yu. I. Novitskii and E. V. Markman, Additional data on radish plants with different orientation of the root creases, *in: Effect of Natural and Weak Artificial Magnetic Fields on Biological Objects,* Belgorod (1973), pp. 92–93.
395. Yu. I. Novitskii, N. I. Kalugin, V. P. Ovcharenko, A. F. Prun, and A. K. Vasil'ev, Experimental test of a possible effect of a magnetic field on orientation of root creases of "Duganskii" radish, *in: Reaction of Biological Systems to Weak Magnetic Fields,* Moscow (1971), pp. 61–64.
396. Yu. I. Novitskii and O. E. Fedorova, Observations of reactions of dark rye seedlings to a weak constant magnetic field, *Materials of 2nd All-Union Congr. Effect of Magnetic Fields on Biological Objects,* Moscow (1969), pp. 162–163.
397. Yu. I. Novitskii, V. Yu. Strekova, and G. A. Tarakanova, Effect of a steady magnetic field on plant growth, *in: Effect of Magnetic Fields on Biological Objects,* Nauka, Moscow (1971), p. 69.
398. Yu. I. Novitskii and M. P. Travkin, Orientation of roots in the geomagnetic field, *in: Materials of Scientific Methodology Conf., Belgorod State Pedagogic Inst.,* Belgorod (1970), p. 73.
399. Yu. I. Novitskii, M. P. Travkin, A. I. Lebedik, and L. K. Ivanova, Orientation of root creases of some varieties of sugar and food beet and the correlation of some economically important features with this effect, *in: Reaction of Biological Systems to Weak Magnetic Fields,* Moscow (1971), p. 68.

400. Yu. I. Novitskii and Yu. A. Kholodov, Hygienic aspects of biological effect of reduced magnetic fields, *in: Hygienic Assessment of Magnetic Fields (Proc. Symp., May 22–23, 1972),* AN SSSR, Moscow (1972), pp. 139–143.
401. A. I. Ol', Indices of disturbance of the earth's magnetic field and their heliogeophysical significance, *in: Correlation between Effects in Earth's Troposphere and Solar Activity,* Gidrometeoizdat, Leningrad (1969), p. 5.
402. A. I. Ol', Manifestations of solar activity in the earth's magnetosphere and ionosphere, *in: Effect of Solar Activity on Earth's Atmosphere and Biosphere,* Nauka, Moscow (1971), p. 104.
403. A. M. Opalinskaya, An attempt to use a chemical test for the prediction of solar activity, *in: Materials of 2nd All-Union Conf. Effect of Magnetic Fields on Biological Objects,* Moscow (1969), p. 168.
404. A. M. Opalinskaya and G. F. Plekhanov, Change in nature of relation between Piccardi test and geomagnetic field in relation to phase of solar activity, season, and time of day, *in: Effect of Natural and Weak Artificial Magnetic Fields on Biological Objects,* Belgorod (1973), pp. 18–20.
405. A. M. Opalinskaya and G. F. Plekhanov, Effect of weak infralow-frequency magnetic field on agglutination reaction, *in: Proc. 3rd All-Union Symp. Effect of Magnetic Fields on Biological Objects,* Kaliningrad State University, Kaliningrad (1975), pp. 75–76.
406. A. I. Osipov and V. P. Desyatov, Mechanism of effect of variations in solar activity on the human organism, *in: Effect of Solar Activity on Earth's Atmosphere and Biosphere,* Nauka, Moscow (1971), p. 204.
407. V. V. Ovchinnikov, G. Z. Galaktionov, R. I. Polonnikov, and L. A. Ershov, Investigation of Orientation of European river eel (*Anguilla anguilla* L.) in migration season, *in: Questions of Fish Behavior, Trudy AtlantNIRO,* No. 36, Kaliningrad (1971), p. 11.
408. L. V. Parkulab, An investigation of the ESR of infectious patients in a space partially shielded from the geomagnetic field, *in: Materials of 2nd All-Union Conf. Effect of Magnetic Fields on Biological Objects,* Moscow (1969), p. 180.
409. L. V. Parkulab, Author's abstract of dissertation: *Effect of Extreme Factors (Magnetic Fields and Shielded Spaces) on Some Components of the Infectious Process,* Kishinev State Med. Inst. (1970).
410. N. N. Pavlov, Rotation of earth, deformation of earth's crust, and solar activity, *in: Rotation and Tidal Deformations of Earth,* No. 1, Naukova Dumka, Kiev (1970), p. 385.
411. R. N. Pavlova, N. I. Muzalevskaya, and V. V. Sokolovskii, Effect of alternating electromagnetic field of infralow frequency and low strength on oxidation of low-molecular-weight thiols and hemoglobin, *in: Human Adaptation System and the External Environment,* Leningrad (1975), pp. 120–122.
412. S. A. Pavlovich, Conjugational *R* transmission in *Escherichia* in experiments with a fluctuating magnetic field, *Proc. 3rd All-Union Symp. Effect of Magnetic Fields on Biological Objects,* Kaliningrad State University, Kaliningrad (1975), pp. 59–60.
413. S. A. Pavlovich, Magnetic field as an ecological factor in variation of microorganisms, *in: Physicomathematical and Biological Problems of Effect of Electromagnetic Fields and Ionization of Air,* Vol. 2, Nauka, Moscow (1975), pp. 123–130.
414. S. A. Pavlovich and V. A. Gumenyuk, Catalase activity of microorganisms cultivated for a long period in amagnetic conditions, *Proc. 3rd All-Union Symp. Effect of Magnetic Fields on Biological Objects,* Kaliningrad (1975), p. 58.
415. S. A. Pavlovich and A. L. Sluvko, Effect of shielding from magnetic field on *Staphylococcus aureus, Proc. 3rd All-Union Symp. Effect of Magnetic Fields on Biological Objects,* Kaliningrad (1975), p. 56.
416. V. M. Pavlyuchenko, M. Yu. Grigor'ev, A. A. Girgor'ev, and E. S. Imshenetskaya, Gonadotropic function of adenohypophysis of sables and effect of some ecological factors on

it, *in: Collection of Scientific Works of Moscow Veterinary Academy,* Vol. 80, Moscow (1975), pp. 102–105.

417. E. I. Petranyuk, V. V. Navrotskii, Yu. A. Azhitskii, *et al.,* Frequency of pulmonary hemoptysis and hemorrhages during passage of atmospheric fronts and changes in earth's magnetic field, *Trudy Krymskogo Med. Inst.* **53,** 71–73 (1973).
418. G. N. Petrova, *Laboratory Assessment of Stability of Residual Magnetization of Rocis,* Izd. AN SSSR (1961).
419. G. N. Petrova, V. V. Bukha, L. N. Gomov, *et al.,* Characteristic features of transitional regimes of the geomagnetic field, *Izv. Akad. Nauk SSSR, Fiz. Zemlya,* No. 6, 53–76 (1972).
420. G. Piccardi, Solar activity and chemical tests, *in: Effect of Solar Activity on the Earth's Atmosphere and Biosphere,* Nauka, Moscow (1971), p. 141.
421. A. T. Platonova, V. V. Bublis, and V. I. Marchenko, Changes in clotting and anticlotting blood systems and solar activity, *in: Adaptation of the Organism to Physical Factors,* NII Éksperim. i Klin. Med., Vilnius (1969), p. 240.
422. A. T. Platonova, Notes on the term "normal values" used in clinical practice, *Issledovaniya po Geomagnetizmu, Aéronomii i Fizike Solntsa,* No. 17, 10 (1971).
423. G. F. Plekhanov, Author's abstract of dissertation: *Perception of "Imperceptible" Signals by Man,* Tomsk State Med. Inst. (1967).
424. G. F. Plekhanov, N. V. Vasil'ev, and A. M. Opalinskaya, Effect of geomagnetic field on course of agglutination reaction, *in: Physicomathematical and Biological Problems of Effect of Electromagnetic Fields and Ionization of Air,* Vol. 2, Nauka, Moscow (1975), pp. 116–117.
425. G. F. Plekhanov and A. M. Opalinskaya, Effect of shielding from geomagnetic field and an artificial magnetic field on fluctuation of Piccardi reaction, *in: Effect of Natural and Weak Artificial Magnetic Fields on Biological Objects,* Belgorod (1973), pp. 21–22.
426. G. F. Plekhanov and A. M. Opalinskaya, Variation of rate of precipitation of bismuth oxychloride in a magnetic field of varying frequency, *in: Reaction of Biological Systems to Weak Magnetic Fields,* Moscow (1971), p. 33.
427. T. A. Plyasova-Bakunina and É. T. Matveeva, Relation between pc1 type fluctuations and geomagnetic storms, *Geomagnetizm i Aéronomiya* **8**(1), 189 (1968).
428. V. I. Pochtarev, *The Earth—A Large Magnet,* Gidrometeoizdat, Leningrad (1974).
429. V. I. Pochtarev, *The Earth's Magnetism and Cosmic Space,* Nauka, Moscow (1966).
430. A. G. Poddubnyi, Some results of long-range observations of the behavior of migrating fish, *in: Bionics,* Nauka, Moscow (1965), p. 255.
431. A. G. Poddubnyi, *Ecological Topography of Fish Populations in Reservoirs,* Nauka, Leningrad (1971).
432. A. K. Podshibyakin, Heliogeomagnetic reasons for the variation of some biomedical tests, *Materials of 2nd Sci. Conf. Problems of Medical Geography,* No. 1, Leningrad (1965).
433. A. K. Podshibyakin, Prognosis of reliability of operation of "man and machine" system and heliogeomagnetic factors, *in: Questions of Bionics,* Nauka, Moscow (1967).
434. A. K. Podshibyakin, Attempts to determine effect of some heliogeomagnetic disturbances on man and animals, *in: The Sun, Electricity, and Life,* MOIP, Moscow (1969), pp. 37–39.
435. A. K. Podshibyakin and V. P. Kolodchenko, Elements of effect and simulation of action of fluctuations in earth's magnetic field on man and animals, *Proc. All-Union Symp. Effect of Artificial Magnetic Field on Living Organisms,* Nauchnyi Sovet po Kibernetike AN SSSR, Baku (1972), pp. 173–174.
436. A. K. Podshibyakin, I. D. Zosimovich, and V. I. Kolesnik, *et al.,* Some prerequisites for predicting deterioration of the operating reliability of the "man–machine" system in relation to heliogeophysical disturbances, *in: Problems of Bionics,* Nauka, Moscow (1973), pp. 205–210.
437. A. K. Podshibyakin and R. V. Smirnov, Group features of bioelectric anticipation of

geomagnetic disturbances by man, *in: Abstr. Papers Commun. 23rd All-Union Sci. Session Sci.-Tech. Soc. Radio-engineering and Electrocommunication,* Moscow (1967), p. 18.

438. A. K. Podshibyakin, I. D. Zosimovich, and V. I. Shakhova, Analysis of effects of heliogeophysical disturbances on cardiovascular diseases. I. Distribution of incidence of myocardial infarct and state of earth's geomagnetic field, *Solnechnye Dannye,* No. 2, 106 (1968).
439. A. K. Podshibyakin and I. D. Zosimovich, An investigation of the correlations between variations of heliomagnetic factors and the frequency of variation of the cardiovascular system (myocardial infarct), *Rev. Infor. o Zakonchennykh n.-i. Rabotakh v In-takh AN USSR,* No. 3, 77 (1969).
440. A. K. Podshibyakin, V. M. Senik, V. P. Kolodchenko, *et al.,* Ecologophysiological effects of geomagnetic disturbances on man and animals, *Abstr. Sci. Papers 11th Session All-Union Physiol. Soc.,* No. 2, Leningrad (1970), p. 423.
441. N. V. Podzorov, Effect of geomagnetic field on germination of conifers, *Izv. Vyssh. Uchebn. Zaved. Lesnoi Zh.,* No. 5, 155 (1970).
442. T. V. Pokrovskaya, *Synoptico-Climatological and Heliogeophysical Long-Range Weather Forecasts,* Gidrometeoizdat, Leningrad (1969), p. 254.
443. T. V. Pokrovskaya, Solar activity and climate, *in: Effect of Solar Activity on Earth's Atmosphere and Biosphere,* Nauka, Moscow (1971), p. 12.
444. G. A. Pospelova and A. N. Zhudin, Sequence of Pliocene-Quaternary deposits of the Priobskii steppe plateau (from paleomagnetic data), *Geologiya i Geofizika,* No. 6, 11–20 (1967).
445. A. S. Presman, *Electromagnetic Fields and Life,* Nauka, Moscow (1968).
446. A. S. Presman, Electromagnetic fields and seasonal migrations of birds, *Vestnik APN "Nauchnaya Mysl'",* No. 8, 40 (1968).
447. A. S. Presman, *Electromagnetic Fields in the Biosphere,* Znaniya, Moscow (1971).
448. A. I. Pronchenko, Effect of meteorological factors and solar activity on results of treatment of elderly and senile patients with coronary thrombosis, *in: Materials of 4th Sci. Conf. Problem "Climate and Cardiovascular Pathology,"* Moscow (1969), p. 94.
449. V. R. Protasov, *Bioelectric Fields in the Life of Fishes,* TsNIITEIRKh, Moscow (1972).
450. V. R. Protasov, *Electrical and Acoustic Fields of Fish,* Nauka, Moscow (1973).
451. V. R. Protasov, M. N. Ungerman, and É. Kh. Vekilov, *Elements of Biogeophysics,* Znanie, Moscow (1975).
452. V. R. Protasov, V. S. Shneer, and G. A. Fonarev, Effect of natural electric fields in sea on behavior and distribution of fish, *Zool. Zh.* **114**(7), 1098–1101 (1975).
453. M. M. Protod'yakonov, *Properties and Electronic Structure of Powdered Materials,* Nauka, Moscow (1969).
454. M. M. Protod'yakonov and I. L. Gerlovin, *Electronic Structure and Physical Properties of Crystals,* Nauka, Moscow (1975).
455. V. V. Protopopov and V. D. Stakanov, Some physiological changes in germinating seeds and seedlings of trees due to electromagnetic environmental factors, *in: 3rd Ural Conf. Physiology and Ecology of Woody Plants,* Ufa (1970), p. 24.
456. Z. E. Pryadchenko and F. V. Sushkov, Effect of geomagnetic field on coagulation of mammalian cells in a constant magnetic field of high strength, *in: Effect of Natural and Weak Artificial Magnetic Fields on Biological Objects,* Belgorod (1973), pp. 142–144.
457. V. P. Pyatkin, Change in state of patients with ischemic heart disease and disturbances of magnetic fields of natural origin, *in: Materials All-Union Symp. "Effect of Artificial Magnetic Fields on Living Organisms,"* Baku (1972), p. 194.
458. V. P. Pyatkin, Disturbances of atmospheric electromagnetic field and earth's magnetic field, and their effect on coronary blood circulation of patients with ischemic heart disease, *in:*

Climato-Medical Problems and Questions of Medical Geography of Siberia, Vol. 1, Tomsk (1974), pp. 158–161.
459. V. P. Pyatkin, Effect of geomagnetic field on state of patients with chronic pneumonia, *in: Physicomathematical and Biological Problems of Effect of Electromagnetic Fields and Ionization of Air,* Vol. 1, Nauka, Moscow (1975), pp. 70–71.
460. V. P. Pyatkin, G. D. Latyshev, and Yu. A. Azhitskii, Geomagnetic disturbances and clinical electrocardiographic changes in patients with ischemic and hypertensive disease, *in: Reaction of Biological Systems to Weak Magnetic Fields,* Moscow (1971), p. 156.
461. V. Raab, Neurohumoral factors in the pathology of the heart and blood vessels, *Terapevticheskii Arkhiv. Med.,* No. 4, 34 (1959).
462. V. Raab, Neural and humoral disorders of blood pressure regulation, *Terapevticheskii Arkhiv. Med.,* No. 4, 30 (1959).
463. P. Rakhno, M. Aksel', L. Sirp, and Kh. Riis, *Population Dynamics of Soil Microorganisms and Nitrogen Compounds in the Soil,* Valgus, Tallin (1971).
464. É. D. Rogacheva, A. P. Dubrov, and N. K. Bulatova, Effect of earth's magnetic field on crystallization of epsomite, *in: Materials of All-Union Symp. ''Biological Effect of Natural and Weak Artificial Fields,''* Belgorod (1973).
465. É. D. Rogacheva, A. P. Dubrov, V. P. Ershov, and N. K. Bulatova, Orienting effect of earth's magnetic field on formation of crystals of epsomite and morenosite, *Dokl. Akad. Nauk SSSR* **221**(1), 84–86 (1975).
466. É. D. Rogacheva, V. P. Ershov, and A. P. Dubrov, Tendency toward oriented growth of epsomite crystals in the earth's magnetic field, *Proc. 3rd All-Union Symp. Effect of Magnetic Fields on Biological Objects,* Kaliningrad State University, Kaliningrad (1975), p. 42.
467. I. I. Rokityanskii and R. V. Shchepetnov, Trend toward 27-day periodicity of short-period variations of the pc and pi type in 1957–1959, *Geomagnitnye Issledovaniya,* No. 6, 110 (1964).
468. Yu. Ross and V. Ross, Spatial orientation of leaves in crops, *in: Photosynthetic Productivity of the Plant Cover,* Tartu (1969), p. 60.
469. M. I. Rozanov, Tree growth curves as a source of information about some heliophysical and geophysical processes, *in: The Sun, Electricity, and Life,* Izd. MGU, Moscow (1972), p. 44.
470. E. D. Rozhdestvenskaya, Effect of geomagnetic disturbances on vegetative nervous system of healthy people and patients with atherosclerosis, *in: Abstr. 4th Interregional Conf. of Ural Therapists,* Sverdlovsk (1972), p. 214.
471. E. D. Rozhdestvenskaya, Effect of geomagnetic disturbances on vegetative nervous system of healthy people and patients with atherosclerosis, *Abstr. 4th Interregional Conf. of Ural Therapists,* Sverdlovsk (1972), pp. 214–218.
472. E. D. Rozhdestvenskaya, Author's abstract of dissertation: *Effect of Heliogeophysical Factors on State of Blood Coagulation and Their Role in the Etiopathogenesis and Prognosis of Thrombotic and Hermorrhagic Complications in Cardiovascular Diseases,* Sverdlovsk State Med. Inst. (1973).
473. E. D. Rozhdestvenskaya, Strength of earth's magnetic field as a factor affecting blood-clotting system of patients with atherosclerosis and ischemic heart disease, *in: Humoral Disturbances in Ischemic Heart Disease,* (S. S. Barats, ed.), Sverdlovsk (1975), pp. 28–34.
474. E. D. Rozhdestvenskaya and K. F. Novikova, Effect of solar activity on fibrinolytic system of blood of patients with cardiovascular diseases, *Solnechnye Dannye,* No. 8, 92 (1968).
475. E. D. Rozhdestvenskaya and K. F. Novikova, Effect of solar activity on blood clotting system and incidence of hemorrhagic complications, *in: Adaptation of Organism to External Factors,* NII Éksperim. i Klin. Med., Vilnius (1969), p. 275.
476. E. D. Rozhdestvenskaya and K. F. Novikova, Effect of solar activity on blood fibrinolytic system, *in: Effect of Solar Activity on the Earth's Atmosphere and Biosphere,* Nauka,

Moscow (1971), p. 193.

477. B. A. Ryvkin, Effect of meteorological and heliophysical factors on course and outcome of cardiovascular diseases, *Klinicheskaya Med.,* No. 6, 86 (1964).
478. B. A. Ryvkin, Author's abstract of dissertation: *Effect of Heliogeophysical and Meteorological Factors on Course and Outcome of Cardiovascular Diseases in Leningrad,* I. P. Pavlov Inst. of Physiology (1966).
479. B. A. Ryvkin, Effect of magnetic disturbances and barometric pressure drops on functional state of human cardiovascular system, *Trudy Leningradskogo Ob. Estestvoispytatelei* **76**(1), 71 (1971).
480. B. A. Ryvkin and F. Z. Ryvkina, Effect of solar and geomagnetic disturbances on cell composition and prothrombin index of blood, *Solnechnye Dannye,* No. 1, 76 (1966).
481. B. A. Ryvkin, F. Z. Ryvkina, B. I. Kobrina, N. A. Parfenova, L. I. Il'ina, N. A. Kostyukhina, M. I. Mel', P. I. Shurpach, and A. N. Babenko, Question of solar-terrestrial correlations in the clinical treatment of cardiovascular diseases, *Solnechnye Dannye,* No. 4, 84 (1967).
482. B. A. Ryvkin, F. Z. Ryvkina, M. Sh, Tastambekova, V. I. Sapozhkov, and A. V. Karavai, Effect of solar activity on dynamics of cardiovascular accidents, *in: Abstr. Papers of Conf. Physiotherapists and Health-Resort Specialists of Central Kazakhstan, Dedicated to 50th Anniversary of Kazakhstan,* Karaganda (1970), p. 155.
483. Yu. D. Safonov, A study of the magnetic field of the heart in the clinical treatment of heart diseases, *in: Proc. 2nd All-Union Conf. of Therapists,* Moscow (1966), p. 427.
484. Yu. D. Safonov and V. B. Pubekh, *The Magnetocardiographic Method of Investigating the Bioelectric Magnetic Field of the Heart,* AN SSSR, Moscow (1966).
485. Yu. D. Safonov, Some aspects of the clinical use of magnetocardiography, *Kardiologiya* **7**(4), 70 (1967).
486. Yu. D. Safonov and V. M. Provotorov, Method of recording the magnetic field of the heart (magnetocardiography), *Byul. Ėksperim. Biol. i Med.* **64,** 1022 (1967).
487. A. R. Sakayan, Hypothesis of electromagnetic orientation of animals, *in: Conf. Effect of Magnetic Fields on Biological Objects,* Scientific Council on the Complex Problem "Cybernetics," Moscow (1966), p. 64.
488. A. R. Sakayan, A study of the orientation of homing pigeons, *in: Questions of Bionics,* Nauka, Moscow (1967), p. 510.
489. A. R. Sakayan, Fish navigation based on earth's magnetic field, *in: Problems of Navigation and Automatic Control,* VINITI, Moscow (1969).
490. I. A. Sapov and A. S. Solodkov, Physiological provision for travel in submarines, *Voenno-med. Zh.*, No. 10, 66 (1970).
491. L. K. Sapozhkov, Author'a abstract of dissertation: *Processing of Biospherological Information by Multidimensional Statistical Methods,* Leningrad Institute of Automatic Control, Leningrad (1973).
492. L. K. Sapozhkov and V. E. Manoilov, Cybernetic aspects of effect of variations of meteorological factors and the geomagnetic field on the dynamics of increase of diseases, *in: Materials of 1st Republican Conf. Use of Mathematical Methods and Computer Techniques in Health Care and Medical Science,* Tbilisi (1971), p. 215.
493. P. V. Savostin, Magnetophysiological effects in plants, *Trudy Moskovskogo Doma Uchenykh,* No. 1, 111 (1937).
494. R. Sh. Sayapova, Effect of nature of distribution of potato plants in field on their growth and development, *in: Collection of Post-Graduate Work of Kazan University, Natural Sciences,* Biology, Kazan (1968), p. 123.
495. R. Sh. Sayapova, Structure of potato crops in relation to position of plants in fields, *in: Materials of 1st Intercollege Sci. Conf. Questions of Agrophytocenology,* Kazan (1969), p. 25.

496. B. I. Sazonov, Disturbance of pressure field as a geophysical index, *Trudy GGO im. A. I. Voeikova* **198,** 107 (1966).
497. B. I. Sazonov, Dual nature of rhythms in geophysical processes, *in: Solar-Atmospheric Relations in Theory of Climate and Weather Forecasting,* (É. R. Mustel', ed.), Gidrometeoizdat, Leningrad (1974), pp. 70–79.
498. K. R. Sedov and N. N. Koroleva, Solar activity and cardiovascular accidents, *Solnechnye Dannye,* No. 11, 83 (1966).
499. E. P. Semenov and A. Ya. Chegodar', Physiological autoantibodies in animals exposed for long periods to a low-frequency electromagnetic field, *in: Effect of Electromagnetic Fields on Biological Objects,* Kharkovskii Gosudarstvennyi Med. Inst., Kharkov (1973), pp. 22–24.
500. V. A. Semykin and M. A. Golubeva, Daily biorhythm of optical activity of plant organisms shielded from and exposed to the local magnetic field, *Proc. 3rd All-Union Symp. Effect of Magnetic Fields on Biological Objects,* Kaliningrad State University, Kaliningrad (1975), p. 183.
501. G. A. Sergeev, Methodological problems of psychotronics, *in: Proc. 2nd Inter. Congr. Psychotronic Research,* Monte Carlo (1975), pp. 9–24.
502. G. A. Sergeev, G. D. Shushkov, and É. G. Gryaznukhin, Methods of recording and statistical treatment of bioplasmograms, *in: Questions of Bioenergetics,* Alma-Ata (1969).
503. N. K. Serov, Some problems of thorough diachronic investigation of processes of human psychic and social life, *in: Effect of Some Cosmic and Geographic Factors on the Earth's Biosphere,* Nauka, Moscow (1973), pp. 26–46.
504. I. I. Shafranovskii, *Symmetry in Nature,* Nedra, Moscow (1968).
505. V. G. Shakhbazov, L. M. Chepel', and V. A. Grabina, Effect on plant growth of vertical magnetic field with different directions of magnetic flux, *in: Effect of Natural and Weak Artificial Magnetic Fields on Biological Objects,* Belgorod (1973), pp. 101–102.
506. N. S. Shcherbinovskii, Relation between terrestrial processes and the rhythm of solar activity, *in: The Desert Schistocercus Locust,* Sel'khozgiz, Moscow (1952).
507. A. É. Shem'i-zade and R. U. Mambetov, Measurements of natural radioactivity of air in mountain conditions, *Atomn. Én.* **36**(1), 61062 (1974).
508. A. É. Shem'i-zade, Increase in natural radioactivity of atmosphere due to geomagnetic storms and possible biological effect of this phenomenon, *in: Physicomathematical and Biological Problems of Effect of Electromagnetic Fields and Ionization of Air,* Vol. 1, Nauka, Moscow (1975), pp. 198–202.
509. R. V. Shepetnov, Author's abstract of dissertation: *Planetary Characteristics of Geomagnetic Micropulsations and Their Use for the Investigation of Circumterrestrial Space,* Inst. of Earth Physics, Academy of Sciences of the USSR, Moscow (1968).
510. A. D. Shevnin and A. P. Dubrov, Geomagnetic variations accompanying motion of artificial earth satellites and their possible biological significance, *in: Effect of Natural and Weak Artificial Magnetic Fields on Biological Objects,* Belgorod (1973), pp. 28–30.
511. I. S. Shnitser, *Angina Pectoris,* Tsentr Inst. Usovershenstv. Vrachei, Moscow (1970).
512. L. N. Shrager, Cytogenetic effect of reduced magnetic fields on right- and left-handed isomers of onion, *Proc. 3rd All-Union Symp. Effect of Magnetic Fields on Biological Objects,* Kaliningrad State University, Kaliningrad (1975), p. 194.
513. H. Strughold and H. B. Hale, Biological and physiological rhythms, *in: Fundamentals of Space Biology and Medicine,* Vol. 2, Book 2, Nauka, Moscow (1975), pp. 139–152.
514. A. V. Shubnikov, *Problems of Dissymmetry of Material Objects,* Izd. AN SSSR, Moscow, (1961).
515. N. A. Shul'ts, Effect of variations of solar activity on numbers of white blood corpuscles, *in: The Earth in the Universe,* Mysl', Moscow (1964), p. 382.
516. N. A. Shul'ts, Author's abstract of dissertation: *Effect of Solar Activity on Incidence of Functional Leucopenias and Relative Lymphocytoses,* Academy of Medical Sciences of the

USSR, Moscow (1967).

517. M. E. Shumakov, Investigation of migrational orientation of passeriforms, *Vestnik LGU, Ser. Biol.,* 106 (1967).
518. A. P. Shushakov, K. F. Novikova, and T. N. Panov, Variations of earth's magnetic field and myocardial infarct, *in: Questions of Physiotherapy and Health-Resort Studies in the Urals,* Sverdlovsk (1964), p. 44.
519. A. P. Shushakov, T. N. Panov, and K. F. Novikova, Variations of earth's magnetic field and myocardial infarct, *Trudy Len. Ob. Estestvoispytatelei* **76**(1), 59 (1971).
520. I. V. Shust and I. M. Kostinik, Effect of a strong constant magnetic field and a hypomagnetic environment on histochemical indices of white rat liver, *Kosmicheskaya Biol. i Aviakosmicheskaya Med.* **9**(6), 19–25 (1975).
521. A. G. Sidorskii, É. A. Sidorskaya, and V. A. Gos'kova, Effect of earth's magnetic field on direction of sexual differentiation of bisexual plants, *Uch. Zap. Gor'kovskogo Gos. Ped. Inst., Ser. Biol.* **122,** 105 (1971).
522. L. V. Sirotina, A. A. Sirotin, and M. P. Travkin, Some features of biological effect of weak magnetic fields, *in: Reaction of Biological Systems to Weak Magnetic Fields,* Moscow (1971), p. 95.
523. L. V. Sirotina, Amylolysis of starch of millet seedlings subjected to weak magnetic fields, *in: Effect of Natural and Weak Artificial Magnetic Fields on Biological Objects,* Belgorod (1973), pp. 106–107.
524. L. V. Sirotina, A. A. Sirotin, M. P. Travkin, and M. F. Trifonova, Pigment production in millet leaves subjected to a magnetic field, *in: Effect of Natural and Weak Artificial Magnetic Fields on Biological Objects,* Belgorod (1973), pp. 108–110.
525. L. V. Sirotina and A. A. Sirotin, Pigment accumulation in millet leaves of different layers after presowing exposure to a magnetic field, *Proc. 3rd All-Union Symp. Effect of Magnetic Fields on Biological Objects,* Kaliningrad State University, Kaliningrad (1975), p. 186.
526. L. V. Sirotina and A. A. Sirotin, Pigment production in ontogenesis of millet after presowing exposure to a magnetic field, *Proc. 3rd All-Union Symp. Effect of Magnetic Fields on Biological Objects,* Kaliningrad State University, Kaliningrad (1975), pp. 193–194.
527. L. M. Slabkii, *Methods of Measuring Ultraweak Fields,* Nauka, Moscow (1974).
528. Ts. N. Slavova, An attempt to demonstrate the role of magnetic storms in the outcome of vascular accidents, *Problems of Bioclimatology and Climatophysiology,* Novosibirsk (1970), p. 364.
529. A. D. Slonim, *Animal Heat and Its Regulation in the Mammalian Organism,* AN SSSR (1952).
530. A. L. Sluvko, Variation of diphtheria corynebacteria cultivated for a long period in a permalloy chamber, *Proc. 3rd All-Union Symp. Effect of Magnetic Fields on Biological Objects,* Kaliningrad State University, Kaliningrad (1975), p. 61.
531. V. V. Sokolovskii, N. I. Muzalevskaya, and R. N. Pavlova, Structural changes in erythrocyte membrane and blood proteins due to a weak alternating electromagnetic field, *Proc. 3rd All-Union Symp. Effect of Magnetic Fields on Biological Objects,* Kaliningrad State University, Kaliningrad (1975), p. 74.
532. A. S. Solodkov, State of basal metabolism in submariners and divers, *Voenno-med. Zh.* No. 5, 60 (1966).
533. A. P. Solomatin, Solar activity and mortality from myocardial infarct and strokes in Western Siberia, *in: Proc. 1st Sci. Conf. Problems of Medical Geography,* Cheboksary, (1975), pp. 77–78.
534. A. P. Solomatin, Geomagnetic activity and its effect on patients with cardiovascular disorders, *Proc. 3rd All-Union Symp. Effect of Magnetic Fields on Biological Objects,* Kaliningrad State University, Kaliningrad (1975), p. 25.

535. A. P. Solomatin, S. A. Kuznetsova, L. M. Nepomnyashchii, and G. S. Pashchenko, Geomagnetic disturbances on earth and cerebral accidents in climato-geographic zones of West Siberia and North Europe, *In: Medico-Biological Aspects of Adaptation Processes,* Novosibirsk (1975), pp. 244–246.
536. G. R. Solov'eva, Some features of the geomagnetic field and artificial magnetic fields, *in: Effect of Natural and Weak Artificial Fields on Biological Objects,* Belgorod (1973), p. 16.
537. M. N. Solov'eva, Relationship between cyclicity of development of earth and the evolutionary process (as exemplified by foraminifera), *in: The Cosmos and the Evolution of Organisms (Proc. Conf.),* Paleontologicheskii Inst. AN SSSR (1974), p. 293–314.
538. A. V. Sosunov, S. A. Petrova, and A. S. Golovatskii, A study of the effect of magnetic field energy on tumor-cell growth in a tissue culture, *in: Proc. Jubilee Sci. Session of Ivan-Franko Med. Inst.,* Kiev (1968), pp. 47–101.
539. A. V. Sosunov, S. A. Petrova, F. O. Borisov, and A. A. Chernov, A study of the clinical-laboratory significance of the ESR of the healthy and sick organism in shielded spaces, *in: Reaction of Biological Systems to Weak Magnetic Fields,* Moscow (1971), p. 153.
540. A. V. Sosunov, B. A. Golubchak, V. Ya. Semkin, and A. V. Mel'nikov, Observations on some biological processes in shielded spaces, *Hygienic Assessment of Magnetic Fields (Proc. Symp. May 20–22, 1972),* AN SSSR, Moscow (1972), pp. 144–146.
541. A. V. Sosunov and L. V. Parkulab, Clinical-laboratory significance of study of ESR in infectious patients in a partially shielded space, *in: Effect of Natural and Weak Artificial Magnetic Fields on Biological Objects,* Belgorod (1973), p. 75.
542. M. G. Spasskaya and N. I. Muzalevskaya, Dynamics of some biological indices in relation to action of weak alternating magnetic fields, *in: Materials of All-Union Symp. "Effect of Magnetic Fields on Living Organisms,"* Baku (1972), p. 90.
543. M. G. Spasskaya and N. I. Muzalevskaya, A morphological assessment of the effect of a weak magnetic field of infralow-frequency on the animal organism, *in: Hygienic Assessment of Magnetic Fields (Proc. Symp. May 20–22, 1972),* AN SSSR, Moscow (1972), pp. 150–156.
544. V. D. Stakanov, Effect of geomagnetic field on germination of tree seeds, *Izv. Sib. Otd. AN SSSR, Ser. Biol.* **10**(2), 130 (1969).
545. V. S. Starchevskii and G. N. Perfil'ev, Solar activity, geomagnetic disturbance, meteorological factors, and myocardial infarct in Simferopol, *Trudy Krymskogo Med. Inst.* **53,** 73–75 (1973).
546. I. G. Stupelis, Ya. Yushenaite, I. Labanauskas, and N. Stakaitite, Activity of earth's magnetic field, time of year, and risk of death from myocardial infarct, *in: Materials of Papers 7th Conf. of Therapists of the Latvian SSR,* Riga (1970), p. 247.
547. I. Stupelis, K. Dapkene, and G. Versatskene, Possible use of monthly prognoses of activity of earth's magnetic field in the prediction of the incidence of myocardial infarct and its prophylaxis, *Proc. Sci. Conf. Med. Faculty of Vilnius State Univ.,* Vilnius (1973), pp. 102–103.
548. F. V. Sushkov, Equivalence of some reactions of tissue culture cells to increase and reduction of magnetic field strength, *in: Physicomathematical and Biological Problems of Effect of Electromagnetic Fields and Ionization of Air,* Vol. 2, Nauka, Moscow (1975), pp. 112–113.
549. K. V. Sudakov and G. D. Antimonii, Central mechanisms of action of electromagnetic fields, *Uspekhi Fiziol. Nauk* **42**(2), 101 (1973).
550. Yu. G. Sulima, Some aspects of biomagnetic reaction of phytosymmetric objects, *in: 2nd Zonal Symp. Bionics,* Abstracts of Papers, Minsk (1967), p. 90.
551. Yu. G. Sulima, Features of reaction of symmetric and dissymmetric grains to meridional orientation in earth's magnetic field, *in: Methods of Selection and Seed-Raising of Corn in Moldavia,* Kishinev (1970).

552. Yu. G. Sulima, *Biosymmetric and Biorhythmic Effects and Features in Crop Plants,* Kishinev (1970), p. 130.
553. L. I. Surkova, Features of winter rest of citrus plants. *Fiziol. Rast.* **9**(5), 609 (1962).
554. L. I. Sverlova, Effect of magnetic reversals on evolution of organic world, *in: The Cosmos and the Evolution of Organisms,* Paleontologicheskii Institut AN SSSR, Moscow (1974), pp. 340–342.
555. A. G. Sytin and A. A. Maksimov, Correlation of mobility of rodents and moles with indices of solar and geomagnetic activity, *in: The Sun, Electricity and Life,* Izd. MGU, Moscow (1972), p. 84.
556. G. A. Tarakanova, Author's abstract of dissertation: *Effect of Steady Magnetic Fields on Growth and Energy Metabolism of Plants,* Inst. of Plant Physiology, Academy of Sciences of the USSR, Moscow (1971).
557. N. A. Temur'yants, Effect of magnetic fields of infralow frequency and low strength on peripheral blood leukocytes, *in: Reaction of Biological Systems to Weak Magnetic Fields,* Moscow (1971), p. 43.
558. N. A. Temur'yants, Author's abstract of dissertation: *Effect of Weak Electromagnetic Fields of Infralow Frequency on the Morphology and Some Metabolic Indices of Peripheral Blood Leukocytes of Animals,* Crimean State Med. Inst., Simferopol' (1972).
559. N. A. Temur'yants, Morphocytochemical study of the biological effectiveness of a weak magnetic field of infralow frequency, *in: Materials of All-Union Symp. Effect of Artificial Magnetic Fields on Living Organisms,* Baku (1972), p. 23.
560. N. A. Temur'yants, Changes in rabbit blood system due to prolonged exposure to a weak electromagnetic field, *in: Physicomathematical and Biological Problems of Effect of Electromagnetic Fields and Ionization of Air,* Vol. 2, Nauka, Moscow (1975), pp. 134–136.
561. N. A. Temur'yants and Yu. S. Krivoshein, Functional activity of mouse peripheral blood leukocytes on exposure to a weak infralow-frequency magnetic field, *Trudy Krymskogo Med. Insta.* **53,** 33–36, Kharkov (1973).
562. N. A. Temur'yants, Yu. S. Krivoshein, and O. G. Tishkin, Effect of a weak field on functional activity of peripheral blood neutrophils, *in: Effect of Natural and Weak Artificial Magnetic Fields on Biological Objects,* Belgorod (1973), pp. 64–65.
563. N. A. Temur'yants, V. Pavlenko, and I. Khondoshko, Effect of a weak magnetic field on respiratory and cardiovascular systems of rabbits, *Proc. 3rd All-Union Symp. Effect of Magnetic Fields on Biological Objects,* Kaliningrad State University, Kaliningrad (1975), pp. 102–103.
564. Yu. V. Tornuev, Author's abstract of candidate's dissertation: *Experimental Investigation of Electrostatic Fields Produced in Air by Bioobjects,* Tomsk (1971).
565. M. P. Travkin, Effect of an anomalous field on crop yield, *in: Materials of 2nd All-Union Conf. Effect of Magnetic Fields on Biological Objects,* Moscow (1969), p. 222.
566. M. P. Travkin, Effect of Kursk Magnetic Anomaly on yield of most important crops, *Materials of Sci. Methodology Conf. Belgorod Pedagogic Inst.,* No. 5 (1969), p. 34.
567. M. P. Travkin, Effect of weak magnetic field on seed germination and orientation of radicles, *Materials of Sci. Methodology Conf. Belgorod Pedagogic Inst.,* No. 5 (1969), p. 28.
568. M. P. Travkin, Effect of weak magnetic field on plant bioelectric potentials, *in: Reaction of Biological Systems to Weak Magnetic Fields,* Moscow (1971), p. 76.
569. M. P. Travkin, Specific effect of magnetic field on biological objects, *in: Reaction of Biological Systems to Weak Magnetic Fields,* Moscow (1971), p. 73.
570. M. P. Travkin, *Life and the Magnetic Field (Materials for a Special Course on Magnetobiology,* Belgorod State Pedagogic Institute) (1971).
571. M. P. Travkin, Effect of a weak magnetic field on the bioelectric potentials of *Tradescantia, Fiziol. Rast.* **19,**(2), 448–450 (1972).

572. M. P. Travkin, Change in bioelectric activity of *Setreasea purpurea* due to a constant and pulsating magnetic field, *Biofizika* **18**(1), 172–173 (1973).
573. M. P. Travkin and N. M. Antipova, Effect of reduced magnetic field on development and fecundity of *Drosophila melanogaster, in: Effect of Natural and Weak Artificial Magnetic Fields on Biological Objects,* Belgorod (1973), p. 82.
574. M. P. Travkin, E. A. Draganets, N. V. Kursevich, A. A. Dirotin, and L. V. Sirotina, Effect of anomalous geomagnetic field of Kursk magnetic anomaly on biological objects, *in: The Cosmos and the Evolution of Organisms (Proc. Conf. Cosmic Factors and the Evolution of the Organic World),* Paleontologicheskii Inst. AN SSSR (1974), pp. 150–153.
575. M. P. Travkin and A. F. Kolchanov, Relicit plants of Belgorod region and the anomalous field of the Kursk magnetic anomaly, *in: The Cosmos and the Evolution of Organisms (Proc. Conf. Cosmic Factors and the Evolution of the Organic World),* Paleontologicheskii Inst. AN SSSR (1974), pp. 154–155.
576. M. P. Travkin and A. M. Kolesnikov, Effect of Kursk magnetic anomaly on incidence of disease in the population, *in: Materials of 2nd All-Union Conf. Effect of Magnetic Fields on Biological Objects,* Moscow (1969), p. 225.
577. M. P. Travkin and A. M. Kolesnikov, Effect of anomalous magnetic field in region of Kursk magnetic anomaly on incidence of disease in population, *in: Hygienic Assessment of Magnetic Fields (Proc. of Symposium on May 22–23, 1972),* AN SSSR, Moscow (1972), pp. 23–25.
578. K. S. Trincher, *Biology and Information,* Nauka, Moscow (1964).
579. K. S. Trincher and A. G. Dudoladov, Thermodynamic and quantum-mechanical bases of theory of magnetization of water, *in: Questions of Theory and Practice of Magnetic Treatment of Water and Aqueous Systems,* Tsvetmetinformatsiya, Moscow (1971), p. 31.
580. S. V. Trofimyuk, Response of peripheral blood of ischemic patients to geomagnetic disturbances, *Proc. 1st Sci. Conf. Young Scientists of Stavropol Med. Inst.,* Stavropol (1974), pp. 26–28.
581. V. A. Troitskaya, Micropulsations of earth's magnetic field and diagnostics of the magnetosphere, *Vestnik AN SSSR* **39**(6), 67 (1969).
582. V. A. Troitskaya and A. V. Gul'el'mi, Geomagnetic pulsations and diagnostics of the magnetosphere, *Vsp. Fiz. Nauk,* No. 3 (1969).
583. Yu. A. Urmantsev, Biosymmetry. Symmetry and dissymmetry of plant florets, *Izv. Akad. Nauk SSSR, Ser. Biol.,* No. 1 (1965).
584. Yu. A. Urmantsev, Properties of D and L modifications of biological objects, *Usp. Sovr. Biol.,* No. 3, 1 (1966).
585. Yu. A. Urmantsev, Isomerism in living nature. I. Theory, *Bot. Zh.,* No. 2, 153 (1970).
586. Yu. A. Urmantsev and A. M. Smirnov, Right-handed and left-handed roots in plants, *Bot. Zh.,* **47,** No. 8, 1073 (1962).
587. R. F. Usmanov, Heliogeophysical factors and current ways of assessing their effect in therapeutic practice, *in: The Sun, Electricity, and Life,* Izd. MGU, Moscow (1972), p. 17.
588. R. F. Usmanov, Role of inhomogeneities of earth's crust in effect of solar activity on atmosphere, *in: Sun-Atmosphere Relations in Theory of Climate and Weather Forecasting,* (É. R. Mustel', ed.), Gidrometeoizdat, Leningrad (1974), pp. 149–160.
589. I. F. Usmanov, Change in orientation of O_2 and NO molecules relative to geomagnetic field as a possible index of effect of solar activity on lower atmosphere, *in: Sun-Atmosphere Relations in Theory of Climate and Weather Forecasting* (É. R. Mustel', ed.), Gidrometeoizdat, Leningrad, (1974), pp. 297–300.
590. I. F. Usmanov, Effect of geomagnetic variations on kinetics of chemical reactions, *in: Physicomathematical and Biological Problems of Effect of Electromagnetic Fields and Ionization of Air,* Nauka, Moscow (1975), pp. 118–123.

591. Yu. A. Urmantsev, *Symmetry of Nature and Nature of Symmetry,* Mysl', Moscow (1974).
592. G. Vaies, *Physics of Galvanomagnetic Semiconducting Devices,* Énergiya, Moscow (1974).
593. U. S. Valeev, Yu. V. Tornuev, and D. F. Rakityanskii, The electric field of the nerve, *Biofizika* **15**(4), 652 (1970).
594. A. S. Vasil'ev and S. I. Gleizer, Change in activity of the eel *Anguilla anguilla* L. in magnetic fields, *Voprosy Ikhtiol.* **13**(2), 381 (1973).
595. A. S. Vasil'ev, S. I. Gleizer, V. A. Sokovishin, L. M. Sokovishina, and V. A. Khodorkovskii, Magnetoreceptive responses in the elver, *Biofizika* **18**(1), 132 (1973).
596. L. L. Vasil'ev, Possible effect of microcurrents circulating in organism on growth and development of tissues and organs, *Trudy Leningradskogo Ob. Estestvoispytatelei* **76**(1), 5 (1971).
597. N. V. Vasil'ev, O. V. Bukharin, P. N. Kuz'min, A. P. Luda, V. I. Polyakov, and G. N. Polyakova, A possible relation between fluctuations of the earth's magnetic field and some factors of human nonspecific immunity, *Proc. 3rd All-Union Symp. Effect of Magnetic Fields on Biological Objects,* Kaliningrad State University, Kaliningrad (1975), pp. 30–31.
598. P. V. Vasilik, Change in physical type of the ancient population of the Ukraine and fluctuations of the earth's magnetic field, *in: Materials of All-Union Symp. Effect of Artificial Magnetic Fields on Living Organisms,* Baku (1972), p. 102.
599. P. V. Vasilik, Western drift of geomagnetic field variations and some features of the manifestation of acceleration in the past, *in: Abstr. Papers of 21st Ukrainian Republic Sci. Tech. Conf.,* Kiev (1972), p. 73.
600. P. V. Vasilik, Effect of magnetic sunspot cycle on particular manifestations of acceleration, *Abstr. Papers of 21st Ukrainian Republic Sci. Tech. Conf.,* Kiev (1972), pp. 67–68.
601. P. V. Vasilik, Acceleration and retardation of ancient population of Ukraine and variations of earth's magnetic field, *in: Mathematical Models in Biology,* Kiev (1972), pp. 102–109.
602. P. V. Vasilik, Intraspecific variation of some animals and variations in intensity of earth's magnetic field, *in: Effect of Natural and Weak Artificial Magnetic Fields on Biological Objects,* Belgorod (1973), pp. 79–81.
603. P. V. Vasilik, Changes in physical development of contemporary man and their relation to geographic characteristics of the geomagnetic field, *in: Effect of Natural and Weak Artificial Magnetic Fields on Biological Objects,* Belgorod (1973), 76–78.
604. P. V. Vasilik, Parallel course of secular variation of some geophysical and biological processes, *in: Medical Cybernetics,* Izd. Inst. Kibernet. AN USSR, Kiev (1974), pp. 17–28.
605. P. V. Vasilik, Effect of environmental factors on rate of individual human development and problem of investigating changes in incidence of disease in population, *Proc. 2nd Ukrainian Republican Symp. Automation of Collection and Processing of Med. Information and the Use of Biotelemetry in Health-Resort Practice,* Izd. Inst. Kibernet. AN USSR, Kiev (1974), pp. 69–71.
606. P. V. Vasilik, Geomagnetic hypothesis of acceleration and some evolutionary processes, *in: The Cosmos and the Evolution of Organisms (Proc. Conf. Cosmic Factors and the Evolution of the Organic World),* Paleontologicheskii Inst. AN SSSR, Moscow (1974), pp. 115–132.
607. P. V. Vasilik, Intensity of earth's magnetic field as a factor affecting course of coronary disease, *Abstr. Ukrainian Republic Conf. Theory and Practice of Development of Automated Med. Information Systems in Health Resorts,* Izd. Inst. Kibernet. AN USSR, Kiev (1975), pp. 21–23.
608. P. V. Vasilik, Reversals of geomagnetic poles and aromorphoses of cryptozoa, *Proc. 3rd All-Union Symp. Effect of Magnetic Fields on Biological Objects,* Kaliningrad State University, Kaliningrad (1975), pp. 28–29.
609. P. V. Vasilik, Construction of a geomagnetic model of faunistic disasters, *Proc. 3rd All-Union Symp. Effect of Magnetic Fields on Biological Objects,* Kaliningrad State University, Kaliningrad (1975), p. 207.

610. P. V. Vasilik, Parallelism of secular course of changes in morphological characters of man and some other mammals in the Holocene, *in: Medical Cybernetics,* Izd. Inst. Kibernet. AN USSR, Kiev (1975), pp. 11–21.
611. P. V. Vasilik, Parallelism of secular course of changes in morphological characters of man and some other mammals during the Wurms glaciation, *Medical Cybernetics,* Izd. Inst. Kibernet. AN USSR (1975), pp. 21–28.
612. P. V. Vasilik, N. V. Vasilik, and V. M. Pomogailo, Acceleration and the earth's magnetic field, *in: 4th Ukrainian Republic Sci. Conf.,* Kiev (1970), p. 145.
613. M. M. Vilenchik, Magnetic effects in biology, *Uspekhi Sovr. Biol.* **63,** 54 (1967).
614. S. A. Vinogradov, V. A. Artishchenko, and A. M. Volynskii, Morphological features of myocardial infarct due to action of magnetic field in an experiment, *in: Effect of Electromagnetic Fields on Biological Objects,* Kharkov State Med. Inst., Kharkov (1973), pp. 37–41.
615. L. I. Vinogradova, É. V. Kordyukov, S. M. Mansurov, and L. G. Mansurova, Polarity of interplanetary magnetic field sectors and frequency of occurrence of vegetative vascular paroxysms due to disturbance of activity of hypothalamic brain structures, Preprint No. 17 (132). Inst. of Terrestrial Magnetism, the Ionosphere, and Radiowave Propagation, Academy of Sciences of the USSR, Moscow (1975).
616. L. I. Vinogradova, É. V. Kordyukov, S. M. Mansurov, and L. G. Mansurova, Polarity of interplanetary magnetic field sectors and frequency of occurrence of vegetative vascular paroxysms due to disturbance of activity of hypothalamic brain structures, *in: Physicomathematical and Biological Problems of Action of Electromagnetic Fields and Ionization of Air,* Vol. 2, Nauka, Moscow (1975), pp. 59–66.
617. L. A. Vitel's, Method of processing geophysical data according to the 27-day calendar, *Solnechnye Dannye,* No. 4 (1955).
618. L. A. Vitel's, Effect of solar activity cycles of different duration on some characteristics of atmospheric processes, *Trudy Vses. Nauchn. Meteorol. Soveshchaniya* **3,** 161 (1963).
619. L. A. Vitel's, Solar-terrestrial correlations in 27-day cycle at different phases of the 11-year cycle, *Trudy GGO im. A. I. Voeikova,* No. 227, 51 (1968).
620. Yu. I. Vitinskii, Morphology of Solar Activity, Nauka, Moscow–Leningrad (1966).
621. V. P. Vizgin, *Development of Relationship between Invariance Principles and Conservation Laws in Classical Physics,* Nauka, Moscow (1972).
622. B. M. Vladimirskii, Disturbance of earth's natural electromagnetic field in ultralow-frequency range, *Trudy Krymskogo Med. Inst.* **53,** 3–7 (1973).
623. B. M. Vladimirskii, Yu. N. Achkasova, and L. V. Monastyrskikh, Disturbances of earth's electromagnetic field and the problem of heliobiological relations, *in: The Sun, Electricity, and Life,* Izd. MGU, Moscow (1972), p. 54.
624. B. M. Vladimirskii, Yu. N. Achkasova, and L. V. Monastyrskikh, Disturbances of earth's electromagnetic field and problem of heliobiological relations, *in: Effect of Some Cosmic and Geophysical Factors on the Earth's Biosphere,* Nauka, Moscow (1973), pp. 195–199.
625. B. M. Vladimirskii and A. M. Volynskii, Changes in cardiac activity of animals due to low-frequency electromagnetic fields, *in: Experimental Cardiology,* Part 1, Vladimir (1970), p. 25.
626. B. M. Vladimirskii and A. M. Volynskii, Effect of short-period fluctuations (SPF) of geomagnetic field of pc1 type on cardiovascular and nervous system of animals, *in: Reaction of Biological Systems to Weak Magnetic Fields,* Moscow (1971), p. 131.
627. B. M. Vladimirskii and A. M. Volynskii, Effect of electromagnetic fields with strength close to that of the natural field on physicochemical and biological systems, *in: Physicomathematical and Biological Problems of Effect of Electromagnetic Fields and Ionization of Air,* Vol. 1, Nauka, Moscow (1975), pp. 126–150.
628. B. M. Vladimirskii, A. M. Volynskii, S. A. Vinogradova, Z. I. Brodovskaya, N. A.

Temur'yants, Yu. N. Achkasova, V. D. Rozenberg, and Zh. D. Chepkova, Experimental investigation of effect of ultralow-frequency electromagnetic fields on homeothermic animals and microorganisms, *in: Effect of Solar Activity on Earth's Atmosphere and Biosphere,* Nauka, Moscow (1971), p. 224.

629. K. S. Voichishin, Determination of informational-prognostic indices of biometeorological phenomena, *in: Extraction and Transmission of Information,* No. 33, Naukova Dumka, Kiev (1972), pp. 16–28.
630. K. S. Voichishin, Author's Abstract of Candidate's Dissertation: *Questions of Statistical Analysis of Unsteady (Rhythmic) Effects in Relation to Geophysical Problems,* Institute of Earth Physics, Academy of Sciences of the USSR, Moscow (1975).
631. K. S. Voichishin and Ya. P. Dragan, A simple stochastic model of natural rhythmic processes, *in: Extraction and Transmission of Information,* No. 29, Naukova Dumka, Kiev (1971), pp. 7–15.
632. K. S. Voichishin and Ya. P. Dragan, Some properties of a stochastic model of a system with periodically varying parameters, *in: Algorithmization of Industrial Processes,* Izd. Inst. Kibernet. AN USSR, Kiev (1972), pp. 11–26.
633. K. S. Voichishin, Nature of variation of rhythm in natural phenomena, *in: Extraction and Transmission of Information,* No. 36, Naukova Dumka, Kiev (1973), pp. 6–9.
634. K. S. Voichishin, Ya. P. Dragan, V. I. Kuksenko, and V. N. Mikhailovskii, Detection of periodicity in structure of natural stochastic processes, *in: Extraction and Transmission of Information,* No. 30, Naukova Dumka, Kiev (1971), pp. 3–15.
635. K. S. Voichishin, Ya. P. Dragan, V. I. Kuksenko, and V. N. Mikhailovskii, *Informational Relations of Bioheliogeophysical Phenomena and Elements of Their Prognosis,* Naukova Dumka, Kiev (1974).
636. K. S. Voichishin and V. N. Mikhailovskii, *On the Transmitter of Biometeorological Information,* Naukova Dumka, Kiev (1966), pp. 172–181.
637. E. Ya. Voitinskii, B. S. Gendel's, B. I. Gol'tsman, N. I. Muzalevskaya, and M. E. Livshits, Effect of a weak magnetic field on electrical activity of brain, *in: The Magnetic Field in Medicine,* Kirgizskii Gos. Med. Inst., Frunze (1974), pp. 21–22.
638. A. G. Volokhonskii, The genetic code and symmetry, *in: Symmetry in Nature,* Leningrad (1971), pp. 371–375.
639. A. G. Volokhonskii, Formal structure of genetic code. *Tsitologiya i Genetika* **6**(6), 487–492 (1971).
640. A. M. Volynskii, Change in cardiac and nervous activity in animals of different age due to an electromagnetic field of low frequency and low strength. Communication I, *Trudy Krymskogo Med. Inst.* **53,** 7–13 (1973).
641. A. M. Volynskii, Analysis of effect of electromagnetic fields of low frequency and strength on nervous system of animals, *Proc. 3rd All-Union Symp. Effect of Magnetic Fields on Biological Objects,* Kaliningrad State University, Kaliningrad (1975), p. 104.
642. A. M. Volynskii, Z. I. Brodovskaya, N. A. Temur'yants, Zh. D. Chepkova, and B. M. Vladimirskii, A study of the effect of low-frequency electromagnetic fields on various systems of the animal organism, *in: Adaptation of the Organism to Physical Factors,* NII Éksperim. i Klin. Med., Vilnius (1969), p. 354.
643. A. M. Volynskii and B. M. Vladimirskii, Modeling of the effect of a magnetic storm on mammals, *Solnechno-zemnaya Fiz.,* No. 1, 294 (1969).
644. Yu. A. Voronov, Role of weak stimuli in nature (with reference to the science of biogeophysics), *Trudy Leningradskogo Ob. Estesvoispytatelei* **76**(1) (1971), p. 44.
645. D. S. Vorontsov, *General Electrophysiology,* Medgiz, Moscow (1961).
646. A. M. Vyalov, Clinico-hygienic and experimental data on the effect of magnetic fields in industrial conditions, *in: Effect of Magnetic Fields on Biological Objects,* Nauka, Moscow (1971), p. 165.

647. V. N. Yagodinskii, Elements of cyclicity of epidemic process, *Zh. Mikrobiol. Épidemiol. Immunobiol.*, No. 11, 33 (1969).
648. V. N. Yagodinskii, Results of comparison of dynamics of epidemic process with change in index of magnetic disturbance, *Issledovaniya po Geomagnetizmu, Aéronomii, i Fizike Solntsa*, No. 17, Nauka, Moscow (1971), p. 31.
649. V. N. Yagodinskii and Yu. V. Aleksandrov, The ''sun-disease'' problem as a component part of medical geography, *in: Geographic Problems of Exploitation of Desert and Mountain Territories of Kazakhstan,* Alma-Ata (1965), p. 238.
650. V. N. Yagodinskii, Z. P. Konovalenko, and I. P. Druzhinin, Relation between epidemic process and solar activity, *in: Effect of Solar Activity on Earth's Atmosphere and Biosphere,* Nauka, Moscow (1971), p. 81.
651. B. M. Yanovskii, *Terrestrial Magnetism,* Leningrad State University (1964).
652. V. V. Yakhontov, *Insect Ecology,* Vysshaya Shkola, Moscow (1969).
653. V. É. Yakobi, Orientation of birds in flight, *in: Mechanisms of Flight and Orientation of Birds,* Nauka, Moscow (1966), p. 64.
654. M. I. Yakovleva, *Physiological Mechanisms of Action of Electromagnetic Fields,* Meditsina, Leningrad (1973).
655. V. Ya. Yurazh, Meteotropic reactions in hypertensive disease and coronary atherosclerosis in relation to air fronts and heliogeophysical factors, *in: Climate and Cardiovascular Pathology,* Meditsina, Moscow (1965), p. 75.
656. Ya. P. Yushenaite, L. Z. Lautsevichus, and G. Ya. Guobis, Meteotropic reactions in patients with rheumatism and the earth's magnetic field, *in: Adaptation of the Organism to Physical Factors,* NII Éksperim. i Klin. Med., Vilnius (1969), p. 245.
657. Ya. P. Yushenaite, I. Stupelis, N. Stanaitite, and Kh. Kibarskis, Aggravation of ischemic heart disease in relation to solar activity and level of earth's magnetic field, *in: Chronic Ischemic Heart Disease (Diagnosis, Treatment, Prophylaxis),* Vilnius (1971), p. 217.
658. V. P. Zabotin, Author's abstract of dissertation: *An Investigation of Low-Frequency Electromagnetic Fields Formed Around Living Organisms,* Leningrad State University (1968).
659. L. V. Zabrodina, Effect of magnetic field of very low strength on blood clotting system, *Issledovaniya po Geomagnetizmu, Aéronomii, and Fizike Solntsa,* No. 17, 68 (1971).
660. L. V. Zabrodina, Effect of magnetic field of very low strength on blood clotting system in experiment, *in: Materials of All-Union Symp. Effect of Artificial Magnetic Fields on Living Organisms,* Baku (1972), p. 163.
661. L. M. Zakharova, Effect of geomagnetic disturbances on brain blood circulation in vascular patients, *Zdravookhranenie Turkmenii,* No. 4, 6–10 (1974).
662. V. P. Zhokhov and E. N. Indeikin, Relation between acute attacks of glaucoma and variations of earth's magnetic field, *in: Effect of Solar Activity on Earth's Atmosphere and Biosphere,* Nauka, Moscow (1971), p. 210.
663. V. A. Zubakov, Faunistic complexes and paleomagnetic scale of Pliocene–Pleistocene in the USSR, *in: The Cosmos and the Evolution of Organisms,* Paleontologicheskii Inst., AN SSSR (1974), pp. 258–262.
664. A. N. Zudin, G. A. Pospelova, and V. N. Saks, The problem of the boundary of the Neogene and Quaternary Periods in the light of paleomagnetic data, *Geologiya i Geofizika,* No. 8, 2–9 (1969).
665. R. A. Abler, Magnet in biological research, *in: Biological Effects of Magnetic Fields,* (M. Barnothy, ed.), Vol. 2, Plenum Press, New York (1969), p. 2.
666. G. Abrami, Correlation between lunar phases and rhythmicities in plant growth under field conditions, *Can. J. Bot.* **50**(11), 2157–2166 (1972).
667. K. Agabab'yan (Agababjan) and P. Vasilik (Wasilik), Evolution durch Erdmagnetismus, *Exact (Stuttgart),* No. 2. 21–25 (1975).
668. W. H. Allen, Bird migration and magnetic meridians, *Science* **108,** 2817, 708 (1948).

669. A. M. Alvarez, Apparent points of contact between the daily course of the magnetic components of the Earth together with certain solar elements and the diastolic pressure of human beings and the total count of their leucocytes, *Puerto Rico J. Public Health Tropical Med.* **10**(3), 388 (1935).
670. K. Antonowicz, Possible superconductivity at room temperature, *Nature* **247**(5440), 358–359 (1974).
671. J. Aschoff, Comparative physiology: diurnal rhythms, *Ann. Rev. Physiol.* **25,** 581 (1963).
672. D. Assman, *Die Wetterfühligkeit des Menschen,* VEB Gustav Fischer-Verlag, Jena (1963).
673. L. Audus, Magnetotropism: a new plant growth response, *Nature* **185**(4707), 132 (1960).
674. L. Audus and J. Wish, Magnetotropism, *in: Biological Effects of Magnetic Fields,* (M. Barnothy, ed.), Vol. 1, Plenum Press, New York (1964), p. 170.
675. E. Bach and L. Schluck, Untersuchungen uber den Einfluss von meteorologischen, ionosphärischen und solaren Faktoren sowie den Mondphasen auf die Auslösung von Eklampsie und Präeklampsie, *Zbl. Gynäkol.* **66**(1), 192 (1942).
676. M. Barnothy (ed.), *Biological Effects of Magnetic Fields,* Vol. 1, Plenum Press, New York (1964).
677. M. Barnothy, Proposed mechanisms for the navigation of migrating birds, *in: Biological Effects of Magnetic Fields,* (M. Barnothy, ed.), Vol. 1, Plenum Press, New York (1964), p. 287.
678. M. Barnothy (ed.), *Biological Effects of Magnetic Fields,* Vol. 2, Plenum Press, New York (1969).
679. F. H. Barnwell, A day-to-day relationship between oxidative metabolism and world wide geomagnetic activity, *Biol. Bull.* **119**(2), 303 (1960).
680. F. H. Barnwell and F. A. Brown, Jr., Responses of planarians and snails, *in: Biological Effects of Magnetic Fields,* Vol. 1, Plenum Press, New York (1964), p. 263.
681. G. Baule and P. McFee, Detection of the magnetic field of the heart, *Am. Heart J.*, **66**(1), 95 (1963).
682. T. E. Bearden, Biofields, orthoframes and orthorotation. Basis for a new science for a new class of phenomena, Preprint, T. Bearden, Redstone Arsenal, Alabama (1975).
683. A. E. Becker, As influencias de phenomenas electro-magneticos sobre os fluxos bio electricos de corpo humano, *Ann. Paulistos de Medicina Cirurgia (Sao Paulo)* **34**(3), 185 (1937).
684. A. E. Becker, Da influência das ridiacoes do subsolo na agricultura, *J. Agronomia (Sao Paulo),* No. 2, 107 (1939).
685. G. Becker, Magnetfeld-Orientierung von Dipteren, *Naturwissenschaften* **50**(21), 664 (1963).
686. G. Becker, Ruheeinstellung nach der Himmelsrichtung, eine Magnetfeld-Orientierung bei Termiten, *Naturwissenschaften,* **50,** 455 (1963).
687. G. Becker, Reaktion von Insekten auf Magnetfelder, elektrische Felder und Atmospherics, *Z. Angew. Entomol.* **54**(1–2), 75 (1964).
688. G. Becker, Zur Magnetfeld-Orientierung von Dipteren, *Z. Vgl. Physiol.* **51**(2), 135 (1965).
689. G. Becker, On the orientation of Diptera according to the geomagnetic field, *in: Proc. 3rd Intern. Biomagnetic Symp., Univ. of Illinois,* Chicago (1966), p. 9.
690. G. Becker, Magnetfeld-Einfluss auf die Galleriebau-Richtung bei Termiten, *Naturwissenschaften* **58**(1), 60 (1971).
691. G. Becker, Aktivitätsschwankungen bei Termiten, ein Phänomen von grundsätzlicher biologischer Bedeutung, *Z. Angew. Entomol.* **72**(3), 273–289 (1972/1973).
692. G. Becker and W. Gerisch, Zusammenhänge zwischen der Frassaktivität von Termiten und solaren Einflussen, *Z. Angew. Entomol.* **73**(4), 365–386 (1973).
693. G. Becker, Einfluss des Magnetfelds auf das Richtungsverhalten von Goldfischen, *Naturwissenschaften* **61**(5), 220–221 (1974).
694. G. Becker and U. Speck, Untersuchungen über die Magnetfeld-Orientierung von Dipteren, *Z. Vgl. Physiol.* **49**(3), 301 (1964).

695. R. O. Becker, Relation between natural magnetic field intensity and the incidence of psychotic disturbances in the human population, *in: Intern. Conf. High Magnetic Field. Final Report (Abstr. No. 5)* (1961).
696. R. O. Becker, The biological effects of magnetic fields, *Med. Electron. Biol. Eng.* **1,** 293 (1963).
697. R. O. Becker, Relationship of geomagnetic environment to human biology, *N.Y. State J. Med.* **63**(15), 2215 (1963).
698. R. O. Becker, The effect of magnetic fields upon the central nervous system *in: Biological Effects of Magnetic Fields* (M. Barnothy, ed.), Vol. 2, Plenum Press, New York (1969), p. 207.
699. R. O. Becker, C. Bachman, and H. Friedman, The direct current control system. A link between environment and organism, *N.Y. State J. Med.* **62**(8), 1169 (1962).
700. D. E. Beischer, Human tolerance to magnetic fields, *Astronautics* **7**(3), 24 (1962).
701. D. E. Beischer, Biological effects of magnetic fields in space travel, *Proc. 12th Intern. Astronautical Congr.,* New York (1963), p. 515.
702. D. E. Beischer, Biological effects of magnetic fields in their relation to space travel, *in: Bioastronautics,* (K. E. Schaefer, ed.), Macmillan, New York (1964), p. 173.
703. D. E. Beischer, Biomagnetics, *Ann. N.Y. Acad. Sci.* **134**(1), 454 (1965).
704. D. E. Beischer, Physiological aspects of a magnetically field-free environment, *Aerospace Med.* **37**(3), 265 (1966).
705. D. E. Beischer, Generation of low magnetic field environment for study of space null magnetic field effects on man, *in: The Future of Bioengineering in Our Daily Lives (K. R. Jackman, ed.), Anaheim. Calif. Inst. Environmental Sci., Annu. Tech. Meeting, Proc. Session C-5, C-6* (1969), pp. Q1–Q5.
706. D. E. Beischer, The null magnetic field as reference for the study of geomagnetic directional effects in animals and man, *Ann. N.Y. Acad. Sci.* **188,** 324–330 (1971).
707. D. E. Beischer and F. F. Miller, Exposure of man to low-intensity magnetic fields, NSAM-823, NASA Order R-39, Pensacola. Florida (1962).
708. D. E. Beischer, F. F. Miller, and J. C. Knepton, Exposure of man to low intensity magnetic fields in a coil system, NASA-NAMI Joint Report, 1018 (1967).
709. F. C. Bellrose, Radar in orientation research, *Proc. 14th Intern. Ornithol. Congr.,* Oxford (1967), p. 281.
710. F. C. Bellrose and R. R. Grabar, A radar study of the flight direction of nocturnal migrants, *Proc. 13th Intern. Ornithol. Congr.,* Ithaca, New York (1963), p. 362.
711. M. H. Benedick and B. Greenberg, The sanguine biological-ecological research program, *IEEE Trans. Comm.* **22**(4), 570 (1974).
712. M. V. L. Bennett, Electrolocation in fish, *Ann. N.Y. Acad. Sci.* **188,** 242–269 (1971).
713. M. F. Bennett and J. Huguenin, Persistent seasonal variations in the diurnal cycles of earthworms, *Z. Vgl. Physiol.* **60**(1), 34 (1968).
714. M. F. Bennett and J. Huguenin, Geomagnetic effects on a circadian difference in reaction time in earthworms, *Z. Vgl. Physiol.* **63**(4), 440 (1969).
715. H. Berg, Eklampsie, akutes Glaukom, traumatische Epilepsie und ihre Auslösung durch geophysikalische Faktoren, *Bioklimat. Beibl.* **10,** 130 (1943).
716. H. Berg, Die Beeinflussung der Häufigkeit epileptischer Auffälle durch Menstruation und geophysikalische Vorgänge. II. Epilepsie und geophysikalische Vorgänge, *Dtsch. Gesundheitswesen* **2**(1), 97 (1947).
717. H. Berg, *Wetter und Krankheiten,* H. Bouvier und Co. Verlag, Bonn (1948).
718. H. Berg, Sonnenflecken, solar Vorgänge und biologisches Geschehen, *Grenzgebiete Med.* **2,** 293 (1949).
719. H. Berg, Kasuistik, Statistik und Experiment in medizin-meteorologischen Grenzgebiet, *Medizin-meteorol. Hefte* **2,** 27 (1950).

720. H. Berg, Solare und magnetische Störungen im Hinblick auf die Problematik medizin-meteorologischer Vorhersagen, *Arch. Phys. Therapie* **6**(1), 216 (1954).
721. H. Berg, *Solare-terrestrische Beziehungen in Meteorologie und Biologie,* Akademie-Verlag, Geest und Portig K.-G., Leipzig (1957).
722. W. A. Berggren, J. D. Phillips, *et al.*, Late Pliocene–Pleistocene stratigraphy in deep sea core from south-central North Atlantic, *Nature* **216**(5112), 253 (1967).
723. D. S. Bhaskara Rao and B. J. Srivastava, Influence of solar and geomagnetic disturbances on road traffic accidents, *Bull. Nat. Geophys. Res. Inst. (India)* **8**(1–2), 32 (1970).
724. K. Birzele, *Sonnenaktivität und Biorhythmus des Menschen,* F. Deuticke, Wien (1966).
725. D. I. Black, Cosmic ray effects and faunal extinctions at geomagnetic field reversals, *Earth Planetary Sci. Lett.* **3**(3), 225–236 (1967).
726. Z. Bochenski, M. Dylewske, J. Gieszczykiewicz, and L. Sych, Homing experiments in birds, XI. Experiments with swallows *Hirundo rustica* L. concerning the influence of Earth magnetism and partial eclipse of the sun on their orientation, *Zesz. Nauk Univ. Jagiellonskiego* **33**(5), 125 (1960).
727. K. Bodman, Von die Wirkung von Pharmaka auf Kurzzeitschätzung und circadiane Periodik von Säugetieren, *Z. Vgl. Physiol.* **68**(2), 276–292 (1970).
728. J. Borsarello and G. Cautini, Bio-cyclergology and astronautics, *Rev. France d'Astronautique* **21** (1967).
729. H. Bortels, Beziehungen zwischen Wetterungsablauf physikalisch–chemischen Reaktionen, biologischen Geschehen und Sonnenaktivität unter besonderer Berücksichtigung eigener mikrobiologischer Versuchergebnisse, *Naturwissenschaften* **38,**(8), 165 (1951).
730. L. Brauner and R. Diemer, Über die geoİektrischen Reaktionvermögen der Inflorescencezachsen von Lupinus polyphyllus, *Planta* **81**(2), 113 (1968).
731. D. Brenner, S. J. Williamson, and L. Kaufman, Visually evoked magnetic fields of the human brain, *Science,* **190,** 480–482 (1975).
732. P. W. Bretsky and D. M. Lorenz, An essay on genetic-adaptive strategies and mass extinctions, *Geol. Soc. Ann. Bull.* **81,** 2449–2456 (1970).
733. R. R. Brewer, E. L. Kanabrocki, L. E. Scheving, and J. E. Pauly, Physiological chronotypes seen in man, *Chronobiologia, Suppl. 1,* 9 (1975).
734. P. Brisotto, Ricerche sul fenomeno dell'orientamento dei piccioni viaggiatori, *Boll. delle malattie dell'orecchio della cola del naso* **55,** 41 (1937).
735. F. A. Brown, Response of animals to pervasive geophysical factors and the biological clock problem, *Cold Spring Harbor Symp. Quant. Biol.* **25,** 57 (1960).
736. F. A. Brown, Responses of the planarian *Dugesia* and the protozoan *Paramecium* to very weak horizontal magnetic fields, *Biol. Bull.* **123**(2), 264 (1962).
737. F. A. Brown, Organismic responsiveness to very weak magnetic field, *Proc. 3rd Intern. Biomagnetic Symp.*, Univ. of Illinois, Chicago (1966), p. 6.
738. F. A. Brown, Jr., Some orientational influence of non-visual, terrestrial electromagnetic fields, *Ann. N.Y. Acad. Sci.* **188,** 224–241 (1971).
739. F. A. Brown, Jr., The clock timing biological rhythms, *Am. Sci.* **60,** 756 (1972).
740. F. A. Brown, F. H. Barnwell, and H. M. Webb, Adaptation of the magnetoreceptive mechanism of mud-snails to geomagnetic strength, *Biol. Bull.* **127**(2), 221 (1964).
741. F. A. Brown, M. F. Bennett, H. M. Webb, and C. L. Ralph, Persistent daily, monthly, and 27-day cycles of activity in oyster and quahog, *J. Exp. Zool.* **131**(2), 235 (1956).
742. F. A. Brown, Jr. and C. S. Chow, Interorganismic and environmental influence through extremely weak electromagnetic field, *Biol. Bull.* **144**(3), 437 (1973).
743. F. A. Brown, Jr. and C. S. Chow, Phase shifting an exogenous variation in hamster activity by uniform daily rotation, *Proc. Soc. Exp. Biol. Med.* **145,** 7–11 (1974).
744. F. A. Brown and C. S. Chow, Differentiation between clockwise and counterclockwise

magnetic rotation by the planarian *Dugesia dorotocephala, Physiol. Zool.* **48**(2), 168–176 (1975).
745. F. A. Brown, Jr., and C. S. Chow, Non-equivalence for bean seeds of clockwise and counterclockwise magnetic motion: a novel terrestrial adaptation? *Biol. Bull.* **148**(3), 370–379 (1975).
746. F. A. Brown, Jr., J. W. Hastings, and J. D. Palmer, *The Biological Clock: Two Views,* Academic Press, New York (1970).
747. F. A. Brown, H. M. Webb, and F. H. Barnwell, A compass directional phenomenon in mud-snails and its relation to magnetism, *Biol. Bull.* **127**(2), 206 (1964).
748. F. A. Brown, Jr., Extrinsic rhythmicality: a reference frame for biological rhythms under so-called constant conditions, *Ann. N.Y. Acad. Sci.* **98,** 775 (1962).
749. F. A. Brown, Jr., How animals respond to magnetism, *Discovery* **24**(11), 18 (1963).
750. F. A. Brown, Jr., A unified theory for biological rhythms, *in: Circadian Clocks* (J. Aschoff, ed.), Amsterdam (1965), p. 231.
751. F. A. Brown, Jr., Endogenous biorhythmicity reviewed with new evidence, *Scientia* **103**(673–674), 245 (1968).
752. F. A. Brown, Jr., A hypothesis for extrinsic timing of circadian rhythms, *Can. J. Bot.* **47**(2), 287 (1969).
753. F. A. Brown, Jr., and Y. H. Park, Phase-shifting a lunar rhythm in planarians by altering the horizontal magnetic vector, *Biol. Bull.* **129**(1), 79 (1965).
754. F. A. Brown, Jr., Y. H. Park, and J. R. Zeno, Diurnal variation in organismic response to very weak gamma radiation, *Nature* **211** 830–833 (1966).
755. T. Bryant, Gas exchange in dry seeds: circadian rhythmicity in absence of DNA replication, transcription and translation, *Science* **178,** 634–636 (1972).
756. D. H. Bulkley, *Biomagnetics and Life,* Rogue Press, Rogue River (1972).
757. O. Burkard, Einige statistische Versuche zur Piccardischen Fällung-Reaktion, *Arch. Meteorol., Geophys., Biokhim., Ser. B* **6**(4), 506 (1955).
758. H. S. Burr, Effect of a severe storm on electric properties of a tree and the Earth, *Science* **124**(3233), 1204 (1956).
759. H. S. Burr, *The Fields of Life,* Ballantine Books, New York (1972).
760. D. E. Busby, Space biomagnetics, *Space Life Sci.,* **1**(1), 23 (1968).
761. J. Carstoiu, Les deux champs de gravitation et propagation des ondes gravitiques, *Compt. Rend., Ser. A* **268**(3), 201 (1969).
762. J. Casamajor, Le mystérieux "sens de l'espace" chez les pigeons voyageurs, *Nature (Paris)* **54**(2748), 366 (1926).
763. J. Casamajor, Le mystérieux "sens de l'espace," *Rev. Sci. (Paris)* **65**(18), 554 (1927).
764. V. B. Chernyshev (Tschernyshev), N. I. Ershova, *et al.,* Influence of electric charges of Earth surface on some soil arthropods, *Pedobiologia* **13**(6), 437–440 (1973).
765. D. N. Chetaev, E. N. Fedorov, S. M. Krylov, V. P. Lependin, *et al.,* On the vertical electric component of the geomagnetic pulsation field, *Planet. Space Sci.* **23,** 311–324 (1975).
766. A. L. Chizhevskii (Tchizhevsky), L'action de l'activité périodique solaire sur la mortalité générale, *in: Traité de Climatologie Biologique et Médicale* (M. Piery, ed.), Vol. 2, Paris (1934), p. 1042.
767. A. L. Chizhevskii (Tchizhevsky), *Les épidémies et les pérturbations électromagnétiques du milieu exterieur,* Paris (1938).
768. P. P. Chuvaev, The effect of magnetic field and rotation of the Earth on germination and shoot growth, *Plant Physiol.* **14,** 456–459 (1967).
769. R. Citron, *Annu. Progr. Rep. 1968.* Center for Shortlived Phenomena, Smithsonian Inst., Cambridge, Mass. (1969).
770. D. Cohen, Magnetic field around the torso: production by electrical activity of the human

heart, *Science* **156**(3775), 652 (1967).
771. D. Cohen, Magnetoencephalography: evidence of magnetic fields produced by alpha-rhythm currents, *Science* **161**(3843), 784 (1968).
772. D. Cohen and L. Chandler, Measurements and simplified interpretations of magnetocardiograms from humans, *Circulation* **39**(3), 395–402 (1969).
773. O. Cohen, Large-volume conventional magnetic shields, *Rev. Phys. Appl*. **5,** 53–58 (1970).
774. M. Comnoiu, Histochemical study of cholesterol in the adrenals after low-tension magnetic fields, *Rev. Roumaine d'Embryol. Cytol., Ser. D* **7**(1), 22 (1970).
775. C. C. Conley, A review of the biological effects of very low magnetic fields, NASA Techn. Note D-5902, Washington, D.C. (1967).
776. C. C. Conley, Effect of near-zero-magnetic field upon biological systems, *in: Biological Effects of Magnetic Fields* (M. Barnothy, ed.), Vol. 2, Plenum Press, New York (1969), p. 29.
777. C. C. Conley, W. J. Mills, and P. A. Cook, Enzymic activity in macrophages from animals exposed to very low magnetic field, *Proc. 3rd Intern. Biomagnetic Symp.,* Univ of Illinois, Chicago (1966), p. 13.
778. F. W. Cope, Evidence from activation energies for superconductive tunneling at physiological temperatures, *Physiol. Chem. Phys.* **3,** 403–410 (1971).
779. F. W. Cope, Biological sensitivity to weak magnetic fields due to biological superconductive Josephson junctions, *Physiol. Chem. Phys.* **5**(3), 173–176 (1973).
780. F. W. Cope, Enhancement by high electric fields of superconduction in organic and biological solids at room temperature and a role in nerve conduction, *Physiol. Chem. Phys.* **6**(5), 405–410 (1974).
781. O. Costa de Beauregard, Note finale, *in: Preuves en Biologie de Transmutation à Faible Energie* (C. L. Kervran, ed.), Maloine, Paris (1975), pp. 283–298.
782. C. Courty, Sur les propriétés magnetiques des enzymes et des aminoacides, *Compt. Rend.* **278**(26), 3383 (1974).
783. A. Cox, Geomagnetic reversals, *Science* **163**(3864), 237–245 (1969).
784. A. Cox, R. R. Doell, and C. B. Dalrymple, Radiometric time-scale for geomagnetic reversals, *Geol. Soc. London* **124,** *Part 1,* 495 (1968).
785. I. K. Crain, Possible direct causal relation between geomagnetic reversal and biological extinctions, *Geol. Soc. Am. Bull.* **82,** 2603–2606 (1971).
786. I. K. Crain and P. L. Crain, New stochastic model for geomagnetic reversals *Nature* **228,** 39–41 (1970).
787. G. Crepaldi and M. Muggeo, Plurichronocorticoid treatment of bronchial asthma and chronic bronchitis. Clinical and endocrino-metabolic evaluation, *Chronobiologia* **1,** *Suppl. 1,* 407–427 (1974).
788. C. Craciun, M. Granescu, E. Morariu, G. Pap, V. Orban, *et al.* La variation du champ geomagnétique facteur de prevision medical-meteorologique, *in: Congr. National de Balneologie,* Thesis, Bucharest (1974), pp. 127–128.
789. M. Curry, *Bioklimatik,* Vol. 1, Riedereau (Ammersee) (1946), p. 110.
790. W. Cyran, Die bioklimatische Reizwirkung erdmagnetischer Störungen, *Medizin-meteorol. Hefte* **1**(5), 89 (1951).
791. A. Daanje, Haben die Vögel einen Sinn für den Erdmagnetismus wie Deklination, Inklination, und Intensität, *Ardea* **25**(1), 107 (1936).
792. A. Daanje, Heimfindeversuche und Erdmagnetismus, *Vogelzug* **12,** 15 (1941).
793. L. Davis, Remarks on the physical basis of bird navigation, *J. Appl. Phys.* **19**(3), 302 (1948).
794. T. A. Davis, Possible geophysical influence of asymmetry in coconut and other plants, *in: FAO Technical Report. Working Party on Coconut Production, Protection and Processing, Second Session, held in Colombo, Ceylon* (1964).

795. A. R. Davis and W. C. Rawls, Jr., *Magnetism and Its Effect on the Living System,* Exposition Press, Hicksville, New York (1974).
796. C. L. Deelder, On the migration of the elver (*Anguilla vulgaris* Turt) at sea, *J. Cons. Perm. Int. Explor. Mer.* **18,** 187–218 (1952).
797. P. J. Deoras, *Termites in the Humid Tropics,* UNESCO, Paris (1962), p. 101.
798. J. Deutsch and O. Zinke, Abschirmung von Messräumen und Messgeräten gegen elektromagnetischen Felder, *Frequenz* **7**(1), 94 (1953).
799. P. Diesing, Steine mit magnetischen Eigenschaften in Magen grosserer Vögel, *Wild und Hund* **73**(10), 232 (1970).
800. S. Dijkgraf and A. J. Kalmijn, Versuche zur biologischen Bedeutung der Lorenzinischen Ampullen bei der Elasmobranchiern, *Z. Vgl. Physiol.* **53**, 187–194 (1968).
801. Th. Dobzhansky and S. Wright, Genetics of a natural population. X. Dispersion rate in *Drosophila pseudoobscura, Genetics* **28**(4), 304 (1943).
802. Th. Dobzhansky, *Genetics of the Evolutionary Process,* Columbia Univ. Press, New York (1970).
803. Th. Dobzhansky, *in: Ecological Genetics and Evolution,* (R. Greed, ed.), Oxford Univ. Press, London (1971).
804. Th. Dobzhansky and B. Wallace, The genetics of homeostasis in Drosophila, *Proc. Nat. Acad. Sci. U.S.A.* **39,** 162 (1963).
805. J. Dorst, *The Migration of Birds,* Houghton Mifflin, Boston (1962).
806. H. B. Dowse and J. D. Palmer, Entrainment of circadian activity rhythms in mice by electrostatic fields, *Nature* **222**(5193), 564 (1969).
807. W. H. Drury and J. C. T. Nisbet, Radar studies of orientation of songbird migrants in Southeastern New England, *Bird Banding* **35**(1), 69 (1964).
808. J. Dubois, L'influence du magnétisme sur l'orientation des colonies microbiennes, *Compt. Rend. Soc. Biol., Ser. 8* **3**(1), 127 (1886).
809. A. P. Dubrov, The effects of natural magnetic and electrical fields on biological rhythms, *in: Abstr. Papers IXth Intern. Congr. Anatomists,* Leningrad (1970), p. 197.
810. A. P. Dubrov, Heliobiology, *Soviet Life,* No. 1, 14 (1972).
811. A. P. Dubrov, Biogravitation and psychotronics, *Impact of Science on Society* **24**(4), 311–319, UNESCO, Paris (1974).
812. A. P. Dubrov, Unknown factors in chronobiology, *Chronobiologia* (1976), in press.
813. A. P. Dubrov, The hypothesis of biogravity as a basis of the specificity of time-space forms in biological systems, *2nd Intern. Congr. Psychotronic Res.,* Monte Carlo, Monaco (1975), pp. 30–33.
814. A. P. Dubrov, Interaction of biological objects with real physical time and space, *Astrologia (Santa Barbara, Calif.)* **1**(2), 27–40 (1975).
815. A. P. Dubrov, Solar-cosmic basis of rhythmicity in the Earth's biosphere, *Rep. Conf. "Solar–Cosmic–Terrestrial Factors in Biosphere,"* Ancona, Italy (1974).
816. S. Düll, Magnetic seeds: A mystery unfolds, *Furrow* **76,** 4–5 (1971).
817. B. Düll, *Wetter und Gesundheit,* Vol. 1, Steinkopf-Verlag, Dresden und Leipzig (1941).
818. T. Düll and B. Düll, Zusammenhänge zwischen Störungen des Erdmagnetismus und Häufungen von Todes fallen, *Dtsch. Med. Wschr.* **61**(3), 95 (1935).
819. T. Düll and B. Düll, Statistik über die Abhängigkeit der Sterblichkeit von geophysikalischen und kosmischen Vörgange, *in: Medizin-Meteorol. Statistik* (F. Linke and D. de Rudder, eds.), Gesammelte Vorträge, Berlin (1936).
820. T. Düll and B. Düll, Erd- und sonnenphysikalische Vorgänge in ihrer Bedeutung für die Krankheits und Todesauslösung, *Nosokomeion* **9,** 103 (1938).
821. T. Düll and B. Düll, Frage solar-aktiver Einflüsse auf die Psyche, *Z. Ges. Neurol. Psychiat.* **162**(3), 495 (1938).

822. T. Düll and B. Düll, Kosmische-physikalische Störungen der Ionosphäre, Troposphäre, und Biosphäre, *Bioklimat. Beibl.* **6**(65), 121 (1939).
823. T. Düll and B. Düll, Neue Untersuchungen über die Beziehung zwischen den Zahl der täglichen Todesfälle und dem magnetischen Störungscharakter, *Bioklimat. Beibl.* **2**(1), 24 (1953).
824. J. W. Dungey, Solar-wind interaction with the magnetosphere: Particle aspects: *In: The Solar Wind,* (R. J. Mackin and M. Neugebauer, eds.), Pergamon Press, Oxford & New York (1966).
825. A. M. Dycus and A. J. Shultz, A survey of the effects of magnetic environments on seed germination and early growth (abstract), *Plant Physiol.* **39**(5), 29 (1964).
826. J. Edmiston, Effect of exclusion of the earth's magnetic field on the germination and growth of seeds of white mustard (*Sinapis alba*), *Biochem. Physiol. Pflanzen* **167**(1), 97–100 (1975).
827. J. E. Eiselein, H. M. Boutell, and M. W. Biggs, Biological effects of magnetic fields—negative results, *Aerospace Med.* **32**(5), 383 (1961).
828. S. T. Emlen, Bird migration: influence of physiological state upon celestial orientation, *Science* **165,** 716 (1969).
829. S. T. Emlen, The influence of magnetic information on the orientation of the indigo bunting, *Passerina cyanea, Animal Behavior* **18,** 215 (1970).
830. S. Fiber, Untersuchungen über die Auslösung des Myokard-infarktes durch meteorologische und solare Faktoren, *Angew. Meteorol.* **1**(5), 9 (1952).
831. H. Fidelsberger, *Astrologie 2000,* Verlag Kremayr und Scheriau, Vienna (1972).
832. H. Fidelsberger, *Sterne und Leben,* Verlag Kremayr und Scheriau, Vienna (1974).
833. M. D. Fiske, Temperature and magnetic dependences of the Josephson tunnelling current, *Rev. Mod. Phys.* **36**(1), 221–222 (1964).
834. R. L. Forward, Common theory of relativity for explorer, *Trans. IRE* **49**(5), 1047 (1961).
835. C. L. Frederick, *Effects of Electromagnetic Energy on Man,* Southwest Res. Inst. (1972).
836. E. H. Frei, Biomagnetics, *Dig. Intern. Conf. Kyoto,* No. 4, 21/1 (1972).
837. H. Friedman and R. Becker, Geomagnetic parameters and psychiatric hospital admissions, *Nature* **200**(4907), 626 (1963).
838. H. Friedman, R. O. Becker, and C. H. Bachman, Psychiatric ward behavior and geophysical parameters, *Nature* **205,** 1050 (1965).
839. H. Friedman, R. O. Becker, and C. H. Bachman, Effect of magnetic field on reaction time performance, *Nature* **213,** 949 (1967).
840. K. von Frisch, *Tanzsprache und Orientierung der Bienen,* Springer-Verlag, Berlin, Heidelberg, New York (1965).
841. H. G. Fromme, Untersuchungen über das Orientierungsvermögen nächtlich ziehender Kleinvögel (*Erithacus rubecula, Sylvia communis*), *Z. Tierpsychol.* **18,** 205 (1961).
842. A. S. Garay and P. Hrasko, Neutral currents in weak interactions and molecular asymmetry, *J. Mol. Evolution* **6**(2), 77–89 (1975).
843. M. Gardner, *The Ambidextrous Universe,* Basic Books, New York and London (1964).
844. M. Gauquelin, *Les Horloges Cosmiques,* Denoel Ed., Paris (1970).
845. J. Gaustier, Les pigeons voyageurs, *Rev. de l'Hypnotisme* **7,** 10 (1893).
846. R. J. Gavalos, D. O. Walter, J. Hamer, and W. R. Adey, Effect of low-level low-frequency electric fields on EEG and behavior in *Macaca menestrina, Brain Res.* **18,** 491 (1970).
847. T. H. Geballe, New superconductors, *Sci. Am.* **225**(5), 22 (1971).
848. J. A. Generelli, N. J. Holter, and W. R. Glasscock, Magnetic fields accompanying transmission of nerve impulses in the frog's sciatic, *J. Psychol.* **52,** 317 (1961).
849. J. A. Generelli, N. J. Holter, and W. R. Glasscock, Further observations on magnetic fields accompanying nerve transmission and tetanus, *J. Psychol.* **57**(1), 201 (1964).
850. V. L. Ginzburg, The problem of high temperature superconductivity. II, *Usp. Fiz. Nauk* **13,** 335–352 (1970).

851. B. Glass and D. Erickson, Geomagnetic reversals and Pleistocene chronology, *Nature* **216,** 437–442 (1967).
852. S. Goldfein, Some evidence for high-temperature superconductivity in cholates, *Physiol. Chem. Phys.* **6,** 261–268 (1974).
853. H. N. B. Gopalan, Cordycepin inhibits induction of puffs by ions in Chironomous salivary gland chromosomes, *Experientia* **29**(6), 724–726 (1973).
854. D. A. Gordon, Sensitivity of the homing pigeon to the magnetic field of the earth, *Science* **108**(2817), 710 (1948).
855. N. D. Gotlieb and W. E. Caldwell, Magnetic field effects on the compass mechanism and activity level of the snail *Helisoma duryiediscus, J. Genetic Psychol.* **11,** First half, 85 (1967).
856. R. Grandpierre and R. Lemaire, La déminéralisation chez les cosmonautes, *Medicine et Hygiene,* No. 796, 1057 (1967).
857. A. E. Green and M. H. Halpern, Response of tissue culture cells to low magnetic fields, *Aerospace Med.* **36**(2), 147 (1965).
858. A. E. Green and M. H. Halpern, Response of tissue culture cells to low magnetic fields, *Aerospace Med.* **37**(3), 251 (1966).
859. B. S. Green and M. Lahov, Crystallization and solid-state reaction as a route to asymmetric synthesis from achiral starting materials, *in: Proc. Intern. Symp. Generation and Amplification of Asymmetry in Chemical Systems* (W. Thiemann, ed.), Julich, BRD (1974), p. 421.
860. R. E. Gribble, The effect of extremely low frequency electromagnetic fields on the circadian biorhythm of common mice, *Chronobiologia* **2,** *Suppl.1,* 24–25 (1975).
861. D. R. Griffin, The sensory basis of bird navigation, *Quart. Rev. Biol.* **19**(1), 21 (1944).
862. D. R. Griffin, Bird navigation, *Biol. Rev.* **27**(4), 359 (1952).
863. D. R. Griffin, *Bird Migration,* The Natural History Press, Garden City, New York (1964).
864. D. R. Griffin and R. J. Hock, Experiments on bird navigation, *Science* **107**(2779), 347 (1948).
865. O. Grünner, Vliv geomagnetickeho klimatickeho komplexu na neuroticke potize lasynskych pacientu (Effect of geomagnetic climatic complex on neurotic troubles of balnear patients), *Fysiat. Reumat. Vestn.* **51**(1), 21–29 (1973).
866. O. Grünner, Vegetativni reaktivita a uroven bdelosti pod vlivem staleho stejnosmevneho magnetickeho pole male intenzity (Vegetative reactivity and vigilance level under the influence of constant direct current of a low intensity magnetic field), *Čas. Lek. Česk* **114**(20), 618–622 (1975).
867. P. I. Gulyaev (Guliaev), Cerebral electromagnetic fields, *Intern. J. Parapsychol.* **7**(4), 399 (1965).
868. A. Hagentorn, Etwas über Reize und Reizleitung, *Munchen. Med. Wschr.* **81**(14), 530 (1934).
869. F. Halberg, Physiological rhythms and bioastronautics, *in: Bioastronautics* (K. E. Schaefer, ed.), Macmillan Co., New York (1964), p. 181.
870. F. Halberg, Chronobiology, *Ann. Rev. Physiol.* **31,** 675 (1969).
871. F. Halberg, W. Nelson, W. Runge, H. Schmitt, G. Pitts, J. Tremor, and O. Reynolds, Plans for orbital study of rat biorhythms, *Space Life Sci.* **4**(2), 437 (1971).
872. F. Halberg, C. Vallbona, L. F. Dietlien, J. A. Rummel, C. E. Berry, G. C. Pitts, and S. A. Nunnely, Circadian circulatory rhythms of man in weightlessness during extraterrestrial flight as well as in bedrest with and without exercise, *Space Life Sci.* **3**(1), 18 (1970).
873. E. J. Hall and J. S. Bedford, Some negative results in the search for a lethal effect of magnetic fields on biological material, *Nature* **203**(4949), 1086 (1964).
874. M. H. Halpern and J. H. Van Dyke, Very low magnetic fields: Biological effects and their implication for space exploration, *Aerospace Med.* **37**(3), 281 (1966).
875. M. H. Halpern and J. H. Van Dyke, Very low magnetic fields: Biological effects and their

implications for space flight, *in: Proc. 37th Annu. Meeting Aerospace Med. Assoc.*, Las Vegas (1966).
876. E. H. Halpern and A. A. Wolf, *in: Cryogenic Engineering* (K. D. Timmerhaus, ed.), Vol. 17, Plenum Press, New York (1972), p. 109.
877. K. C. Hamner, Endogenous rhythms in controlled environmentals, *in: Environmental Control of Plant Growth* (R. Evans, ed.), Academic Press, New York (1963), pp. 215–232.
878. C. G. A. Harrison, Evolutionary processes and reversals of the Earth's magnetic field, *Nature* **217**(5123), 46–47 (1968).
879. C. G. A. Harrison and B. M. Funnell, Relationships of paleomagnetic reversals and micropaleontology in two late Cenozoic cores from the Pacific Ocean, *Nature* **204,** 566–569 (1964).
880. G. A. Harrison, J. S. Weiner, J. M. Tanner, and N. A. Barnicott, *Human Biology,* Clarendon Press, Oxford (1964).
881. S. S. Hatai, I. Kokubo, and N. Abe, The Earth currents in relation to the response of catfish, *Proc. Imp. Acad. Japan* **8,** 478–481 (1932).
882. C. B. Hartfield and M. J. Camp, Mass extinctions correlated with galactic events, *Geol. Soc. Am. Bull.* **81,** 911–914 (1970).
883. E. Hartmann, *Krankheit als Standortproblem,* Karl F. Haug Verl., Heidelberg (1967).
884. J. D. Hays, Faunal extinctions and reversals of the earth's magnetic field, *Geol. Soc. Am. Bull.* **82**(9), 2433–2447 (1971).
885. J. D. Hays, The stratigraphy and evolutionary trends of Radiolaria in North Pacific deep-sea sediments, *Geol. Soc. Am. Mem.* **126,** 185–218 (1970).
886. J. D. Hays and N. D. Opdyke, Antarctic Radiolaria, magnetic reversals and climatic change, *Science* **158,** 1001–1011 (1967).
887. J. D. Hays, T. Saito, N. D. Opdyke, and L. H. Burckle, Pliocene–Pleistocene sediments of the equatorial Pacific, their paleomagnetic, biostratigraphic and climatic record, *Geol. Soc. Am. Bull.* **80**(8), 1481 (1969).
888. H. Heckert, Lunationrhythmen des Menschlichen Organismus, *in: Methodisches und Ergebnisse (Probleme der Bioklimatologie),* Vol. 7, Akademie-Verlag, Leipzig (1961).
889. W. Hellpach, *Die Geopsychischen Erscheinungen,* F. Enke, Stuttgart (1950).
890. H. P. Heppner, Electric field in the magnetosphere, *in: Proc. Joint COSPAR, IAGA, URSI Symp. Critical Problems of Magnetospheric Physics, Madrid, May 11–13, 1972,* Madrid (1972), pp. 107–115.
891. T. Herron, Phase modulation of geomagnetic micropulsation, *Nature* **207** (1965).
892. K. Hoffman, Zur synchronisation biologischer Rhythmen, *Verh. Dtsch. Zool. Ges.* **64,** 266–273 (1970), Table 1.
893. K. Hoffman, Biological clocks in animal orientation and other functions, *in: Proc. Intern. Symp. Circadian Rhythmicity,* Wageningen (1971), pp. 175–205.
894. T. Hoshizaki and K. C. Hamner, An unusual stem bending response of *Xanthium pensylvanicum* to horizontal rotation, *Plant Physiol.* **37,** 453–459 (1962).
895. T. Hoshizaki, B. H. Carpenter, and K. C. Hamner, The interaction of plant hormones and rotation around a horizontal axis on the growth and flowering of *Xanthium pensylvanicum, Planta* **61,** 178–186 (1964).
896. T. Hoshizaki, W. R. Adey, and K. C. Hamner, Growth responses of barley seedling to simulated weightlessness induced by two-axis rotation, *Planta* **69**(3), 218–229 (1966).
897. T. Hoshizaki, Effect of net zero gravity on the circadian leaf movements of pinto beans, *Paper for Symp. Gravity and the Organism,* September, 1967, New York.
898. T. Hoshizaki, Photoperiodism and circadian rhythmus: an hypothesis, *Bioscience* **24**(7), 407 (1974).
899. L. R. C. Howard, Subjective variables in electro-physiological recording, *Acta Biotheoretica* **17,** Part 4, 194 (1967).

900. M. Innamorati and C. Lenzi Crillini, L'ipotesi dell'effetto del campo geomagnetico e la variabilitá tra le ripetizioni in prove di germinazione ed accrescimento in Triticum, *Giorn. Bot. Ital.* **106**(6), 301–338 (1972).
901. M. Innamorati and A. B. Godi, Mancanza di effetto di campi magnetici deboli sull'accrescimento delle plantule di Triticum, *Giorn. Bot. Ital.* **108**(1–2), 27–53 (1974).
902. E. Irving, *Paleomagnetism,* Wiley, New York (1964).
903. I. Itoh, Tsujioko, and T. Saito, Blood clotting time under metal cover, *Intern. J. Bioclimat. Biometeorol., Sect. A* **3**(5) (1959).
904. L. G. Jacchia, Solar plasma velocity, exospheric temperature, and geomagnetic activity, *J. Geophys. Res.* **70**(4), 385 (1965).
905. L. G. Jacchia and J. W. Slowely, The shape and location of the diurnal bulge in the upper atmosphere, Year Book of the Smithsonian Institute (1966).
906. E. Jahn, Über Ursachen der Massenvermehrungen forstschädlicher Insekten, *Anz. Schädlingskde* **43,** 145–151 (1970).
907. E. Jahn, Hinweise zur Auswirkung biophysikalischer Unweltverhaltnisse auf forstschädliche Insekten, untersucht inbesondere an *Lymantria monacha* L., *Anz. Schadlingskde* **46,** 37–43 (1973).
908. E. Jonas, *Cosmobiological Birth Control,* ESPress, Inc., Washington, D.C. (1975).
909. R. L. Jones, Response of growing plants to a uniform daily rotation, *Nature* **185,** 775 (1960).
910. B. D. Josephson, Weakly coupled superconductors, *in: Superconductivity* (R. D. Parks, ed.), Vol. 1, Marcel Dekker, New York (1969), pp. 423–447.
911. E. Kanabrocki, L. E. Scheving, F. Halberg, R. L. Brewer, and T. J. Bird, Circadian variations in presumably healthy men under conditions of peace-time Army Reserve Unit training, *Space Life Sci.* **4,** 258–270 (1973).
912. V. P. Kaznacheev (Kaznacheyev), On relation of physiological and biological mechanisms of human adaptation, *Report to 3rd Intern. Symp. Circumpolar Health, July 8–11,* Yellowknife, Canada (1974), p. 6.
913. W. T. Keeton, Orientation by pigeons: Is the sun necessary? *Science* **165,** 922 (1969).
914. W. T. Keeton, Magnets interfere with pigeon homing, *Proc. Nat. Acad. Sci. U.S.A.* **68,** 102 (1971).
915. W. T. Keeton, Effect of magnets on pigeon homing, *in: Animal Orientation and Navigation,* (S. R. Galler, K. Schmidt-Koenig, G. J. Jacobs, and R. E. Belville, eds.), NASA, Washington (1972), pp. 579–594.
916. W. T. Keeton, The orientational and navigational basis of homing in birds, *in: Advances in the Study of Behavior,* (D. S. Lehrman, J. S. Rossenblatt, R. A. Hinde, and E. Shaw, eds.), Vol. 5, Academic Press, New York (1974), pp. 47–132.
917. W. T. Keeton, T. S. Larkin, and D. M. Windsor, Normal fluctuations in the earth's magnetic field influence pigeon orientation, *J. Comp. Phys.* **95,** 95–103 (1975).
918. J. P. Kennett and N. D. Watkins, Geomagnetic polarity change, volcanic maxima and faunal extinction in the South Pacific, *Nature* **227**(5261), 930 (1970).
919. C. L. Kervran, *Transmutation Naturelles,* Maloine, Paris (1966).
920. C. L. Kervran, *Preuves Relatives à l'Existence des Transmutations Biologiques,* Maloine, Paris (1970).
921. C. L. Kervran, *Les Transmutations Biologiques en Agronomie,* Maloine, Paris (1970).
922. C. L. Kervran, *Preuves en Biologie de Transmutations à Faible Energie,* Maloine, Paris (1975).
923. A. N. Khramov and D. M. Pecherskii, Mesozoic paleomagnetic scale of the USSR, *Nature* **244**(5417), 499–501 (1973).
924. I. H. Kimm, Orientation of cockchafers, *Nature* **188**(4744), 69 (1960).
925. J. W. King and H. Kohl, Upper atmospheric winds and ionospheric drifts caused by neutral and pressure gradients, *Nature* **206**(4985), 699 (1965).

926. K. E. Klein, H. Bruner, H. Holtmann, H. Rehme, J. Stolze, W. D. Steinhoff, and H. M. Wegmann, Circadian rhythm of pilot's efficiency and effects of multiple time zone travel, *Aerospace Med.* **41,** 125 (1970).
927. H. N. Kluijver, Ergebnisse eines Versuches über das Heimfindevermögen von Staren, *Ardea* **24,** 227 (1935).
928. H. König, Messungen zur Abschirmung atmosphärische Störungen durch metallische Hüllen, *Medizin-Meteorol. Hefte* **2**(9), 32 (1954).
929. H. König and F. Ankermüller, Über den Einfluss besonders neiderfrequenter elektrischer Vorgänge in der Atmosphäre auf den Menschen, *Naturwissenschaften* **47**(21), 486 (1960).
930. H. König, Biological effects of extremely low frequency electrical phenomena in the atmosphere, *J. Interdisc. Cycle Res.* **2**(3), 317–323 (1971).
931. H. L. König, Relations between ELF and VLF signals, *Biometeorology* **6,** Part 1, 121 (1975) (*Suppl. to Int. J. Biometeorology* **19,** 1975).
932. H. L. König, *Unsichtbar Umwelt—Der Mensch in Spielfeld elektromagnetischer Kräfte*, Heinz Moos Verlag, Munich (1975).
933. T. Kopecek, Effects of microelectromagnetic fields on plant growth, *Acta Univ. Agric. Fac. Agron.* **20**(2), 199–210 (1972).
934. J. Kopp, Tierkrankheiten und geophysikalische Bodenreize, *Schweiz. Arch. Tierheilkunde* **96**(1), 33 (1954).
935. J. Kotleba, J. Bielek, I. Glos, and J. Barta, O možnom vplyve magnetickeho pola země na spanok člověka, *Česk. Fysiol.* **22**(5), 459–460 (1973).
936. G. Kramer, Über Richtungstendenzen bei der nächtlichen Zugunruhe gekäftiger Vögel, *in: Ornithologie als Biologische Wissenchaft,* Heidelberg (1949), p. 269.
937. L. Kadik and G. Biczo, Evidence from activation energies for superconductive tunneling in biological systems at physiological temperatures, *Comment. Physiol. Chem. Phys.* **4**(5), 495 (1972).
938. L. Ladik, G. Biczo, and J. Redly, Possibility of superconductive-type enhanced conductivity in DNA at room temperature, *Phys. Rev.* **188,** 710–715 (1969).
939. D. R. Lambe and M. E. Mayer, Sensitivity of the pigeon to changes in the magnetic field (ability of pigeon to discriminate changes in earth's magnetic field for use in navigation), *Psychonomic Sci.*, No. 5, 349 (1960).
940. M. M. Lamotte, The influence of magnets and habituation to magnets on inexperienced homing pigeons, *J. Comp. Physiol.* **89**(4), 379 (1974).
941. S. Lang, Influence of an electric field of 10 Hz on the metabolism of lipids and the water electrolyte balance, *Biometerology* **6,** Part 1, 121–122 (1975) (*Int. J. Biometeorology* **19,** *Suppl.*).
942. R. Lauterbach, Biogeophyikalische Umweltfaktoren, Vortrag 14.9.1972 in der Akad. D. Wiss. der DDR (Klass. Mensch und Umwelt, 1972).
943. J. A. Lear, A primitive human guidance system, *New Scientist* **13**(273), 316 (1962).
944. H. V. Leitner and H. W. Ludwig, Therapy with 10 cps pulses, *Biometeorology* **6,** Part 1, 122–123 (1975) (*Int. J. Biometeorology* **19,** *Suppl.*).
945. M. Lenzi, On the possibility of biological effects of the magnetic field at the cellular level, *Riv. Med. Aeronaut. Speciale* **29**(1), 24 (1966).
946. B. Leszcynski, Wplyw amplitudy wahań natężenia pola magnetyczkego ziemi na wypadkowši ryzy pracy w swietle własnych badan (The effect of amplitudes of fluctuations of magnetic field intensity on the frequency of accidents at work in the light of original investigations), *Wiadom. Lek.* **26**(2), 149–153 (1973).
947. W. C. Levengood, Factors influencing biomagnetic environments during the solar cycle, *Nature* **205**(4970), 465 (1970).
948. M. Lezzi, Induction eines Ecdyson-aktivierbaren Puff in isolierten Zellkernen von Chironomus durch KC1, *Exp. Cell Res.* **43**(3), 571 (1966).

949. M. Lezzi and L. T. Gilbert, Differential effects of K^+ and Na^+ on specific band of isolated polytene chromosomes of *Chironomus tentans*, *J. Cell Sci.* **6**(3), 615–627 (1970).
950. L. M. Libber, Extremely low frequency electromagnetic radiation biological research, *Bioscience* **20**(21), 1169 (1970).
951. M. Lindauer and H. Martin, Die Schwereorientierung der Bienen unter dem Einfluss des Erdmagnetfeldes, *Z. Vgl. Physiol.* **60**(3), 219 (1968).
952. H. W. Lissmann, On the function and evolution of electric organs in fish, *J. Exp. Biol.* **35**(1), 156 (1958).
953. H. W. Lissmann and K. E. Machin, The mechanism of object location in *Gymnarchus niloticus* and similar fish, *J. Exp. Biol.* **35**(4), 451 (1958).
954. W. A. Little, Possibility of synthesizing an organic superconductor, *Phys. Rev.* **134**(6A), 1416 (1964).
955. W. A. Little, Superconductivity at room temperature, *Sci. Am.* **212**(2), 21–27 (1965).
956. J. De Lorge, Operant behavior of rhesus monkeys in the presence of extremely low frequency–low intensity magnetic electric fields, Experiment 1, NAMRL-1155 (1972); Experiment 2, NAMRL-1196 (1973), Pensacola, Florida, U.S.A.
957. J. De Lorge, Do extremely low frequency electromagnetic fields influence behavior in monkeys? *Biometeorology* **6,** Part 1, p. 123 (1975) (*Int. J. Biometeorology* **19,** *Suppl.*).
958. H. W. Ludwig, Der Einfluss von elektromagnetischen Tiefstfrequenz-Wechselfeldern auf höhere Organismem, *Biomedizin. Technik* **16**(2) (1971).
959. H. W. Ludwig, M. A. Persinger, and K. P. Ossenkopp, Physiologische Wirkung elektromagnetische Wellen bei tiefen Frequenzen. I. Historischer Überblick und gegenwartiger Stand der Forschung, *Arch. Meteorol. Geophys. Bioklimatol.* **21**(1), 99 (1973).
960. E. L. McBridge and A. E. Comer, Effect of magnetic fluctuations on bean rhythms, *Chronobiologia* **2** Suppl. 1, 44–45 (1975).
961. J. D. McCleave, S. A. Rommel, and C. L. Cathcart, Weak electric and magnetic fields in fish orientation, *Ann. N.Y. Acad. Sci.* **188,** 270–281 (1971).
962. D. L. McDonald, Orientation of homing pigeons: Direct sight of sun not necessary, *J. Exp. Zool.* **191**(2), 161–167 (1975).
963. R. M. McDowall, Lunar rhythms in aquatic animals, *Tuatura* **17**(3), 133–144 (1969).
964. M. W. McElhinny, Geomagnetic reversal during the Phanerozoic, *Science* **172,** 157–159 (1971).
965. M. Makklin and A. L. Barnetz, Control of differentiation by calcium and sodium ions in *Hydra pseudoligatis,* Exp. Cell Res. **44**(2–3), 665 (1966).
966. J. Malek, J. Gleich, and V. Maly, Characteristics of the daily rhythm of menstruation and labor, *Ann. N.Y. Acad. Sci.* **98**(4), 1042 (1962).
967. K. Marha, *Elektromagneticke Pole a Životni Potsredi,* SZdn. Prague (1968).
968. R. Markson, Tree potentials and external factors, *in: The Fields of Life* (H. S. Burr, ed.), Ballantine Books, New York (1972), pp. 186–206.
969. H. M. Martin and M. Lindauer, Orientierung im Erdmagnetfeld, *Fortschr. Zool.* **21,** 211–228 (1973).
970. R. Martini, Der Einfluss der Sonnentätigkeit auf die Häufung von Unfällen, *Zbl. Arbeitsmed. Arbeitsschutz (Darmstadt),* No. 2, 98 (1952).
971. J. P. Marton, Conjectures on superconductivity and cancer, *Physiol. Chem. Phys.* **5**(3), 259–270 (1973).
972. P. Masoero, S. Maletto, and V. Beljin, Hipoteza o novoij ekologiji, *Stočarstvo* **20**(1–2), 29 (1966).
973. G. V. T. Matthews, The experimental investigation of navigation in homing pigeons, *J. Exp. Biol.* **28,** 508 (1951).
974. G. V. T. Matthews, An investigation of homing ability in two species of gulls, *Ibis* **94,** 243 (1952).

975. G. V. T. Matthews, The problems of bird orientation, *Sci. News,* No. 23, 46 (1952).
976. G. V. T. Matthews, *Bird Navigation,* Cambridge Univ. Press (1955).
978. G. V. T. Matthews, *Bird Navigation,* 2nd ed., Cambridge Univ. Press (1968).
979. C. Maurain, Les propriétés magnétiques et éléctriques terrestres et la faculté d'orientation du pigeon voyageur, *Nature (Paris)* **54**(2728), 44 (1926).
980. E. S. Maxey and J. B. Beal, The electrophysiology of acupuncture: How terrestrial electric and magnetic fields influence air-ion energy exchanges through acupuncture points, *Biometeorology* **6,** Part 1, 124 (1975) (*Int. J. Biometeorology* **19,** *Suppl.*).
981. E. S. Maxey, Critical aspects of human versus terrestrial electromagnetic symbiosis, Report presented to U.S. National Committee of Int. Union of Radio Science Meeting, Boulder, Colorado, October 20–23 (1975), pp. 1–9.
982. E. Mayr, *Animal Species and Evolution,* Oxford University Press, London (1965).
983. P. Melchior, *The Earth Tides,* Pergamon Press, Oxford (1966).
984. F. W. Merkel, Orientation behavior of birds in Kramer cages under different physical cues, *Ann. N.Y. Acad. Sci.* **188,** 282–294 (1971).
985. F. W. Merkel, H. G. Fromme, and W. Wiltschko, Nichtvisuelles Orientierungsvermögen bei nächtlich zugunruhigen Rotkehlchen, *Vogelwarte,* No. 22, 168 (1964).
986. F. W. Merkel and W. Wiltschko, Magnetismus und Richtungsfinden zugunruhiger Rotkehlchen (*Erithacus rubecula*), *Vogelwarte,* No. 23, 71 (1965).
987. F. W. Merkel and W. Wiltschko, Nächtliche Zugunruhe und Zugorientierung bei Kleinvögeln, *Verh. Dtsch. Zool. Ges.,* 356 (1966).
988. M. E. Meyer and D. R. Lambe, Sensitivity of the pigeon to changes in the magnetic field, *Psychonomic Sci.* **5,** 349 (1966).
989. M. J. Meyer, Étude par le méthode des regressions multiples du déterminisme d'un comportement d'évitement chez la souris, *Compt. Rend.* **270,** Sér. D, No. 16, 2012 (1970).
990. A. T. Middendorf, Die Isepipetsen Russlands. Grundlagen zur Erforschung der Zugzeiten und Zugrichtungen der Vögel Russlands, *Mem. Acad. Sci. St.-Petersbourg, Ser. VI* **8,** 1–143 (1855).
991. S. G. Miles, Laboratory experiments on the orientation of the adult American eel, *J. Fish Res. Board of Canada* **25,** 10 (1968).
992. L. Miro, Champs magnétiques et biologie, *Rev. Méd. Aéronaut* **7**(3), 171 (1968).
993. L. Miro, G. Detour, and A. Pfister, Réalization et action biologiques des ambiances hypomagnétiques, *Rev. Méd. Aéronaut.* **8**(32), 175 (1969).
994. L. Miro, G. Detour, A. Pfister, R. Kaiser, and R. Grandpierre, Action biologiques des ambiances hypomagnetiques, *Presse Therm. Climat.* **107**(1), 32 (1970).
995. S. K. Mitra, *The Upper Atmosphere,* The Asiatic Society, Calcutta (1952).
996. *Molecular Asymmetry in Biology,* Vols. 1 and 2, Pergamon Press, New York (1969–1970).
997. F. R. Moore, Influence of solar and geomagnetic stimuli on the migratory orientations of herring gull chicks, *Auk* **92**(4), 655–664 (1975).
998. S. Moore, *Biological Clocks and Patterns,* Criterion Books, New York (1967).
999. H. Moriyama, *Studies on X Agent,* Igaku Shoin Ltd., Tokyo (1970).
1000. H. Moriyama, *Challenge to Einstein's Theory of Reactivity (Further Studies on X-agent),* Igaku Shoin Ltd., Tokyo (1975).
1001. L. Mortberg, Celestial asymmetry as a possible cause of small chemical asymmetry and autocatalytic amplification of the latter, *in: Proc. Intern. Symp. Generation and Amplification of Asymmetry in Chemical Systems* (W. Thiemann, ed.), Julich, BRD (1974), p. 109.
1002. I. R. Mose and G. Fischer, Das elektrostatische Feld als klima-faktor, *Wien Med. Wschr.* **125**(1–3), 30–35 (1975).
1003. H. Motoyama, *Chakri, Nadi of Yoga and Meridians, Points of Acupuncture,* Institute of Religious Psychology, Tokyo (1972).

1004. F. S. Mozer, W. D. Gonzales, F. Bogot, M. C. Kelley, and S. Schutz, High-latitude electric fields and the three-dimensional interaction between the interplanetary and terrestrial magnetic field, *J. Geophys. Res.* **79,** 56 (1974).
1005. W. H. Munk and G. J. F. MacDonald, *The Rotation of the Earth,* Cambridge Univ. Press, London and New York (1964).
1006. L. E. Murr, Plant physiology in simulated geoelectric and geomagnetic fields, *Adv. Frontiers Plant Sci.,* **15,** 97–120 (1966).
1007. R. W. Murray, The response of the ampullae of Lorenzini of elasmobranches to electrical stimulation, *J. Exp. Biol.* **39,** 119–128 (1962).
1008. T. Nagata, *Rock Magnetism,* Maruzen Comp. Ltd., Tokyo (1961).
1009. D. Neumann, Die Kombination verschiedener endogener Rhythmen bei der zeitlichen Programmierung von Entwicklung und Verhalten, *Oecologia* **3,** 166–183 (1969).
1010. J. R. Neville, An experimental study of magnetic factors possibly concerned with bird navigation, *Dissert. Abstr.* **15,** 1885 (1955).
1011. N. D. Newell, Crises in the history of life, *Sci. Am.* **208,** 76–92 (1963).
1012. A. N. Nicholson, Sleep patterns in the aerospace environment, *Proc. Roy. Soc. Med.* **65**(2), 192 (1972).
1013. J. C. T. Nisbet and W. H. Drury, Orientation of spring migrants studied by radar, *Bird Banding* **38,** 173 (1967).
1014. A. Nishida, Geomagnetic D2 fluctuations and associated magnetospheric phenomena, *J. Geophys. Res.* **73**(5), 1795–1804 (1968).
1015. F. Nixon, *Born to Be Magnetic,* Vols. 1 and 2, Magnetic Publishers, Chemainus, British Columbia (1971).
1016. H. P. Noey and W. A. Bonner, On the origin of molecular "handedness" in living systems, *J. Mol. Evolut.* **6**(2), 91–98 (1975).
1017. J. Novák, Über den Einfluss exogener Faktoren auf das Entstehen und den Verlauf einiger klinischer Krankheitsformen, *Dermatol. Wschr.* **138**(39), 1059 (1958).
1018. J. Novák, Vliv zmen geomagnetickych na vznik a prübeh nekterych chorob, *in: Sbornik Praci II Bioklimatologicke Konference v Lublicich* (1958).
1019. J. Novák, Pokus o prevenci professionálnich a lékových exantémů účinkem modifikovaného geomagnetického pole, *Prakt. Lékar* **41**(11), 509 (1961).
1020. J. Novák, Therapieversuch bei einigen Dermatosen durch experimentelle Akklimatisation, *Dermatol. Wschr.* **13**(41), 372 (1964).
1021. J. Novák and L. Valek, Attempt at demonstrating the effect of a weak magnetic field on *Taraxacum officinale, Biol. Plantarum* **7**(6), 469 (1965).
1022. H. T. Odum, The bird navigation controversy, *Auk* **65**(4), 584 (1948).
1023. J. M. Olivereau and M. Bousquet, Les effets psycho-physiologique de l'ionisation atmosphérique, *Concours Med.* **96**(3), 385–393 (1974).
1024. D. E. Olson, The evidence for auroral effects on atmospheric electricity, *Pure Appl. Geophys.* **84**(1), 118–138 (1971).
1025. N. D. Opdyke, B. Glass, J. D. Hays, and J. Foster, Paleomagnetic study of Antarctic deep-sea cores, *Science* **154**(3747), 349 (1966).
1026. A. R. Orgel and S. C. Smith, Test on the magnetic theory of homing, *Science* **120,** 890 (1954).
1027. I. Ormeny, Dynamische Richtung der medizinischen Klimatologie, *Arch. Phys. Therap.* **22**(2), 71–83 (1970).
1028. K. P. Ossenkopp and M. D. Ossenkopp, Open-field behavior in juvenile rats exposed to an ELF rotating magnetic field: Differential sex effect, *Biometeorology* **6,** Part 1, 125 (1975) (*Int. J. Biometeorology* **19,** *Suppl.*).
1029. K. P. Ossenkopp, W. Koltek, and M. A. Persinger, Prenatal exposure to extremely low

frequency low intensity rotating magnetic field and increase in thyroid and testicle weights in rats, *Development. Psychobiol.* **5,** 275–285 (1972).

1030. J. D. Palmer, The effect of weak magnetic fields on the spatial orientation and locomotory activity responses of *Volvox aureus, Diss. Abstr.* **23**(9), 3093 (1963).

1031. J. D. Palmer, Organismic spatial orientation in very weak spatial magnetic fields, *Nature* **198**(4885), 1061 (1963).

1032. J. D. Palmer, Geomagnetism and animal orientation, *Nat. Hist.* **76,** 54 (1967).

1033. S. Panek, Zagadniene sezonowej zmiennosci we wzrastanii organizmu czlowieka, *Zeszyty Naukowe Uniw. Jagiellonskiego, Prace Zool.,* **5**(33), 6 (1960).

1034. B. J. Patton and J. Fitsch, Design of a room-size magnet shield, *J. Geophys. Res.* **67**(3), 1117 (1962).

1035. M. Peju, L'avenir de l'humanité dépend d'une boussole, *Sci. Vie* **110**(588), p. 58 (1966).

1036. A. C. Perdeck, Does navigation without visual clues exist in robins, *Ardea* **51,** 91–104 (1963).

1037. M. A. Persinger, Open-field behavior in rats exposed prenatally to a low intensity low frequency rotating magnetic field, *Development. Psychol.* **2**(3), 168–171 (1969).

1038. M. Persinger, Prenatal exposure to an ELF rotating magnetic field, ambulatory behavior, and lunar distance at birth: a correlation, *Psychol. Rep.* **28,** 435–438 (1971).

1039. M. A. Persinger, Possible cardiac driving by an external rotating magnetic field, *Int. J. Biometeorol.* **17**(3), 263 (1973).

1040. M. A. Persinger and W. S. Foster, ELF rotating magnetic fields: Prenatal exposure and adult behavior, *Arch. Meteor. Geophys. Bioklimatol., Ser. B* **18,** 363 (1970).

1041. M. A. Persinger, Geophysical models for parapsychological experiences, *Psychoenergetic Systems,* Vol. 1 (1975), pp. 1–12.

1042. M. A. Persinger, G. B. Glavin, and K. P. Ossenkopp, Physiological changes in adult rats exposed to an ELF rotating magnetic field, *Int. J. Biometeor.* **16**(2), 163–172 (1972).

1043. M. A. Persinger and J. T. Janes, Significant correlations between human anxiety scores and perinatal geomagnetic activity, *Biometeorology* **6,** Part 1, 126 (1975) (*Int. J. Biometeorology* **19,** *Suppl.*).

1044. M. A. Persinger and G. F. Lafrenière, Relative hypertrophy of rat thyroid following ten day exposures to an ELF magnetic field: determining intensity thresholds, *Biometeorology* **6,** Part 1, 126–127 (1975) (*Int. J. Biometeorology* **19,** *Suppl.*).

1045. M. A. Persinger, H. W. Ludwig, and K. P. Ossenkopp, Psychophysiological effect of extremely low frequency electromagnetic fields: a review, *Perceptual and Motor Skills* **36,** 1131 (1973).

1046. M. A. Persinger and K. P. Ossenkopp, Some behavioral effects of pre- and neo-natal exposure to an ELF rotating magnetic field, *Int. J. Biometeorology* **17,** 217–220 (1973).

1047. M. A. Persinger and J. J. Pear, Prenatal exposure to an ELF rotating magnetic field and subsequent increase in conditioned suppression, *Development. Psychobiol.* **5,** 269 (1972).

1048. H. E. Petschek, The mechanism for reconnection of geomagnetic and interplanetary field lines, *in: The Solar Wind,* (R. J. Mackin and M. Neugebauer, eds.), Pergamon Press, Oxford and New York (1966).

1049. G. Piccardi, *The Chemical Basis of Medical Climatology,* C. C Thomas, Springfield (1962).

1050. G. Piccardi and M. Danti, "Wirkung einer Metall-Abschirmung auf biologischer Test," Wetter und Leben, No. 7, 152 (1972).

1051. G. Piccardi, D. Senatra, M. R. Voegelin, and L. Corti, Bioelectric rhythms in human blood, *in: Proceedings of Second ICRI Symposium on Interdisciplinary Cycle Research, Noordwijk, Netherlands* (1970).

1052. H. D. Picton, The responses of *Drosophila melanogaster to weak electromagnetic fields, Diss. Abstr.* **25**(4), 6727 (1965).

1053. H. D. Picton, Some responses of Drosophila to weak magnetic and electrostatic fields, *Nature* **211**(5046), 303 (1966).
1054. A. A. Pilla, Electrochemical information transfer at living ell membranes, *Ann. N.Y. Acad. Sci.* **238,** 249 (1974).
1055. U. J. Pittman, Growth reaction and magnetotropism in roots of winter wheat, *Can. J. Plant Sci.* **42,** 430 (1962).
1056. U. J. Pittman, Effects of magnetism on seedling growth of cereal plants, *in: Biomed. Sci. Instrum.* **1**(4), 117 (1963).
1057. U. J. Pittman, Magnetism and plant growth: effect on germination and eartly growth of cereal seeds, *Can. J. Plant Sci.* **43**(4), 513 (1963).
1058. U. J. Pittman, Magnetism and plant growth. II. Effect on root growth of cereals, *Can. J. Plant Sci.* **44**(3), 283 (1964).
1059. U. J. Pittman, Biomagnetic responses in Kharkov 22 MC winter wheat, *Can. J. Plant Sci.* **47**(4), 389 (1967).
1060. U. J. Pittman, Biomagnetism—A mysterious plant growth factor, *Can. Agr.* **13**(3), 14 (1968).
1061. U. J. Pittman, Magnetism and plant growth, *Crop Soils* **20**(8), 8 (1968).
1062. H. C. Regnart, The generation of electric currents by water moving in a magnetic field, *Proc. Univ. Durham Phil. Soc.* **8,** 291–300 (1932).
1063. U. J. Pittman, Magnetotropic responses in roots of wild oats, *Can. J. Plant Sci.* **50**(3), 350 (1970).
1064. U. J. Pittman, Biomagnetic responses in germinating malting barley, *Can. J. Plant Sci.* **51,** 64–65 (1971).
1065. U. J. Pittman, Biomagnetic responses in potatoes, *Can. J. Plant Sci.* **52**(5), 727–733 (1972).
1066. U. J. Pittman and T. H. Anstey, Magnetic treatment and seed orientation of single harvest snap-beans (*Phaseolus vulgaris* L.), *Proc. Am. Hort. Soc.* **91,** 310 (1967).
1067. R. Plonsey, *Bioelectric Phenomena,* McGraw-Hill, New York (1969).
1068. A. D. Pokorny and R. B. Mefferd, Jr., Geomagnetic fluctuations and disturbed behavior, *J. Nerv. Mental Disease* **143**(2), 140 (1966).
1069. M. Poumailloux and R. Viart, Corrélation possible entre l'incidence des infarctus du myocarde et l'augmentation des activités solaire et géomagnétique, *Bull. Acad. Nat. Méd.* **143**(7–8), 167 (1959).
1070. M. Poumailloux, Influence du géomagnétisme d'origine cosmique sur l'organisme humain, *Sem. Hôpit.* **45**(39), 2107 (1969).
1071. A. S. Presman (Pressman), Zugvögel und magnetische Felder, *Bild Wissenscha.* **6**(3), 235 (1969).
1072. A. S. Presman, Akzeleration und elektromagnetische Felder der Biosphäre, *Moderne Medizin,* **2**(4), 224–228 (1972).
1073. J. S. J. Puig, El sol y la tuberculosis, *Public. popul. del Observ. de San Miguel, Buenos Aires, Ser. F,* No. 1, 20 (1935).
1074. M. Puma, Richerche sperimentali sui campi magnetici in biologia, *Riv. Biol.* **54**(4), 541 (1952).
1075. S. C. Ratner *et al.,* Magnetic fields and orienting movements in molluscs, *J. Comp. Phys. Psych.* **65,** 365 (1968).
1076. L. J. Ravitz, History, measurement and applicability of periodic changes in the electromagnetic field in health and disease, *Ann. N.Y. Acad. Sci.* **98,** 1144–1201 (1962).
1077. J. Regnault, L'orientation des animaux et les influences magnétiques, *Rev. Pathol. Comp. (Paris)* **19**(157), 4 (1919).
1078. A. Reille, Essai de mise en évidence d'une sensibilité du pigeon au champ magnétique l'aide d'un conditionnement noticeptif, *J. Physiol. (Paris)* **30,** 85 (1968).

1079. R. Reiter, Biometeorologische Indikatoren vor grossräumiger und prognostichen Bedeutung, *Ber. Dtsch. Wetterdien. in der US-Zone* **35,** 325 (1952).
1080. R. Reiter, Verkehrungsfallziffer und Reaktionszeit unter dem Einfluss verschiedener meteorologischer, kosmischer und luftelektrischer Faktoren, *Meteorol. Rundschau* **5,**(1/2), 14 (1952).
1081. R. Reiter, Gibt es einen direkten Kausalzusammenhang zwischen raschen elektrischen order elektromagnetischen Feldänderungen und biometeorologischen Reaktionen, *Medizin-meteorol. Hefte* **2**(9), 35 (1954).
1082. R. Reiter, Nachweis der biologischen Wirksamkeit elektromagnetischer Wechselfelder niedriger Frequenz, *Naturwissenschaft* **44**(1), 22 (1954).
1083. R. Reiter, Umweltseninflüsse auf die Reaktionszeit des gesunden Menschen, *Munchen. Med. Wschr.* **96**(1032), 1067 (1954).
1084. R. Reiter, *Meteorobiologie und Elektrizität der Atmosphäre,* Akademie-Verlag, Leipzig (1960).
1085. R. Reiter and J. Kampik, Neue Ergebnisse der Klimatologie und Biophysik, *in: Beiträge zur Wirkung Meteorologischer, Kosmischer und Geophysikalischer Reize,* Verlag die Egge, Nuremberg (1948).
1086. M. Renner, Über eine weiteres Versetzungsexperiment zur Analyse des Zeitsinnes und der Sonnenorientierung der Honigbiene, *Z. Vgl. Physiol.* **42,** 449 (1959).
1087. C. Reynoud, The laws of orientation among animals, *Rev. Deux Mondes (Paris),* No. 146, 380 (1898).
1088. W. Riper and E. R. Kalmbach, Homing not hindered by wing magnets, *Science* **115,** 577 (1952).
1089. Y. A. Rocard, A biological effect of weak magnetic gradient, *in: Proc. 2nd Intern. Biomagnetic Symp.* (1963), p. 3.
1090. Y. A. Rocard, Actions of a very weak magnetic gradient: The reflex of the dowser, *in: Biological Effects of Magnetic Fields,* (M. Barnothy, ed.), Vol. 1, Plenum Press, New York (1964).
1091. A. Rochon-Duvigneaud and C. Maurain, Enquête sur l'orientation du pigeon voyageur et son mécanisme, *Nature (Paris)* **51,** 232 (1923).
1092. N. B. Romensky, Influence d'activité solaire et de différents facteurs météorologiques sur les crises cardiovasculaires, *Bull Soc. Méd. Paris* **1**(13) (1967).
1093. S. A. Rommel and J. D. McCleave, Oceanic electric fields: Perception by American eels? *Science* **176**(4040), 1233 (1972).
1094. S. A. Rommel, Jr., and J. D. McLeave, Prediction of oceanic electric fields in relation to fish migration, *J. Cons. Perm. Int. Explor. Mer.* **35,** 27–31 (1973).
1095. S. A. Rommel, Jr., and J. D. McCleave, Sensitivity of American eels (*Anguilla rostrata*) and Atlantic salmon (*Salmo salar*) to weak electric and magnetic fields, *J. Fish. Res. Bd. Can.* **30,** 657–663 (1973).
1096. A. C. Rose-Innes and E. H. Rhoderick, *Introduction to Superconductivity,* Pergamon Press, Oxford (1969).
1097. A. Rothen, The major influence of a magnetic field on a nickel surface used for an immunoelectroadosorption assay, *Physiol. Chem. Phys.* **4**(1), 61–69 (1972).
1098. A. Rothen, Circadian activity of nickel-coated glass slide used for carrying out immunologic reactions at a liquid–solid interface, *Biophys. J.* **14,** 987–989 (1974).
1099. K. H. Roth-Lutra, Der Wandel des anthropologischen Typus bei den Europiden vom Jungpaläolithikum bis ins zweite vorchristliche Jahrtausend, *Homo* **20**(3), 174–185 (1969).
1100. W. Rowan, *The Riddle of Migration,* Williams and Wilkins, Baltimore (1931).
1101. W. F. Royce, L. S. Smith, and A. C. Hartt, Models of oceanic migrations of Pacific salmon and comments on guidance mechanism, *U.S. Fish. Bull.* **66,** 441–462 (1968).

1102. W. F. Ruddiman, Pleistocene sedimentation in the equatorial Atlantic: stratigraphy and faunal paleoclimatology, *Geol. Soc. Am. Bull.* **82**(2), 283–302 (1971).
1103. J. Rummel, E. Sallin, and H. Lipscomb, Circadian rhythms in simulated and manned orbital space flight, *Rass. Neurol. Veg.*, **21**(1–2), 57 (1967).
1104. D. R. Russel and H. G. Hendrick, Preference of mice to consume food and water in an environment of high magnetic field, *in: Biological Effects of Magnetic Fields,* (M. Barnothy ed.), Vol. 2, Plenum Press, New York (1969). pp. 233–240.
1105. C. Sagan, Is the early evolution of life related to the development of the earth's core? *Nature* **206** (4983), 448 (1965).
1106. T. Saito, Statistical studies of three types of geomagnetic continuous pulsations, *Sci. Reps. Tohoku Univ., Ser. 5, Geophys.* **14**(3) (1962).
1107. G. Sardou and M. Faure, Les Tâches solaires et la pathologie humaine, *Presse Méd.*, No. 18, 283 (1927).
1108. J. F. Saunders, Biochemical dimensions of space biology, *Space Life Sci.* **1**(1), 10 (1968).
1109. D. B. O. Savile, Bird navigation in homing and in migration, *Science* **107**(2788), 596 (1948).
1110. K. E. Schaefer (ed), *Man's Dependence on the Earthly Atmosphere,* Macmillan Co., New York (1962).
1111. H. Schaltegger, Versuch zu einer allgemeinen Theorie der chemischelektrischen Informationsübertragung im tierischen Organismus, *Chimia* **7,** 191 (1966).
1112. F. Schneider, Der experimentelle Nachweis einer magnetischen und elektrischen Orientierung der Maikäfers, *Verh. Schweiz. Naturwiss. Ges.* 132 (1960).
1113. F. Schneider, Beeinflussung der Aktvität des Maikäfers durch Veränderung der gegenseitiger Lage magnetischer und elektrischer Felder, *Mitt. Schweiz. Entomol. Ges.* **34**(4), 223 (1961).
1114. F. Schneider, Orientierung und Aktivität des Maikäfers unter dem Einfluss richtungsvariabler kunstlicher elektrischer Felder und weiterer ultraoptischer Bezugssysteme, *Mitt. Schweiz. Entomol. Ges.* **36**(1/2), (1963).
1115. F. Schneider, Systematische Variationen in der elektrischen magnetischen und geographisch-ultraoptischen orientierung des Maikäfers, *Viertel. Naturforsch. Ges. Zür.* **108**(4), 373 (1963).
1116. F. Schneider, Deviationsrhythmen in Bezug auf Künstliche magnetische felder, *Mitt. Schweiz. Entomol. Ges.* **47**(1/2), 1–14 (1974).
1117. K. Schreiber, An unusual tropism of feeder roots in sugar beets and its possible effect on fertilizer response, *Can. J. Plant Sci.* **38**(1), 124 (1958).
1118. A. Schultz, J. Smith, and A. Dycus, Effect on early plant growth of nulled and directional magnetic field environment, *Proc. 3rd Intern. Biomagnetic Symp.*, Univ. of Illinois, Chicago (1966), p. 67.
1119. W. C. Schumacher, A preliminary study of a physical basis of bird navigation, *J. Appl. Phys.* **20,** 123 (1949).
1120. W. Sedlak, Ochrona srodowiska czloweka w zakresie niejonizujacego promieniowania, *Wiadom. Exologiczne* **19**(3), 222 (1973).
1121. W. Sedlak, The electromagnetic nature of life, *in: 2nd Intern. Congr. Psychotronic Res.*, Monte-Carlo (1975), pp. 77–83.
1122, 1123. H. Seipel and R. D. Morrow, The magnetic field accompanying neuronal activity *J. Wash. Acad. Sci.* **50**(6), 1 (1960).
1124. A. B. Severny, J. M. Wilcox, P. H. Scherrer, and D. S. Colburn, Comparison of the mean photospheric magnetic field and the interplanetary magnetic field, *Solar Phys.* **15,** 3–14 (1970).
1125. T. K. Shires, H. C. Pitot, and S. A. Kaufman, The membrane: A functional hypothesis for the translation and regulation of genetic expression, *in: Biomembranes,* Vol. 5, Plenum Press, New York and London (1974), pp. 81–145.

1126. P. V. Siegel, S. J. Gerathewohl, and S. R. Mohler, Time-zone effects, *Science* **164,** 1249 (1969).
1127. J. F. Simpson, Evolutionary pulsations and geomagnetic polarity, *Geol. Soc. Am. Bull.* **77**(2), 197 (1966).
1128. J. F. Simpson, Evolutionary pulsations and geomagnetic polarity *in: Mechanisms of Mutations and Inductive Factors,* Prague (1966), p. 183.
1129. F. D. Sisler and E. Sentfle, Concerning the possible influence of the Earth's magnetic field in geomicrobiological processes in the hydrosphere, *in: Bacteriol. Proc.* **36** (1961).
1130. F. D. Sisler and E. Sentfle, Possible influence of the Earth's magnetic field in geomicrobiological processes in the hydrosphere, *Proc. Am. Soc. Microbiol.* **62,** (1962).
1131. J. Slepian, Physical basis of bird navigation, *J. Appl. Phys.* **19**(3), 306 (1948).
1132. H. W. Smith, Geophysical effects on pupal weights of an insect during retrospective world interval July 26–August, 1972. Report presented (not published) in Kyoto to IAGA Meeting 1973, *Agric. Res. Paper No. 958,* Dept. of Entomology, University of Idaho.
1133. A. Sollberger, *Biological Rhythm Research,* Academic Press, New York (1964).
1134. L. Solymar, *Superconducting Tunneling and Applications,* Wiley, New York (1972).
1135. W. E. Southern, Ph. D. Dissertation: *The Role of Environmental Factors in Ring-Billed and Herring Gull Orientation,* Cornell University (1967).
1136. W. E. Southern, Gull orientation by magnetic cues: A hypothesis revisited, *Ann. N.Y. Acad. Sci.* **188,** 295 (1971).
1137. W. E. Southern, Orientation behavior of ring-billed gull chicks and fledglings, *Condor* **71**(4), 418–425 (1969).
1138. W. E. Southern, Influence of disturbances in the earth's magnetic field on ring-billed gull orientation, *Condor* **74**(1), 102–105 (1972).
1139. W. E. Southern, Magnets disrupt the orientation of juvenile ring-billed gulls, *Bioscience* **22**(8), 476 (1972).
1140. W. E. Southern, Orientation of gull chicks exposed to Project Sanguine's electromagnetic fields, *Science* **189**(4197), 143–145 (1975).
1141. F. R. Spiegel and W. T. Joines, A semiclassical theory for nerve excitation by a low intensity electromagnetic field, *Bull. Math. Biol.* **35**(5–6), 591 (1973).
1142. B. J. Srivastava and Bhaskara Rao, Unusual magnetic activity during 4–10 VIII 1972 and some of its biological consequences, *Ind. J. Radio Space Phys.* **3,** 384–390 (1974).
1143. H. D. Stalov and A. G. W. Cameron, Variations of geomagnetic activity with lunar phase, *J. Geophys. Res.* **69,** 4975–4982 (1964).
1144. H. B. Steen and P. Oftedall, Lack of effect of constant magnetic fields on Drosophila egg hatching time, *Experientia* **23**(10), 814 (1967).
1145. R. A. Stratbucker *et al.,* The magnetocardiogram—a new approach to the fields surrounding the heart, *IEEE Trans. Biomed. Electronics,* **10,** 145–149 (1963).
1146. E. Stresemann, Haben die Vögel einen Ortsinn?, *Ardea,* **24,** 213 (1935).
1147. Th. Struller and E. Schröder, Beeinflussen tellurische Faktoren den Geburtszeitpunkt auf Dauer bzw. Weheneintritt und Stärke? *Medizin-meteorol. Hefte,* **6,** 31 (1951).
1148. A. M. Stutz, Effect of weak magnetic fields on gerbil spontaneous activity, *Ann. N.Y. Acad. Sci.,* **188,** 312–323 (1971).
1149. L. Svalgaard, Sector structure of the interplanetary magnetic field and daily variation of geomagnetic field at high latitude, Danish Meteorol. Inst. Geophys. Papers R-6 (1968).
1150. L. Svalgaard, Polar magnetic variations and their relationship with the interplanetary magnetic sector structure, *J. Geophys. Res.,* **78,** 2064–2078 (1973).
1151. B. M. Sweeney, *Rhythmic Phenomena in Plants,* Academic Press, New York (1969).
1152. R. Tager, Gehören erdmagnetische Störungen zu den Uraschen des Wehenbeginns?, Dissertation, Göttingen (1950).
1153. M. Takata, Über eine neue biologisch wirksame Komponente der Sonnenstrahlung. Beitrag

zu einer experimentellen Grundlage der Heliobiologie, *Arch. Meteorol. Geophys. Bioklimat., Ser. B,* **2,** No. 5, 586 (1951).
1154. L. Talkington, Bird navigation and geomagnetism, *Am. Zool.,* No. 7, 199 (1967).
1155. T. R. Tegenkampf, Mutagenic effect of magnetic fields in *Drosophila melanogaster, in: Biological Effects of Magnetic Fields* (M. Barnothy, ed.), Vol. 2, Plenum Press, New York (1969), p. 189.
1156. E. D. Terracini and F. A. Brown, Jr., Periodisms in mouse "spontaneous" activity synchronized with major geophysical cycles, *Physiol. Zool.* **35**(1), 27 (1962).
1157. F. W. Tesch, Homing of eels, *Marine Biol.* **1**(1), 2 (1967).
1158. F. W. Tesch, Heimfindevermögen von Aalen nach Beeinträctigung des Geruchssinnes nach Adaptation oder nach Verpflanzung in ein nachbar Ästuar, *Marine Biol.* **6**(2), 148 (1970).
1159. F. W. Tesch, Influence of geomagnetism and salinity on the directional choice of eels, *Helgoland, Wiss. Meeresuntersuch.* **26**(3–4), 382–395 (1974).
1160. A. Thauzies, L'orientation, *Rev. Sci. (Paris), Ser. 4* **9,** Part 1, 392 (1898).
1161. A. Thauzies, L'orientation lointaine, *in: 4th Intern. Congr. Psychology* (1910), p. 263.
1162. A. Thauzies, L'orientation lointaine des pigeons voyageurs, *Rev. Sci. (Paris)* **31,** 805 (1913).
1163. E. Thellier and O. Thellier, Sur l'intensité du champ magnétique terrestre dans le passé historique et geologique, *Ann. Geophys.* **15**(3), 285–377, (1959).
1164. A. L. Thompson, *Problems of Bird Migration,* Witherby, London (1926).
1165. T. W. Tibbitts, T. Hoshizaki, and D. K. Alford, Moment responses of Phaseolus leaves in relation to diurnal and other magnetic fluctuations, *Bioscience* **23**(8), 479–484 (1973).
1166. W. A. Tiller, The light source in high-voltage photographing, *in: 2nd Western Hemisphere Conf. Kirlian Photog.,* Gordon and Breach, New York (1974).
1167. W. A. Tiller, Some physical network characteristics of acupuncture points and meridians, *in: Proc. Acad. Parapsychol. Med.,* Los Altos (1972).
1168. P. Tompkins and C. Bird, *The Secret Life of Plants,* Harper and Row, New York (1973).
1169. W. M. Thornton, Electric perception by deep sea fish, *Proc. Univ. Durham Phil. Soc.* **8,** 301–312 (1932).
1170. H. Tremel, *Erdmagnetismus und Eklampsie,* Dissertation, Munich (1951).
1171. K. S. Trincher, Thermodynamischen Modell der metabolisierenden Zelle, *Ideen Exacten Wissen (Stuttgart),* No. 1, 13–17 (1973).
1172. K. S. Trincher, The nonapplicability of the entropy concent in living systems, *in: Unity through Diversity* (W. Gray and N. D. Rizzo, eds.), Part 1, Gordon and Breach, New York, London, and Paris (1973), pp. 315–340.
1173. K. S. Trincher, Information and biological thermodynamics, *in: Entropy and Information in Science and Philosophy.* (L. Kubat and J. Zeman, eds.), Akademia Publ. House, Prague, and Elsevier, Amsterdam (1975), pp. 105–123.
1174. K. S. Trincher and A. Dudoladov, Spin-lattice interaction of water and protein membranes in cell metabolism, *J. Theoret. Biol.* **34,** 557 (1972).
1175. V. A. Troitskaya, Rapid variations of the electromagnetic field of the Earth, *Res. Geophys.* **1,** 485 (1964).
1176. S. W. Tromp, *Psychical Physics,* Elsevier, New York and Amsterdam (1949).
1177. S. W. Tromp, Seasonal and yearly fluctuations in meteorologically induced electromagnetic wave patterns in the atmosphere (period 1958–1968) and their possible biological significance, *J. Interdisc. Cycle Res.* **1**(2), 193–199 (1970).
1178. S. W. Tromp, *Medical Biometeorology,* Elsevier, London (1963).
1179. R. J. Uffen, Influence of the earth's core on the origin and evolution of life, *Nature* **198**(4976), 143 (1963).
1180. M. Valentinuzzi, Recent progresos en magnetobiologia, I, *Ciencia Invest.* **16**(7–8), 250 (1960).
1181. J. H. Van Dyke and M. H. Halpern, Observations on selected life processes in null magnetic

field, *Anat. Rec.* **151**(3), 480 (1965).

1182. W. Van Riper and E. R. Kalmbach, Homing not hindered by wing magnets, *Science* **115**(2995), 577 (1952).

1183. R. H. Varian, Remarks on "A preliminary study of a physical basis of bird navigation," *J. Appl. Phys.* **19**(3), 306 (1948).

1184. Ph. D. Veneziano, The effect of low intensity magneto-static fields on the growth and orientation of the early embryo of *Gallus domesticus, Diss. Abstr.* **25**(7), 4319 (1965).

1185. R. M. Verfaillie, Correlation between the rate of growth of rice seedlings and the *P* indices of the chemical test of Piccardi. A solar hypothesis, *Int. J. Biometeorology,* **13**(12), 113 (1969).

1186. C. Viguier, Le sens de l'orientation et ses organes chez les animaux et chez l'homme, *Rev. Phil.* **14**(7), 1 (1882).

1187. C. J. Waddington, Paleomagnetic field reversals and cosmic radiation, *Science* **158,** 913–918 (1967).

1188. C. Walcott, The homing of pigeons, *Am. Sci.* **62,** 542 (1974).

1189. C. Walcott and R. P. Green, Orientation of homing pigeons altered by a change in the direction of an applied magnetic field, *Science* **184,** 180–182 (1974).

1190. H. G. von Wallraff, Nichtvisuelle orientierung zugunruhiger Rotkehlchen *Erithacus rubecula, Z. Tierpsychol.* **30,** 374–382 (1972).

1191. F. Ward and R. Shapiro, Solar, geomagnetic and meteorological periodicities, *Ann. N.Y. Acad. Sci.* **95,** 200–224 (1961).

1192. L. H. Warner, The present status of the problem of orientation and homing by birds, *Quart. Rev. Biol.* **6**(2), 208 (1931).

1193. G. W. de la Warr and D. Baker, *Biomagnetism,* Delaware Laboratories, Oxford (1967).

1194. N. D. Watkins and H. G. Goodell, Geomagnetic polarity changes and faunal extinction in the Southern Ocean, *Science* **156,** 1083 (1967).

1195. T. H. Waterman, Animal navigation in the sea, *Gunma J. Med. Sci.* **8**(3), 243–262 (1959).

1196. R. Wehner and Th. Labhart, Perception of geomagnetic field in the fly *Drosophila melanogaster, Experientia* **26**(9), 967 (1970).

1197. R. Werman, CNS cellular level: membranes, *Ann. Rev. Physiol.* **34,** 337–374 (1972).

1198. J. A. West and J. G. Toonder, *The Case for Astrology,* Penguin Books, Baltimore (1973).

1199. R. Wever, Über Beeinflussung der circadian Periodik des Menschen durch schwache elektromagnetische Felder, *Z. Vgl. Physiol.* **56**(2), 111 (1967).

1200. R. Wever, Einfluss schwacher elektromagnetischer Felder auf die circadiane Periodik des Menschen, *Naturwissenschaften* **55**(1), 29 (1968).

1201. R. Wever, Gesetzmässigkeiten der circadianen Periodik des Menschen geprüft an der Wirkung eines schwachen elektrischen Wechselfeldes, *Pfügers Arch.* **302**(2), 97 (1968).

1202. R. Wever, The effect of electric fields on circadian rhythmicity in man, *in: Proc. 12th COSPAR Plenary Meeting,* Prague (1969).

1203. R. Wever, The effect of electric fields on circadian rhythms in men, *Life Sci. Space Res.* **8,** 177–187 (1970).

1204. R. Wever, Die circadiane Periodik des Menschen als Indikator fur die biologische Wirkung elektromagnetischer Felder, *Z. Phys. Med.* **2,** 439–471 (1971).

1205. R. Wever, Influence of electric fields on some parameters of circadian rhythm in man, *in: Biochronometry* (M. Menaker, ed.), Nat. Acad. Sci., Washington (1971), pp. 117–132.

1206. R. Wever, Mutual relations between different physiological functions in circadian rhythms in man, *J. Interdisc. Cycle Res.* **3,** 253–265 (1972).

1207. R. Wever, Human circadian rhythms under the influence of weak electric fields and the different aspects of these studies, *Int. J. Biometeorology* **17**(3), 227–232 (1973).

1208. R. Wever, Circadian rhythm in human performance, *in: Drugs, Sleep and Performance* (S. A. Lewis, ed.), (1973).

1209. H. Weyl, *Symmetry,* Princeton Univ. Press, Princeton (1952).
1210. J. M. Wilcox, The interplanetary magnetic field. Solar origin and terrestrial effects, *Space Sci. Rev.* **8,** 258–328 (1968).
1211. J. M. Wilcox, Inferring the interplanetary magnetic field by observing the polar geomagnetic field, *Rev. Geophys. Space Phys.* **10,** 1003–1013 (1972).
1212. J. M. Wilcox and N. F. Ness, Quasi-stationary corotating structure in the interplanetary medium, *J. Geophys. Res.* **70,** 5793–5805 (1965).
1213. D. H. Wilkinson, Some physical principles of bird orientation, *Proc. Linn. Soc. London* **160,** 94 (1949).
1214. C. B. Williams, *Insect Migration,* Macmillan Co., New York (1958).
1215. W. Wiltschko, Über den Einfluss statischer Magnetfelder auf die Zugorientierung der Rotkehlchen (*Erithacus rubecula*), *Z. Tierpsychol.* **25,** 537 (1968).
1216. W. Wiltschko, Einige Parameter des Magnetkompass der Rotkehlchen, *Verh. Dtsch. Zool. Ges.* **65,** 281–285 (1972).
1217. W. Wiltschko, The influence of magnetic total intensity and inclination on direction chosen by migrating European robins, *Animal Orientation and Navigation,* NASA SP-262, U.S. Gov. Print. Off., Washington (1973), p. 569.
1218. W. Wiltschko, Der Magnetkompass der Gartengrasmücke (*Sylvia borin*), *J. Ornithol.* **115**(1), 1–7 (1974).
1219. W. Wiltschko, H. Hock, and F. W. Merkel, Outdoor experiments with migrating European robins in artificial magnetic fields, *Z. Tierpsychol.* **29,** 409–415 (1971).
1220. W. Wiltschko and F. W. Merkel, Orientierung zugunruhiger Rotkehlchen im statischen Magnetfeld, *Verh. Dtsch. Zool. Ges.* 537 (1968).
1221. W. Wiltschko and R. Wiltschko, Magnetic compass of European robins, *Science* **176,** 62–64 (1972).
1222. W. Wiltschko and R. Wiltschko, The interaction of stars and magnetic field in the orientation system of night migrating birds. I. Autumn experiments with European warblers (genus *Sylvia*), *Z. Tierpsychol.* **37,** 337–355 (1975).
1223. K. Wodzicki, W. Puchalski, and H. Liche, Untersuchungen über die Orientierung und Geschwindigkeit des Fluges bei Vögeln, *J. Ornithol.* **87**(1), 99 (1939).
1224. A. E. Woodcock and M. B. Wilkins, The geoelectric effect in plant shoots, I. The characteristics of the effect, *J. Exp. Bot.* **20**(62), 156 (1969).
1225. D. G. Wooley and U. J. Pittman, P_{32} detection of geomagnetotropism in winter wheat roots, *Agronomy J.* **58**(6), 561 (1966).
1226. H. L. Yeagley, A preliminary study of a physical basis of bird navigation, *J. Appl. Phys.* **18**(2), 1035 (1947).
1227. H. L. Yeagley, A preliminary study of a physical basis of bird navigation, II., *J. Appl. Phys.* **22,** 746 (1951).
1228. M. A. Zimmerman and J. D. McCleave, Orientation of elvers of American eels (*Anguilla rostrata*) in weak magnetic and electric fields, *Helgoland Meeresuntersuch.* **27**(2), 175–189 (1975).

Index